Circuit Design of Digital Computers

CIRCUIT DESIGN OF DIGITAL COMPUTERS

Joseph K. Hawkins

Robot Research
La Jolla, Calif.

John Wiley & Sons, Inc.

New York . London . Sydney . Toronto

Library of Congress Catalog Card Number: 68-30911
SBN 471 36272 7
Printed in the United States of America

Preface

Numerically by far the most widely used electronic element in the world today is the digital circuit unit. Tens of thousands are present in typical general-purpose computers, and the computer itself has become as omnipresent as the oscilloscope and the typewriter. Yet the fundamentals of digital circuit behavior have remained curiously fragmented. The dozen or so linear and nonlinear components making up a typical digital unit assembled discretely or formed on a semiconductor chip have played havoc with our best efforts at comprehensive analysis. The result has been a piecemeal approach: static stability equations in one case; approximate switching time calculations under isolated conditions; and fan-in and fan-out limitations for a particular circuit type. In addition, magnetic decision and memory elements have been regarded as existing in a different world, having little or no common basis for comparison with semiconductor or other devices that implement digital functions.

It is my intention in this book to bring some degree or order and unity to the subject of digital circuit performance. For this purpose the traditional figure-of-merit in the linear electronics field, gain-bandwidth, has been extended in concept to encompass nonlinear circuits. The equally opposed variables of power dissipation and lifetime or reliability are also brought explicitly into the performance picture. The result is a conceptual framework within which the designer can see and evaluate the trade-offs he must inevitably make in selecting a particular circuit implementation.

To achieve the necessary depth of treatment a great deal of breadth has been sacrificed. Circuits composed of tunnel diodes, field-effect transistors, parametric amplifiers, and cryogenics, to name a few, have been deliberately omitted to make room for detailed circuit performance analyses on various forms of the two most illustrative and widespread components in use today; the bipolar transistor and the magnetic component.

Much of the material presented here was developed and used in a one-semester graduate engineering course given for five years. Students were expected to be acquainted with elementary digital logic and basic electronic properties of materials and to understand transient circuit analysis. The course was one of a series starting with an introductory semester that covered Boolean logic, basic circuit types, machine organization, and programming. Other one-semester courses in the series treated computer logic and systems design.

The material in this book begins by dealing with the universal circuit requirements for performing digital logic and memory. It then develops the necessary dc and transient large-signal equivalent circuits for both semiconductor and magnetic components used in analyzing circuit performance and shows how these components can be arranged into over-all systems. Interrelationships among fan-in, fan-out, switching speed, power, and reliability are explicitly derived for a representative selection of transistor and magnetic elements in a way that permits unifying performance comparisons. The transistor circuit types treated include resistor-transistor logic (RTL), diode-transistor logic (DTL), direct coupled transistor logic (DCTL), and current-mode logic (CML). Saturating and nonsaturating cases are considered in depth, together with coupling networks with and without storage. In the magnetic field both series and parallel forms of logic and decision elements are treated to bring out similarities in performance criteria. The often-overlooked switching and storage criteria in digital memory elements such as flip-flops, particularly within over-all system constraints, are explored in detail. The different flip-flops derivable from decision elements are developed and the same unifying performance criteria applied to each in turn. The mass storage area is, of course, dominated by magnetic materials, and two chapters are devoted to prime examples of space and time-select memories, the magnetic core coincident current memory, and the magnetic surface recording storage unit. Performance criteria for these media are again developed and evaluated in terms of basic material properties.

It has been my intention in this book to concentrate on principles rather than specific technology. The hope is that these principles will be extended to specific new cases as the technology develops and will not be soon obsolete. I have also tried to unify both the approach to circuits based on semiconductors and magnetic materials and the great diversity of circuit forms within each broad field. The book thus establishes a frame of reference for the student and illustrates and rounds out his understanding of each of the major categories of circuits he has heard about in earlier courses, namely, decision elements, memory elements, and space and time-select storage.

I am grateful to many, in particular, Montgomery Phister, Jr., and

Willis H. Ware, who encouraged this project. I also owe a great debt of gratitude to three patient and plucky ladies who typed, compiled, and helped to edit the successive versions of the manuscript: Mrs. Helen Hawkins, Mrs. Joan Wisenand, and Mrs. Connie Moon. My wife and children also deserve an expression of appreciation for their forbearance.

JOSEPH K. HAWKINS

La Jolla, California
August 1968

Contents

ix

Circuit Design of Digital Computers

Chapter 1

Digital Circuit Specifications and Component Properties

Digital computer circuits, like all computing devices, represent mathematical symbols or the values of physical quantities by means of other physical quantities that can be more readily manipulated. The analog circuit that solves the equation

$$f(t) = ax + b\dot{x} + c\ddot{x}$$

represents the value of the symbol x by means of values of voltage, current, or some other physical quantity. The symbol x may in turn represent the value of the displacement, velocity, or rotational behavior of some element in a mechanical system. The digital computer circuit that calculates the Boolean function

$$f(x_1, x_2) = \overline{x_1 \oplus x_2} \tag{1.1}$$

similarly represents the value of the symbol x_1 by means of a voltage, current, or some other physical quantity. The symbol x_1 may in turn represent the truth value of some statement about the physical world such as "the presence of an alligator" or "a number not equal to zero."

Like all computing devices, digital computer circuits perform the operations specified by the defining function on the physical quantities. The analog circuit performs the multiplications implied by the symbol juxtapositions ax, $b\dot{x}$, and $c\ddot{x}$, and adds. The digital computer circuit likewise mechanizes the EXCLUSIVE-OR operation corresponding to the symbol $\oplus$ and complements. The similarity of digital computer circuits to other computing devices ends here, however, because of differences in the nature of the functions mechanized.

1

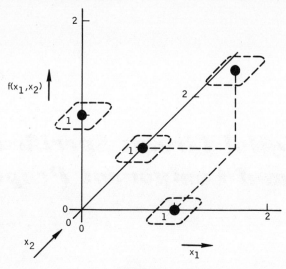

Figure 1.1

Digital computer circuits are characterized by at least two special properties possessed by the functions that they are required to mechanize. The
first of these is nonlinearity. The great majority of networks for which formal
analysis and synthesis techniques have been developed are characterized
by the fact that they obey the laws of superposition. Such networks are
termed *linear*. The only networks of interest to the digital circuit designer do

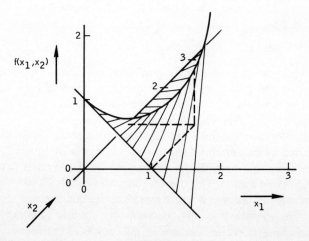

Figure 1.2

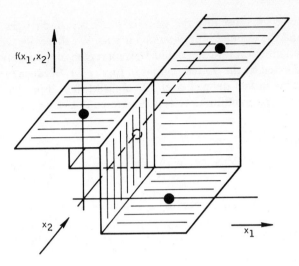

Figure 1.3

not obey the laws of superposition; they are *nonlinear*. In most cases satisfactory mathematical techniques for handling nonlinear equations do not exist. The job of the circuit designer is therefore difficult. A large part of the job consists of making appropriate simplifying assumptions, solving specific nonlinear equations when feasible, and developing useful components that can be made to fit those fragmentary parts of nonlinear circuit theory that are known.

The second important characteristic of digital computer circuits is that the functions that they are called upon to mechanize are defined only at discrete values of the input variables. The algebraic expression

$$f(x_1, x_2) = 1 - x_1 - x_2 + 2x_1x_2 \tag{1.2}$$

is nonlinear, but it is not discrete if its value is defined over some continuous range of the variables x_1 and x_2. The Boolean expression (1.1) is defined to have values only at the points corresponding to $(x_1 = 0, x_2 = 0)$, $(x_1 = 0, x_2 = 1)$, $(x_1 = 1, x_2 = 0)$, and $(x_1 = 1, x_2 = 1)$. Although the function (1.1) is in principal not defined for any values of the variables $x_i \neq 0, 1$, we shall see that some compromise must be made for practical circuit design. The plot of (1.1) is given in Figure 1.1, and the plot of (1.2) is given in Figure 1.2. The functions yield identical values at points $(0, 0)$, $(0, 1)$, $(1, 0)$, and $(1, 1)$. It is clear intuitively, however, that the surface of Figure 1.2 is a poor way of mechanizing (1.1). In any reasonable circuit the variables can be expected to vary somewhat about their nominal discrete values. The result, in Figure

1.2, would be serious, because the value of the function changes rapidly in regions around the defined binary-valued inputs. We therefore tacitly imply that a Boolean function is in fact defined over a region in the vicinity of input points, as indicated by the dotted regions in Figure 1.1. Because the behavior of real circuits is in fact defined over all values of their input variables, the ideal function corresponding to (1.1) is illustrated in Figure 1.3.

Basic Units

The types of circuit element with which this text deals are defined as follows:

Decision element. A circuit whose state (value of physical quantity representing a symbol; the output) is dependent, according to the rule-following characteristics of the device, on the states of external circuits (inputs) at the same time.

The concept of simultaneity ("at the same time") requires greater precision. In the computer environment, "simultaneous" simply means that the functional relationships governing the output of a decision element are only permitted to contain variables corresponding to inputs; that is, prior states of the element itself are not involved. Furthermore, it is implied that the output of a decision element will assume the proper state before that state is detected or employed in any way as an input to subsequent elements. These restrictions are closely related to the rate of change and manner of detecting the state of a logic element, thus introducing the synchronous and asynchronous concepts.

Synchronous system. A system in which the inputs to decision elements change state between externally defined (and usually evenly spaced in time) periods.

Asynchronous system. A system in which the inputs to decision elements change state continuously until a specified operation is complete.

Although the above definitions are far from rigorous, there is little point in belaboring them. It will be seen later that the same circuit can in many cases be arranged to operate in either a synchronous or an asynchronous mode. As a simple example consider the function (1.1) with the addition of a term that is true only at specified times; that is, let

$$f = \overline{(x_1 \oplus x_2)} \cdot x_3.$$

If x_3 is a timing signal, common to all elements in a system, these elements are operating in synchronism.

Memory element. A circuit whose state is dependent upon the states of external circuits at *previous* times.

The classical concept distinguishing between decision and memory elements can be illustrated by means of Figure 1.4, showing first the ideal states that are to be produced by some logic element as a function of a set of inputs. The circuit in this figure is designed to represent one of two symbols, 0 or 1, corresponding to lower and upper levels. Circuit output is shown as a function of time. If the output is sampled or detected at times represented by pulses on a clock (a), the circuit indeed satisfactorily represents the desired function of the inputs. If, however, the sampling repetition rate is increased to that of another clock (b), the circuit output no longer satisfies the desired function at all times; that is, its output is not the given function of the inputs *at the same time.* The output in the example is the desired function of the inputs during the previous clock period. The circuit may therefore be classified as a memory element.

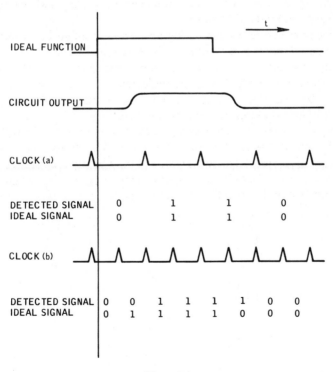

Figure 1.4

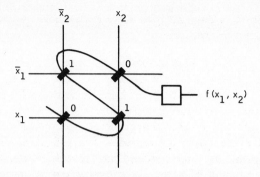

Figure 1.5

We shall see in later chapters that from a circuit standpoint the decision element is the basic building block of digital computers, and that the distinction between decision elements and memory elements is largely one of convenience and conformance to the habits of logic designers. In considering the property called memory, we must give careful thought to the kind of memory being discussed. Every decision element displays at least two (and potentially three) kinds of memory.

First, memory resides in the wiring and circuit configuration itself. This may seem trivial, but it makes difficult the drawing of a distinct line between decision and memory elements. For example, consider the four-bit memory plane of Figure 1.5, with storage as indicated. As long as the stored bits remain unchanged, this unit, as a black box, is indistinguishable from a conventional gate that implements the function (1.1).

Memory also resides in the inherent delay time associated with the propagation of a signal through any physical device. This feature has been employed in practical systems to convert what would be regarded as an ordinary pulse amplifier or a delay-type memory element, as in Figure 1.4, into a set-reset flip-flop of the so-called dynamic type.

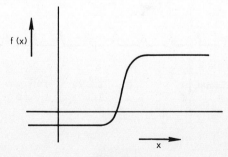

Figure 1.6

Finally, every decision element possesses the potential for becoming a memory element because of its nonlinear characteristic. The simplest decision element characteristic is illustrated in Figure 1.6. Such a characteristic can always be converted to that of a memory element by introducing positive feedback greater than some critical value. The reverse is also true. Devices that inherently exhibit memory characteristics can, by appropriate feedback and/or selection of signal ranges, be converted to decision elements.

Because decision elements can be arranged in networks to exhibit memory, we are led to a more useful working definition of both types.

Decision element. A circuit that exhibits a single-valued output as a function of the inputs over the range of defined input values.

Characteristics illustrating this definition are given in Figure 1.7. The discrete values at which the input x is defined are marked along the ordinate, and the corresponding values of $f(x)$ are marked by a cross. The above

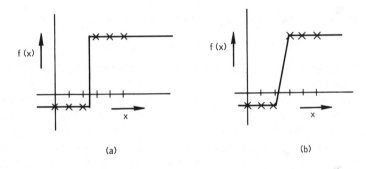

(a)   (b)

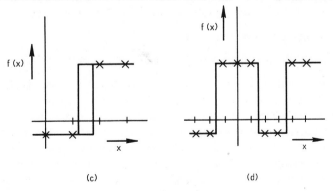

(c)   (d)

Figure 1.7

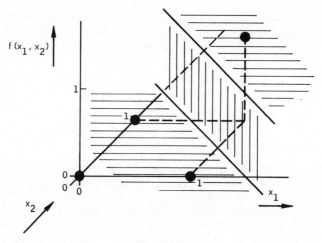

Figure 1.8

definition can be expressed symbolically, with time included, by the form

$$f(t < t + 1) = f[x_1(t), x_2(t), \ldots, x_N(t)].$$

This simply says that the output is a function only of inputs during the period of definition.

The characteristics of Figure 1.7 imply that the input, although discrete, is more than binary-valued. This is merely for convenience, because it is difficult to represent functions of more than a few variables in graphical form. A possible decision element characteristic for two binary variables is illustrated in Figure 1.8. It is single-valued and mechanizes the function, $f(x_1, x_2) = x_1 \cdot x_2$.

Memory element. A circuit that possesses a double-valued output as a function of the inputs, over the range of defined input values.

Symbolically, it may be written as

$$f(t + 1) = f[x_1(t), x_2(t), \ldots, x_N(t), f(t)].$$

This simply says that the output at $t + 1$ includes a term that is dependent on the same output at time t, in addition to input variables.

We shall not attempt to make a precise circuit distinction between a memory element and what we normally regard as a storage element or device. The storage element, in the form of magnetic core memories and surface recording techniques, will be considered in subsequent chapters. The behavior of such devices is distinguished from that of memory elements less by functional definitions than by operation and use in complete systems.

Decision Element Specifications

Three techniques are used to represent decision and memory element behavior. The first of these is the Boolean equation. Equations of the form of (1.1) will be used to represent functional relationships according to the usual conventions of Boolean notation. The symbols that will be employed most frequently are as follows:

$\cdot$ = AND. Example: $f(x_1, x_2, x_3) = x_1 \cdot x_2 \cdot x_3$, which is read "$f$ equals x_1 AND x_2 AND x_3." The AND symbol $\cdot$ will be omitted wherever the meaning is clear, in which case the function may be written $f(x_1, x_2, x_3) = x_1 x_2 x_3$.

$+$ = OR (inclusive). Example: $f(x_1, x_2, x_3) = x_1 + x_2 + x_3$, which is read "$f$ equals x_1 OR x_2 OR x_3." The OR symbol $+$ will never be omitted, and possible confusion with the algebraic "plus" symbol will be avoided by context or appropriate comment.

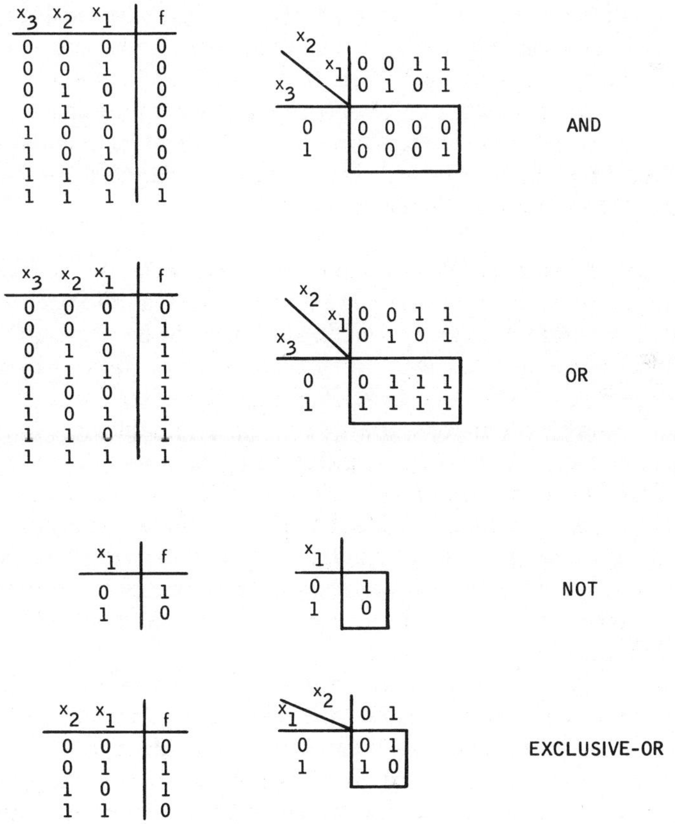

x_3	x_2	x_1	f
0	0	0	0
0	0	1	0
0	1	0	0
0	1	1	0
1	0	0	0
1	0	1	0
1	1	0	0
1	1	1	1

AND

x_3	x_2	x_1	f
0	0	0	0
0	0	1	1
0	1	0	1
0	1	1	1
1	0	0	1
1	0	1	1
1	1	0	1
1	1	1	1

OR

x_1	f
0	1
1	0

NOT

x_2	x_1	f
0	0	0
0	1	1
1	0	1
1	1	0

EXCLUSIVE-OR

Figure 1.9

$\overline{}$ = NOT (or inhibit, complement, or invert). Example: $f(x_1) = \overline{x_1}$, which is read "f equals x_1 BAR" or "f equals NOT x_1."

$\oplus$ = EXCLUSIVE-OR. Example: $f(x_1, x_2) = x_1 \oplus x_2$, which is read "$f$ equals the EXCLUSIVE-OR of x_1 and x_2."

The second method of specifying decision and memory element behavior is by truth tables. This is simply a listing of function values for all values of the input variables, and the list may be arranged in a number of ways. The two that are employed are the truth table listing in ascending binary order of input configurations and the two-dimensional layout of input configurations, as in the Veitch diagram. These methods are illustrated in Figure 1.9 for the Boolean functions defined above.

An additional symbol is employed whenever the function value is indeterminate or the corresponding input configuration never occurs. This is represented by the symbol X and is often called a "don't care" condition.

A third method of specifying decision and memory element behavior is by an $(N + 1)$-dimensional plot, the variables occupying N dimensions and the function value occupying the $(N + 1)$st dimension. Lines and perspective drawings of planes can be used when $N + 1$ is 2 and 3 respectively. When $N + 1 = 4$, it is convenient to plot only input configuration space and label function values by appropriate symbols. On such plots the solid dot ● represents a 1 and the empty dot ○, a zero. These conventions are illustrated in Figure 1.10 for the same functions.

So far only a handful of Boolean functions have been discussed and illustrated. It cannot be overemphasized that these represent an extremely small fraction of the total number of functions that could be specified. Because each variable can take on two values, there are two ways of picking the value of the first variable, two ways of picking the second, and so on, for a total of 2^N possible input configurations in the truth table. For every input configuration there are two ways of picking the function value, and therefore 2^{2^N} possible functions that can be specified. It is desirable to reduce the complexity of this great wealth of possible functions by classifying groups of them according to some criterion of similarity. It is important from a circuit standpoint to make the criterion correspond to simple circuit operations so that the members of a group of functions can all be mechanized by similar circuit arrangements.

One way of specifying the classes of Boolean function is to say that any member of the class can be derived from any other by permuting the input variables and/or complementing them. This is equivalent to saying that we have a circuit with N inputs that realizes a given function, together with a set of N signals and their complements, and it is desired to connect circuit inputs to the available signals in all possible ways that yield different functions. For example, if we have a circuit that mechanizes the AND function, together with $2N$ signals corresponding to N input variables and their complements, it is possible to connect the first circuit input to any one of $2N$ points, the second to any one of $2N - 2$ points, etc. The result is $2^N(N!)$ ways of connecting circuit inputs to signals. Although this number increases rapidly with increasing N, the fraction $2^N(N!)/2^{2^N}$ becomes vanishingly small with increasing N. This means that although the number of members of

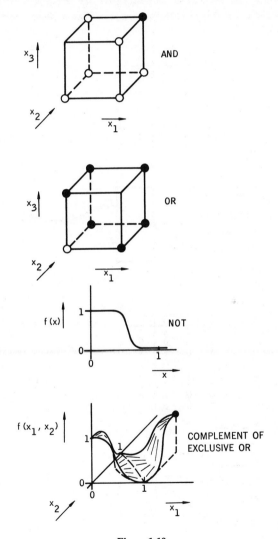

Figure 1.10

each function class defined in this manner becomes large with increasing N, the number of classes is also increasing rapidly with N.

If we regard the complement of a function as belonging to the same class as the function itself, even further simplification exists. This is equivalent to saying that not only are input variables available in both uncomplemented and complemented form, but also that the circuit provides both normal and complemented outputs, and we are free to select between them. Viewed from this standpoint, the AND and OR functions are members of the same class and can be mechanized by identical circuits. This is illustrated in Figure 1.11.

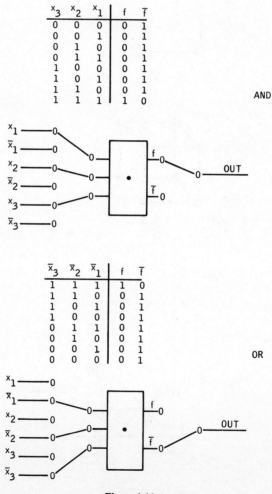

x_3	x_2	x_1	f	$\bar{f}$
0	0	0	0	1
0	0	1	0	1
0	1	0	0	1
0	1	1	0	1
1	0	0	0	1
1	0	1	0	1
1	1	0	0	1
1	1	1	1	0

AND

$\bar{x}_3$	$\bar{x}_2$	$\bar{x}_1$	f	$\bar{f}$
1	1	1	1	0
1	1	0	0	1
1	0	1	0	1
1	0	0	0	1
0	1	1	0	1
0	1	0	0	1
0	0	1	0	1
0	0	0	0	1

OR

Figure 1.11

The possibility of simplifying the specification of decision elements leads to something of a dilemma when considering standard practice and nomenclature. First, many useful circuits have been devised that do not provide both uncomplemented and complemented outputs. Second, different names have been given to many specific functions that are members of the same class, as in the AND and OR case above. In view of these practices an exhaustive listing of possible decision element functions is perhaps the best way to avoid confusion. Of course, this is limited by available paper and patience.

TABLE 1.1

Input	Output Functions			
	0	1	2	3
0	0	0	1	1
1	0	1	0	1

Possible one-input decision elements are listed in Table 1.1. According to the above class membership criterion, only two classes exist, namely, (0 and 3) and (1 and 2). The former is in fact a function of one fewer variables, that is, zero input variables, because it is not a function of the input at all. Function 1 is a pulse amplifier, and function 2 is the NOT, or inverting pulse amplifier. Possible two-input decision elements are listed in Table 1.2.

TABLE 1.2

Inputs		Output Functions															
2	1	0	1	2	3	4	5	6	7	8	9	10	11	12	13	14	15
0	0	0	0	0	0	0	0	0	0	1	1	1	1	1	1	1	1
0	1	0	0	0	0	1	1	1	1	0	0	0	0	1	1	1	1
1	0	0	0	1	1	0	0	1	1	0	0	1	1	0	0	1	1
1	1	0	1	0	1	0	1	0	1	0	1	0	1	0	1	0	1

There are four classes of function. Their members are the following:

Class	Member Function Numbers	Number of Variables
1	0, 15	0
2	3, 5, 10, 12	1
3	1, 2, 4, 7, 8, 11, 13, 14	2
4	6, 9	2

The last column above simply indicates that the first two classes are really functions of two and one fewer variables, respectively, than the number of available inputs, and may therefore be disregarded. Certain members of classes 3 and 4 have been given names. They are the following:

Function Number	Name	Symbol
1	AND	·
6	EXCLUSIVE-OR	⊕
8	Pierce	↓
14	Sheffer stroke	|

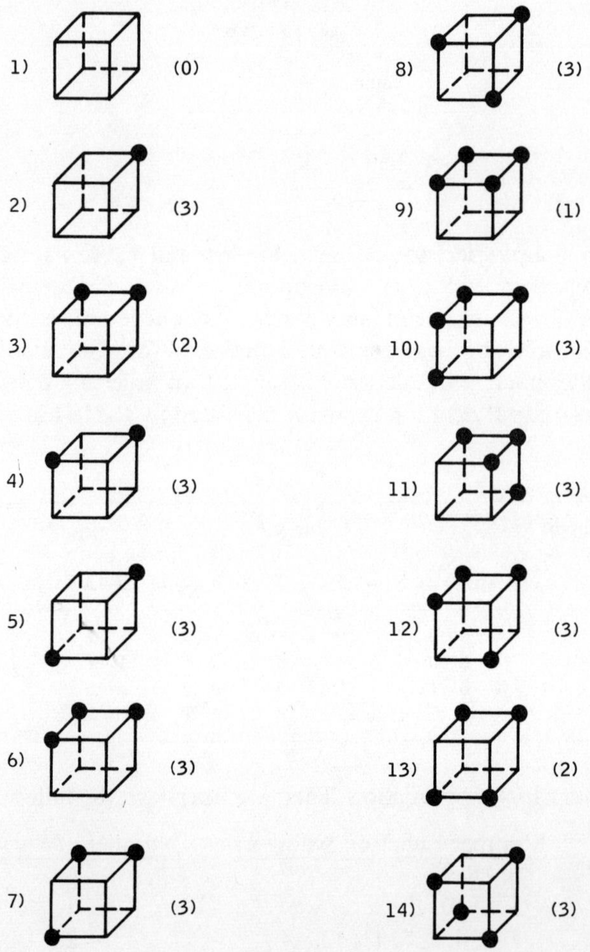

1) (0)

2) (3)

3) (2)

4) (3)

5) (3)

6) (3)

7) (3)

8) (3)

9) (1)

10) (3)

11) (3)

12) (3)

13) (2)

14) (3)

Figure 1.12

From a circuit standpoint there is little point in such special names. Even supposing that the black box mechanizing the function does not provide a complemented output, it is clear that function 9, for example, can be obtained with a $\oplus$ circuit by reconnecting one of the circuit inputs to the complement of one of the input variables. Similarly functions 1, 2, 4, and 8 can be obtained with a single circuit by reconnecting one or both of the inputs. The same holds true of functions 7, 11, 13, and 14.

Possible three-input decision element functions are partially listed in Table 1.3, in which the AND (function 1) and the OR (function 127) are entered among others. The number of functions is too large for exhaustive listing at this point, and it is helpful to convert to a geometrical representation of function types in N-dimensional space. The classes of three-input functions are illustrated in Figure 1.12. They are listed in ascending order of the number of minterms in the positive or uncomplemented form. If m is the number of minterms in the form given, the complement of the function of course has $2^N - m$ minterms. The number to the left of each box is an arbitrary function number. The number in parentheses to the right is the number of variables of which it is a function. Thus functions 3 and 13 are independent of one variable, and function 9 is a function of only one variable.

TABLE 1.3

Inputs			Output Functions								
3	2	1	0	1	2	23	105	127	128	254	255
0	0	0	0	0	0	0	0	0	1	1	1
0	0	1	0	0	0	0	1	1	0	1	1
0	1	0	0	0	0	0	1	1	0	1	1
0	1	1	0	0	0	1	0	1	0	1	1
1	0	0	0	0	0	0	1	1	0	1	1
1	0	1	0	0	0	1	0	1	0	1	1
1	1	0	0	0	1	1	0	1	0	1	1
1	1	1	0	1	0	1	1	1	0	0	1

Some comments on these functions are in order. Number 2 is the standard AND of three variables. When complemented and rearranged, it becomes the standard OR. Function 3 is the AND of two variables, represented in three-variable space. Function 13 is the $\oplus$ of two variables. Function 11 is the standard carry function for the ith bit of an addition operation. Its logic is

$$C_i = A_i B_i + A_i C_{i-1} + B_i C_{i-1},$$

in which A and B are the two words being added, and C is the carry bit.

Function 14 is the standard sum operation whose logic is given by

$$S_i = A_i \oplus B_i \oplus C_{i-1}.$$

Other functions listed have no special name, although they can of course be represented by logic expressions. Number 4, for example, is the function $f = \overline{(x_i \oplus x_2)} \cdot x_3$ previously mentioned.

Memory Element Specifications

The logic specifications for all possible memory elements representing functions of N input variables can be listed exhaustively in much the same manner as the foregoing decision element tabulations. Additional freedom is available in specifying the state of the memory element output, however, because the element may be a function not only of the state of its inputs, but also of its own state at the previous time period. This is not possible with the decision element, because by definition it possesses no "memory" of its previous state and can therefore only be a function of present inputs. In accordance with standard practice the output state of a memory element will be designated Q. The value of Q is, of course, restricted to binary values. There are various ways of specifying a memory element function. Perhaps the simplest is to regard Q as an additional input to the element. The truth table of all N-input memory elements therefore contains $N + 1$ columns in the "input" section, 2^{N+1} rows of input combinations, and $2^{2^{N+1}}$ columns in the "output" section.

The possible one-input memory elements are listed in Table 1.4.

TABLE 1.4

(t)		$Q(t + 1)$																
Q	x	0	1	2	3	4	5	6	7	8	9	10	11	12	13	14	15
0	0	0	0	0	0	0	0	0	0	1	1	1	1	1	1	1	1
0	1	0	0	0	0	1	1	1	1	0	0	0	0	1	1	1	1
1	0	0	0	1	1	0	0	1	1	0	0	1	1	0	0	1	1
1	1	0	1	0	1	0	1	0	1	0	1	0	1	0	1	0	1

As in the case of decision elements certain possible memory elements are not particularly useful. Column 0 represents a circuit always in the 0 state; column 15 a circuit always in the 1 state. Column 3 represents a circuit always in the same state, whatever the state may be. Column 12 represents a 1,0 pattern generator, that is, a circuit which changes state at each time period under consideration, regardless of the input. Other columns represent

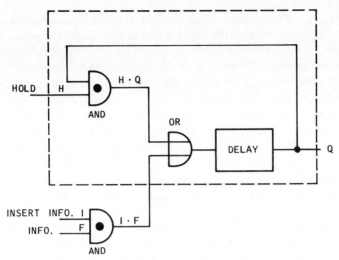

Figure 1.13

circuits that in more subtle ways are difficult to employ in practical logic design. Columns 1 and 2 represent circuits that, once set to the 0 state, will remain there indefinitely. The only statement that can be made about the information content of such circuits is that either they started out in the 0 state or at least once some time in the past received a 1 or 0 input, respectively. The reverse holds true for columns 13 and 14.

One of the most widely used one-input memory elements is function 5. This is commonly called the *delay element*. Its output at one time period is a reproduction of its input at the previous time period. An even more useful element, the complement of the delay, is represented by the inverting delay element of column 10. It mechanizes the NOT operation, one time period delayed. This is not only crucial for the synthesis of arbitrary logic functions but is also, conveniently, the operation performed by many of the amplifying components of interest to digital circuit designers. In combination with appropriate decision elements information may be inserted into a delay element and remain indefinitely or be replaced. Referring to Figure 1.13, for example, as long as control signal H is 1 and control signal I is 0, the output will continue to represent the same value Q from time period to time period. If control signal H is 0 and control signal I is 1, the new value represented by line F will be inserted. If both H and I are 0, the information content of the delay element will be cleared to 0. If both H and I are 1, the value of Q will depend both on past history and input information.

An even more useful one-input memory element is function 6, usually called the *trigger element*. The presence of a 1 at its input reverses the state of the element. The absence of an input 1 leaves its state unchanged. Its complement with respect to the input is function 9. The value of this element also lies in its ability to invert the input signal. It thereby mechanizes the crucial NOT operation. Many logic functions can be mechanized by means of passive networks alone, that is, with power-loss components. Except for the transformer, however, signal inversion can be achieved only by means of active components. It is more than simple convenience and signal attenuation, therefore, that enjoins us to use amplifiers, such as represented by the trigger memory element, at strategic points throughout a computing network. We shall see, in addition, that the fundamental flip-flop design problem can be reduced to the design of the trigger as the basic building block.

Heretofore the output of a logic element, whether decision- or memory-type, has been regarded as capable of assuming only one of two values, symbolized as 0 and 1. In practical circuits, however, the output may be indeterminate under certain input conditions or combinations. One of two things may be meant by the term "indeterminate." It may be that the circuit is not designed to respond in a specific way to some input combination (that is, the element's output will assuredly be either 0 or 1, but which one it will be is dependent on chance transients, particular component tolerances at the time, or relative signal amplitudes). It is not necessarily directly dependent on the input configuration. On the other hand, "indeterminate" may be used as the equivalent of "don't care." This is a term describing the output of a logic element when it is known that the corresponding input combination will never occur. It is, of course, up to the logic designer to ensure the meeting of this constraint.

In the design of memory elements the practice of including indeterminate conditions freely has arisen. This is partly because of intuitive concepts of useful logic functions and partly a result of the purpose for which basic pulse circuits such as the Eccles–Jordan flip-flop were designed. The indeterminate output is commonly represented by the symbol X. This extra degree of freedom makes the number of possible N-input memory element functions that may be written grow to an unwieldy $3^{2^{N+1}}$. Rather than attempt to list or discuss these exhaustively, only the memory elements in common use are treated in this text.

The logic specification of the RS (reset-set) flip-flop is listed in Table 1.5. Operation of this flip-flop is as follows: in the absence of inputs, the element retains its previous state. A pulse on the set line drives the element to the 1 state; a pulse on the reset line drives the element to the 0 state. Simultaneous inputs are forbidden or cause indeterminate circuit operation.

Logic specifications of the JK flip-flop are listed in Table 1.6. The only difference between this element and the RS unit is that simultaneous inputs result in definite triggering of the circuit rather than in indeterminate

TABLE 1.5

(t)			(t + 1)
R	S	Q	Q
0	0	0	0
0	0	1	1
0	1	0	1
0	1	1	1
1	0	0	0
1	0	1	0
1	1	0	X
1	1	1	X

operation. In terms of practical circuit design it will be seen that in many cases involving complete systems of decision and memory elements in which certain freedom of interconnection exists, there is very little difference between the *RS* and *JK* memory elements.

The only three-input memory element that will be considered in this text is the *RST* (reset-set-trigger) flip-flop whose specifications are listed in Table 1.7. This element is simply an *RS* circuit with a separate trigger input. Note that the simultaneous occurrence of more than one input pulse is forbidden. In practice the *RST* flip-flop functions can in many cases also be obtained from the basic *RS* or *JK* circuit by means of minor modifications.

The memory elements specified above represent only a very small sample of the possibilities. A large number of other types of element may be obtained without basic circuit modification simply by including certain decision functions within the "black box" regarding as constituting the memory element. Such respecification is primarily a matter of conceptual convenience,

TABLE 1.6

(t)			(t + 1)
K	J	Q	Q
0	0	0	0
0	0	1	1
0	1	0	1
0	1	1	1
1	0	0	0
1	0	1	0
1	1	0	1
1	1	1	0

TABLE 1.7

(t)				(t + 1)
R	S	T	Q	Q
0	0	0	0	0
0	0	0	1	1
0	0	1	0	1
0	0	1	1	0
0	1	0	0	1
0	1	0	1	1
0	1	1	0	X
0	1	1	1	X
1	0	0	0	0
1	0	0	1	0
1	0	1	0	X
1	0	1	1	X
1	1	0	0	X
1	1	1	0	X
1	1	1	1	X

the truth table depending on where the line is drawn between units, that is, to which portions of the over-all logic the label "element" is applied.

Gain and Bandwidth

In addition to functional specification, both decision and memory elements possess a characteristic signal gain or amplification whose magnitude is of great importance to the logic designer. The designer wishes to know how many times a given term (decision or memory element output signal) may appear in the logic equations and similarly, how many terms (output signals) may appear at the input to a given decision or memory element. These quantities are usually called *fan-out*, or *logic gain*, and *fan-in*, respectively. It will be seen that precise definitions can depend upon the logic structure under consideration. For purposes of this text, the definition of logic amplification or fan-out can be stated as follows:

Logic gain. The number of memory element inputs which can be driven by a single memory element output through the type of decision element structure involved in the set of circuits. For purposes of notation maximum fan-in or, more generally, simply the number of inputs entering a decision element structure, will be designated by N. Maximum fan-out or, more generally, the number of loads on a given element, will be designated by n.

Fan-in is limited by the attenuation characteristics of that portion of a decision element which we shall term the *mapping function*, defined in the

regarded as proportional to some effective bandwidth, we shall define the *gain-bandwidth* of digital circuits as the product of fan-out, or logic gain, and inverse switching time. Thus

$$\text{GB} \equiv \frac{n}{T}.$$

We shall see that expressions for this quantity can be derived in terms of component parameters, and gain bandwidth will therefore be used as a criterion for the selection of suitable components upon which to base a given circuit function design.

Decision Component Properties

Having specified what behavior we expect from decision elements, it is useful to ask in general terms what component properties we must look for in order to mechanize such behavior.

We have seen that the over-all decision element characteristic must resemble that of Figures 1.7 or 1.8. In the case of Figure 1.7 the decision element output $f(x)$ is depicted as being a function of some discrete, multi-valued variable, x. This implies that various configurations of the multiple two-valued input variables have been mapped in some way onto x, and that the decision function $f(x)$ is then applied. Alternatively, the plot of Figure 1.8 implies that mapping and decision are performed in a single, indivisible component having the characteristic illustrated. It is difficult, however, to find single components that behave in the latter manner. The almost universal practice of mechanizing decision elements in a two-stage process has therefore arisen. In the first stage multivariable input configurations are mapped into a single-variable signal in a number of possible ways that we will examine. In the second stage a decision function is applied to the single signal. This process is illustrated in Figure 1.14. Formally we can define circuit functions as follows:

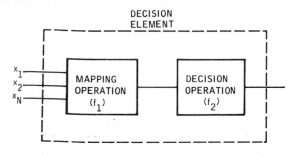

Figure 1.14

following section, and by the amplification of that portion of a de
element which we shall term the *decision function*, also subsequently d
We shall see that in some types of decision element there exists a tr
between N and n, whereas in others N and n are virtually indep
The latter case corresponds to the use of diodes in the decision
and is of predominant importance at the present time. Therefore
frequently act as if n is a function only of the parameters of the acti
ponents in a circuit. This simplification has the advantage of
some measure of the over-all logic usefulness of circuit compor
any case, however, given a decision element fully loaded (e.g., by f
memory elements), the maximum number of inputs for correct sigr
mission may be calculated. Because the particular characteristi
components employed in the decision element determine the att
this calculation has real meaning only for given component prope
discussed in the following section and performed in detail in si
chapters.

Fan-out may be regarded as a measure of the number of logic
that a given circuit represents; that is, n is inversely related to
component count in a system in the following sense: If the fan-ou
building block is inadequate for system requirements, that circu
paralleled by additional similar circuits. Although such additio
are frequently disguised in the complexities of the logic, the total
count is nevertheless a decreasing function of n, other factors
constant. If we take the count to be inversely proportional to n,
given system is proportional to $(1/n)$ (unit cost/component).

Over-all system effectiveness is measured by the number of cc
of a given kind that can be carried out for a given cost, that
putations/unit cost. The system cost per unit time may generall
be proportional to total system cost divided by its lifetime. I
lifetimes of two systems being compared are similar, the rem
we must evaluate is the number of computations that can be ca
unit time. A good measure of this factor is the switching time
basic components employed in the system, because we are l
delay in propagating logic signals through the system. We
take as a reasonable measure of system effectiveness the relati

$$\frac{\text{Computations}}{\text{Unit cost}} = \left(\frac{\text{Components}}{\text{Unit cost}}\right)(\text{Lifetime})(n)\left(\frac{1}{T}\right)$$

In this text we shall not attempt to evaluate the first two facto
expression. The last two, however, can be directly relate
electrical parameters of components employed in circuit de
therefore be optimized as part of the design procedure. Si

Mapping operation. A circuit operation that transforms the configurations of N binary-valued input variables into discrete values of a single output variable.

Decision operation. A circuit operation that transforms the discrete (but not necessarily binary) values of an input variable into binary values of an output variable.

Considering these black boxes in reverse order, we see that the only components of interest for the decision operation f_2 are those exhibiting a single-threshold input-output relationship defined qualitatively by Figure 1.6 or its complement.

This characteristic may be described by a familiar phrase, namely, that the component displays amplification. This may seem contradictory to the usual notion of a linear amplifier.

In practice, however, no such thing as a linear amplifier exists. The output of any physical device is always limited on the upper end of the curve by such considerations as power limits or destruction of the device. If the device is not inherently constrained to operate below such levels by its own gain characteristic, clamping circuits are commonly inserted to prevent it from entering such a region. On the lower end of the characteristic there is always some input signal level below which no amplification occurs. This may lie in the thermal noise level in some devices, but normally there exists an "active" region below which the input of the device can be quiescently biased.

The term "amplification" generally means power gain greater than unity somewhere along the characteristic. Is this really necessary? We can certainly conceive of a device with the necessary shape to its characteristic, which, however, only attenuates. Is such a device a candidate for decision element mechanization? The answer is a qualified no. Because a decision element output must drive the inputs to other elements, power gain must be provided somewhere in the system. It is convenient in most cases to furnish that gain in the box f_2. Some notable exceptions exist, for example, in core memories, in which individual cores exhibit decision characteristics and attenuation. Gain in this case is provided on a shared basis by the read-amplifier-write-amplifier loop.

The job of the mapping operation f_1 is to convert input configurations to some set of values in a signal space such that a single decision level will separate the configurations into two categories, according to the requirements of the truth table. This can always be done, provided f_1 is allowed the greatest possible generality. However, certain types of function, particularly those involving products or their equivalent, are expensive to mechanize.

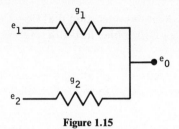

Figure 1.15

Therefore the practice has arisen of restricting f_1 to be the sum of simple functions of the input variables; that is, f_1 is of the form

$$f_1 = g_0(x_0) + g_1(x_1) + g_2(x_2) + \cdots + g_N(x_N). \tag{1.3}$$

Two types of function will be considered in this text. The first type is the linear mapping operation, in which $g(x_i) = a_i x_i$, the coefficient a_i being a constant. The second type is the exponential diode function, in which

$$g(x_i) = I_0(e^{x_i/E_T} - 1), \tag{1.4}$$

where

$\quad g(x_i) =$ current through the ith diode, in amperes,
$\quad\quad I_0 =$ reverse current, in amperes,
$\quad\quad x_i =$ voltage across the ith diode, in volts,
$\quad\quad E_T = kT/q$, in volts,
$\quad\quad\; k =$ Boltzmann's constant, 8.6×10^{-5} eV/°K,
$\quad\quad\; T =$ temperature, in degrees Kelvin,
$\quad\quad\; q =$ electronic charge, 1.6×10^{-19} C.

Consider first the linear case. Assume the coefficients of the input variables are equal and correspond to some conductance g. The circuit is shown in

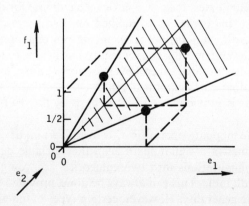

Figure 1.16

Figure 1.15. In this case,

$$f_1 = e_0 = \frac{g_1 e_1 + g_2 e_2}{g_1 + g_2}$$

$$= \tfrac{1}{2}(e_1 + e_2) \qquad \text{if } g_1 = g_2 = g. \tag{1.5}$$

The plot of f_1 as a function of e_1 and e_2 is shown in Figure 1.16. It can be seen that this function maps input configurations as follows:

$$(0, 0) \rightarrow 0$$

$$(0, 1) \text{ and } (1, 0) \rightarrow \tfrac{1}{2}$$

$$(1, 1) \rightarrow 1.$$

Because two input configurations are mapped into a single point, the decision operation which follows f_1 cannot distinguish between them. The only Boolean functions which can thereby be derived are (not including the null functions) $x_1 + x_2$, and $x_1 \cdot x_2$, depending on whether the decision level is placed between 0 and $\tfrac{1}{2}$ or between $\tfrac{1}{2}$ and 1.

By letting $g_1 \neq g_2$, we can tilt the plane of Figure 1.16 in any manner consistent with positive conductances. It is convenient at this point to shift to a two-dimensional representation of the input configurations and plot lines of constant f_1, defined by

$$f_1 = \left(\frac{g_1}{g_1 + g_2}\right) e_1 + \left(\frac{g_2}{g_1 + g_2}\right) e_2 + E_d$$

$$= a_0 + a_1 e_1 + a_2 e_2, \tag{1.6}$$

in which E_d may be regarded as the input level at which the decision operation changes state. Solving for one of the variables, say e_2, and letting $f_1 = 0$ be the value at which the decision changes,

$$e_2 = -\frac{a_0}{a_2} - \frac{a_1}{a_2} e_1. \tag{1.7}$$

Therefore lines of constant $f_1 = 0$, at which the decision element changes state, must have negative slopes on the (e_1, e_2) plane, provided $g > 0$, as shown in Figure 1.17. This restriction may be overcome by employing other types of summing devices. For example, in an ideal multiple-winding transformer, the applied ampere-turns must sum to zero, or

$$\sum_i N_i I_i = 0. \tag{1.8}$$

Because N can be either positive or negative (reversed winding), plots such

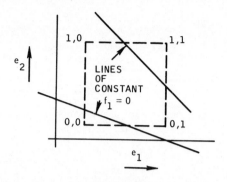

Figure 1.17

as Figure 1.18 can be obtained, from which the Boolean function $x_2 \cdot \bar{x}_1$ could be mechanized by an appropriate decision operation.

With three inputs, surfaces of constant sum become planes in three-dimensional space. With both positive and negative coefficients available, functions such as $(i_1 + i_2)\bar{i}_3$ can be mechanized, as illustrated in Figure 1.19.

The above illustrations indicate the effect of restricting the mapping operation f_1 in the decision element to be a sum of simple functions. Referring to the list of three-input function classes in Figure 1.12, we see that only certain functions can be obtained by passing a single plane through the function space in such a way that the 1 configurations are separated from the 0 configurations. For example, there is no way of passing a plane through function 4 in such a way that the 1 and 0 vertices are separated. The same holds true for functions 5, 7, 8, 10, 12, 13, and 14. Such functions must be obtained by cascading single-plane decision elements.

For the case in which the mapping operation is the sum of exponential terms, as in (1.4), it is convenient to approximate the diode characteristic by

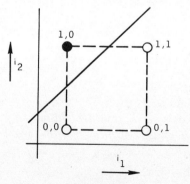

Figure 1.18

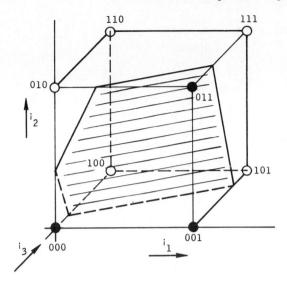

Figure 1.19

the perfect diode (Figure 1.20), based on voltage swings much larger than E_T. Thus the conductance $g \to \infty$ for positive voltage, and $g \to 0$ for negative voltage. Lines of constant mapping-function values e_0, for the two configurations of Figure 1.21a and b, are therefore plotted in Figure 1.22a and b, respectively. Two observations are pertinent. First, all of the functions $g_i(x_i)$ of (1.3) are identical (assuming identical diodes). Second, all conductances are positive. Therefore only one function class can be obtained, the AND and OR being members of the same class. Of the function classes of three variables represented in Figure 1.12, for example, only type 2 can be mechanized with a single decision element of the diode type. (Classes 3 and 9, which

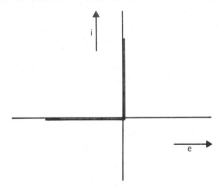

Figure 1.20

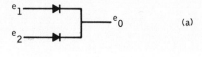

(a)

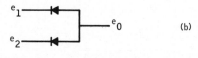

(b)

Figure 1.21

can also be mechanized, are functions of two variables and one variable, respectively).

The introduction of diode nonlinearity in the mapping operation, however, accomplishes two very useful and important results. The first is that it is now possible to cascade mapping operations without intervening decision operations. This is not the case with linear mapping because if one of the variables is itself a linear function of other variables, it can simply be replaced by those variables with appropriately scaled coefficients. We know, in fact, that any function whatever can be mechanized in two levels of diode mapping functions, provided both inputs and their complements are available. This mechanization corresponds to the sum-of-products (AND-OR) or product-of-sums (OR-AND) forms of logic representation. With linear mapping functions two layers of complete decision elements (including the decision operation) are required (for example, by letting the layers be AND's and OR's, respectively). We see, therefore, that the nonlinearity of the diode can be employed to obtain the effects of a passive decision operation. This fact is frequently used to great advantage, particularly in those cases in which component economy is of paramount importance. Conversely, in some cases, such as direct coupled transistor logic, active components are connected to provide both nonlinear mapping and decision operations.

The second beneficial effect of nonlinear mapping is the reduction of signal attenuation. Comparison of Figures 1.18 and 1.22 shows that the decision plane(s) corresponding to a diode mapping function can be placed in all cases equidistant between points in input signal space, whereas the decision planes corresponding to linear mapping more closely approach points in signal space with increasing N. If signal levels are 0 and 1 and all variables can be in error by an equal maximum amount Δ, the worst-case limit is $\Delta \leq 1/2N$ for an N-input AND mechanized with ideal (zero-tolerance) linear-mapping components. With perfect diode mapping, $\Delta \leq \frac{1}{2}$, independent of N.

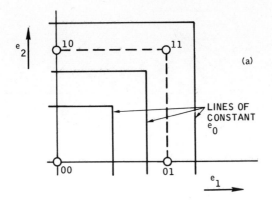

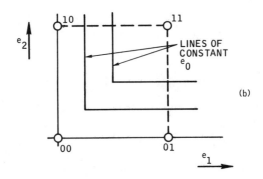

Figure 1.22

These advantages of diode mapping are, in general, paid for in terms of efficiency. Each element employing diode mapping can only mechanize one minterm of a function, and we may therefore look upon these circuits as ones which pick up each minterm of a multiminterm function individually. This restriction does not hold for linear mapping. For example, the structure of Figure 1.23 can mechanize any function of up to $N = 4$, if the boxes contain linear-input decision elements. For diode mapping, the worst function of four inputs is the mod-2 sum, $f = x_1 \oplus x_2 \oplus x_3 \oplus x_4$. This requires 40 diodes in a two-level AND-OR, or 30 diodes in a four-level structure of the form $(x_1 \oplus x_2) \oplus (x_3 \oplus x_4)$. The comparison is not so unfavorable, however, when the types of functions normally occurring in computing systems are considered, together with the limitation on N for linear mapping. We shall see that linear mapping appears further disadvantageous when decision element switching times are considered. Because of the complexity of these trade-offs in circuit design, the minimization problem is by no means

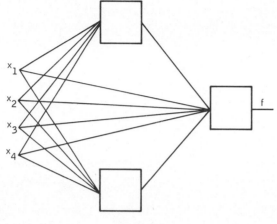

Figure 1.23

solved for the general case. Standard examples of both types of mapping function will be considered in this text.

Memory Component Properties

The electronic components suitable for realizing memory element circuit requirements must exhibit two major properties: (a) amplification, with or without time delay or energy storage properties, and (b) energy storage, either transient or permanent. A variety of components meet these require-ments. Included are components with amplification based upon the principle of charge control, such as vacuum tubes and transistors, and components whose amplification is based upon variable-impedance effects, such as inductor and capacitor parametric amplifiers. Energy storage components may exhibit permanent hysteresis; that is, they may be two-valued and there-fore retain a permanent effect after removal of the input causing that effect, as in ferromagnetic and ferroelectric materials. The magnetic-core shift register is an example of the use of permanent hysteresis. Some components can also be made to exhibit amplification, but they are based on hysteresis for their storage properties. This is the case with the magnetic shift register. Energy may also be stored, as a time function, in linear single-valued com-ponents. Examples that have been employed in digital computers include capacitor storage (both individual capacitors and bulk capacitor), sonic delay lines (e.g., mercury, quartz), and magnetostrictive delay lines.

In general, because energy is required to detect the state of a memory element and components exhibiting hysteresis alone cannot indefinitely supply energy, some form of amplification is required. Compatibility with decision element signal levels is also necessary. Therefore in this text complete

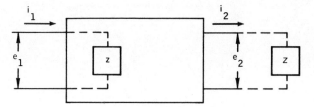

Figure 1.24

memory elements will be regarded as including all necessary amplification and will be assumed to be furnished with some source of power in the appropriate form. Such power is commonly termed *holding power*, implying that its presence is necessary for the element to retain its state *and* supply an indicative output signal. Removal of the holding power will result in either the loss of stored information or the absence of an output signal, or both. Those portions of a computer in which amplification and holding power (readout drive) are shared among a number of components exhibiting memory properties will be referred to as the *storage* of the machine, and the individual components will be referred to as *storage elements.*

When amplification, readout, and a power source are made inherent properties of memory elements, the required behavior of the four-terminal network representing the element may be specified in a number of useful and illuminating ways, without regard to the particular type of component involved. First of all, it must be possible to arrange the circuit into some equivalent of the open-loop four-terminal network of Figure 1.24. In order for the four-terminal equivalent to perform as a memory element it must exhibit open-loop voltage or current gain over some portion of the operating region when loaded by the same impedance one sees by looking into the input terminals; that is, the network must display power gain. For the network to be useful in a real computer system, the gain requirement is more stringent. For the memory element to drive n decision elements (fan-out of n), the circuit must exhibit gain when the load impedance is $1/n$ times that seen by looking into the input terminals. This simply means that the memory element must be capable of driving n other memory elements when the attenuation (or gain) of intervening decision elements (usually, but not necessarily, passive networks) is taken into account.

It is common practice to depict the open-loop gain requirement of memory elements by means of the normalized S-shaped transfer characteristic illustrated in Figure 1.25. The power gain relationship G between output and input power is

$$G = \frac{P_2}{P_1} = \frac{e_2 i_2}{e_1 i_1}.$$

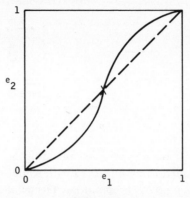

Figure 1.25

For a fan-out of n, with equal input impedances everywhere, the currents required to drive all elements into similar states are equal, and $i_2 = ni_1$. Then, with G_v = voltage gain,

$$G = n\left(\frac{e_2}{e_1}\right) = nG_v.$$

In the normalized amplifier input-output characteristic illustrated in Figure 1.25, e_2/e_1 (or $\Delta e_2/\Delta e_1$) is greater than 1 over some portion of the curve. If the loop is now closed by connecting output to input without attenuation, forcing $e_2 = e_1$, it can be seen from Figure 1.25 that only two stable points on the normalized transfer characteristic satisfy such a condition, namely, $e_1 = e_2 = 0$ and $e_1 = e_2 = 1$. The third point at which $e_2 = e_1$, intermediate between the others, is unstable; that is, a small change in e_1 causes a larger change in e_2, which, under closed-loop conditions of $e_1 = e_2$ or $\Delta e_1 = \Delta e_2$ causes a further change in e_1, etc.

   In general, signal levels associated with the input and output of a particular gain-exhibiting component will not be the same. Therefore, consider the device with the arbitrary input-output characteristic illustrated in Figure 1.26. It is desired to arrange this device with positive feedback to exhibit two-stable-state behavior. The standard arrangement is illustrated in Figure 1.27, in which e_1, e_2, and x are regarded as voltages and g_1, g_2, and g_x are regarded as conductances. From Figure 1.27 we can write

$$x = \frac{e_1 g_1 + e_2 g_2}{g_1 + g_2 + g_x}$$

$$= \gamma e_1 + \beta e_2$$

$$= \gamma e_1 + \beta f(x), \tag{1.9}$$

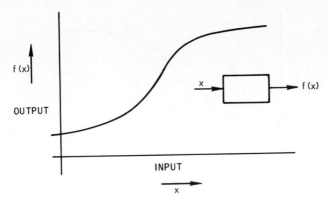

Figure 1.26

in which

$$\gamma \equiv \frac{g_1}{g_1 + g_2 + g_x}$$

$$\beta \equiv \frac{g_2}{g_1 + g_2 + g_x}.$$

Solving (1.9) for $f(x)$,

$$f(x) = \frac{x - \gamma e_1}{\beta} = h(x) \qquad (1.10)$$

If we now plot values of $h(x)$ versus x on the characteristic of Figure 1.26, the intersections $f(x) = h(x)$ constitute solutions to the equations. The $h(x)$ functions are families of lines having slopes $1/\beta$ and intersecting the x axis at values of $x = \gamma e_1$. The criterion for two-state stability is now apparent. The slope $1/\beta$ must be less than the maximum slope of $f(x)$ in order that some member of the $h(x)$ family (corresponding to some value of e_1) will intersect the characteristic at more than one point; that is,

$$\frac{1}{\beta} < f'(x)_{\text{max}} = \frac{df(x)}{dx}\bigg|_{\text{max}} \qquad (1.11)$$

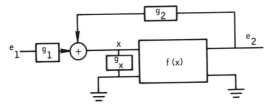

Figure 1.27

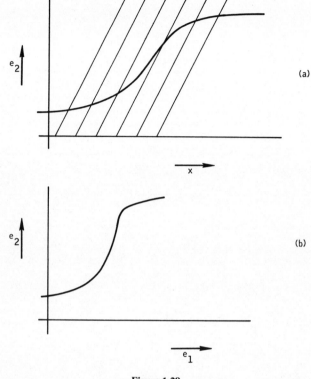

Figure 1.28

or

$$\beta > \frac{1}{f'(x)_{\text{max}}}.$$

A case in which $\beta < 1/f'(x)_{\text{max}}$ is constructed and replotted in Figure 1.28a and b. The case in which $\beta > 1/f'(x)_{\text{max}}$ is similarly plotted in Figure 1.29a and b.

Another common way of stating the criteria for bistable action is that the device must exhibit a negative-resistance region somewhere in its characteristic. This statement can be related to the foregoing manner of representation in the following way. For simplicity, let $x = e_1$, as in Figure 1.30. Then for an infinite input impedance amplifier,

$$i_1 = \beta(x - e_2)$$
$$= \beta[x - f(x)]. \tag{1.12}$$

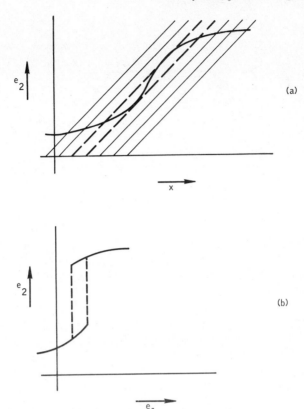

(a)

(b)

Figure 1.29

Taking the derivative and letting the dynamic input resistance $R_1 = de_1/di_1$,

$$R_1 = \frac{de_1}{di_1} = \frac{1}{\beta[1 - f'(x)]}. \tag{1.13}$$

Assuming $f'(x) > 0$ everywhere, as in Figure 1.26, R_1 becomes negative

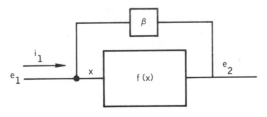

Figure 1.30

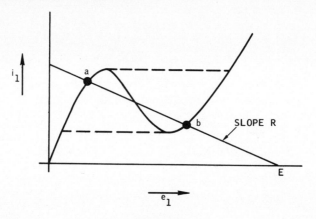

Figure 1.31

wherever $f'(x) > 1$. The result is the typical plot of Figure 1.31. If the input e_1 is replaced by a voltage source E and a series resistor R, then $e_1 = E - i_1R$, and two stable states a and b may be obtained. The resistance R is now equivalent to the conductance ratio γ of the preceding discussion. This is simply, therefore, an equivalent way of saying that provided $f'(x)_{max} > 1$, a value of β and values of E and R (or γe_1) can be found such that the device is bistable.

The feedback plots of Figures 1.28 and 1.29 yield some insight into the response time that can be expected from a given device. Assume for simplicity that β is a unit-time-delay element, and that all other circuit components act instantaneously. At time $t = 0$, apply a fixed input, say $e_1 = E_1$. Then $x = \gamma e_1 = \gamma E_1 = a$, on Figure 1.32, and $e_2 = f(\gamma E_1) = f(a)$. At time $t = 1$, x becomes $\gamma E_1 + \beta f(\gamma E_1) = a + \beta f(a) = b$. This point is obtained simply by projecting the point $f(a)$ horizontally to intersect $h(x) = (x + \gamma E_1)/\beta$,

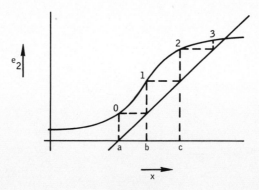

Figure 1.32

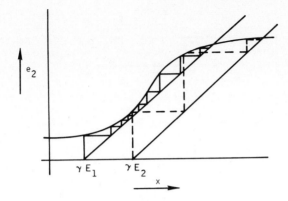

Figure 1.33

because the slope of $h(x)$ is $1/\beta$. The corresponding output at time $t = 1$ is $e_2 = f(b)$, as shown on Figure 1.32. The system thus converges to its final value by a series of steps. Although an infinite time is theoretically required for convergence, for all practical purposes stability is reached when the steps become sufficiently small. This representation indicates qualitatively the importance of overdrive in obtaining fast switching times, a problem we will discuss in more detail in connection with specific devices. However, it can be seen in Figure 1.33 that a signal E_1 just sufficient to switch the circuit to its opposite state may require a large number of steps (delay times) to

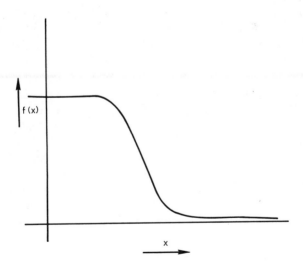

Figure 1.34

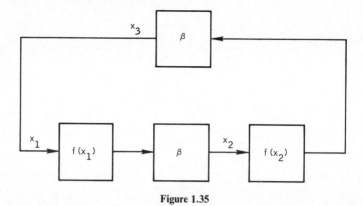

Figure 1.35

traverse portions of the characteristic, whereas a larger signal E_2 can force completion of the transient in a few steps.

These difficulties in dealing with the response times of digital circuits are compounded when we note that the situation depicted in Figures 1.26 and 1.27 is seldom so simple in practice. More commonly, we are dealing with basic components that inherently invert, or furnish negative gain as illustrated in Figure 1.34, rather than furnishing the positive gain implicit in the foregoing discussions. From a system standpoint this is indeed a fortunate circumstance, because some version of the NOT function, or signal negation, is necessary if components are to be universal in the sense of being capable of mechanizing any logic function whatever. In obtaining positive feedback for the construction of memory elements, the analysis is complicated by the necessity of cascading two such components as illustrated in Figure 1.35 in which identical components $f(x)$ and coupling networks β are assumed. The problem of bistability is now as follows. There must be two values of, say, x_1 such that $\beta[f(x_1)] = x_2 \neq x_1$ and $\beta[f(x_2)] = x_3 = x_1$, completing the loop. This is illustrated by means of the graphical device of Figure 1.36 in which $\beta[f(x_1)]$ versus x_1 is plotted on the right and $\beta[f(x_2)]$ versus $x_2 = \beta[f(x_1)]$ is plotted on the left so as to make the x_2 and $\beta[f(x_1)]$ scales coincide. Note that aside from the 90° rotation the two curves are identical (for identical components and coupling networks). The stability criterion is that the distances a and b be equal. This can be seen more readily if we rotate the left half of Figure 1.36 about the vertical axis, producing the equivalent plot of Figure 1.37. The stable points are now readily apparent, namely a and b, whereas c represents a point which satisfies $x_1 = x_3$ but is dynamically unstable. The coupling network can generally be reduced to a bias plus linear attenuation, in which case we can let

$$x_2 = x_0 + \beta f(x_1),$$

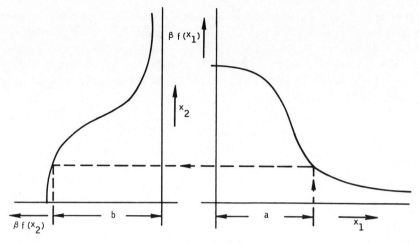

Figure 1.36

in which x_0 and β can be arbitrarily selected (with $\beta < 1$ in practice). Varying the values of x_0 and β corresponds to translating without scale change and changing scale without translating, respectively, for the two curves of Figure 1.37 with respect to their appropriate coordinates.

The foregoing represents a graphical approach to the solution of the circuit equations of a bistable element, and we could, of course, equally well attack

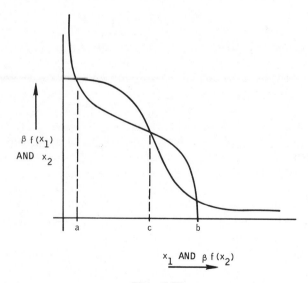

Figure 1.37

the equations analytically. In this case the question is, if we open up the loop between x_3 and x_1 in Figure 1.35, does a change in x_1 produce a greater change in x_3? If so, the circuit is unstable until an operating point is found at which this is no longer true. In order to find dx_3/dx_1 we can simply differentiate the functions representing the separate elements in the network and multiply results; that is,

$$\frac{dx_2}{dx_1} = \frac{dx_2}{d[f(x_1)]} \cdot \frac{d[f(x_1)]}{dx_1} = \beta f'(x_1).$$

Completing the loop, the criterion $dx_3/dx_1 > 1$ is therefore met if

$$\beta^2 [f'(x)]^2 > 1,$$

or simply

$$|f'(x)| > \frac{1}{\beta}$$

somewhere in the characteristic. Of course, this is the same gain criterion as before, with the exception that we have specifically allowed for either positive- or negative-gain components.

So far we have considered only requirements for a single memory element, including some slight attention to time-dependent behavior. It is worthwhile even at this early point, however, to consider some of the constraints imposed upon component behavior by general-purpose systems consisting of more than one element, particularly with respect to timing problems. It turns out, as we shall see in later chapters, that component behavior and system requirements impose certain constraints on the interconnection of elements. This makes it possible to regard most systems as operating in a two-phase or cycle mode; that is, elements may be divided into two (or more) groups. During one time period, say P_1, one group of elements is remaining (nearly) static, while the other group is in the process of storing energy (switching). During the subsequent time period P_2 the roles of the two groups are reversed. The two groups of elements are not always (in fact, are seldom) of the same kind. Thus one group may be active (include gain) while the other may be entirely passive. Various degrees of dissimilarity of the two groups of elements are illustrated in Figure 1.38. In case (a) active memory flip-flops drive only passive memory elements, commonly capacitive or inductive energy storage components. Connections are not allowed between the outputs of one flip-flop and the inputs of another. System operation may be considered to be as follows. During P_1 all flip-flops change state, according to their logic rules, with little change in the energy stored in passive devices. During P_2 flip-flops remain fixed, while passive elements charge (or discharge) to the appropriate energy state, again according to logic rules. Thus logic is performed on each half cycle. In case (b) active amplifiers replace the passive components of case (a). Energy storage in this case can be dependent solely upon the parameters of the active elements. Again, flip-flops are only allowed to drive amplifiers, and vice versa. Both of the above cases are unsymmetrical in the sense that, as usually arranged, an element of one kind drives only one element of the other kind, on one or other half cycle. In case (c) amplifiers have been converted to bistable elements. The elements are nevertheless still divided into two groups (1) and (2), with interconnections allowed only between opposite types. The symmetry may remove some constraints, however, permitting multiple logic connections at either level.

Some components, particularly those such as transistors operating down to dc, appear to fall more naturally into use in systems such as cases (a) and

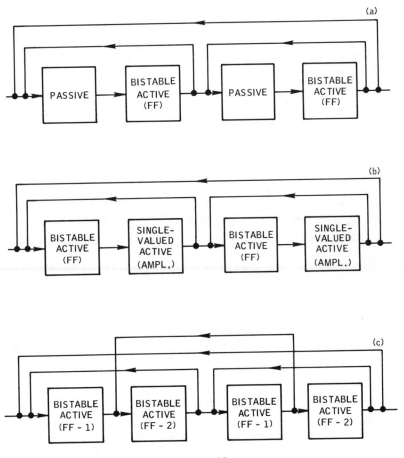

Figure 1.38

(b). Other components, such as magnetic amplifiers requiring some form of ac carrier power source, can only be operated in systems like case (c). Even though they do not differ greatly in basic principles, circuit design of the various types of system calls for somewhat different design approaches. All three are considered in this text.

REFERENCES

[1] M. Phister, Jr., *Logical Design of Digital Computers*, Wiley, New York, 1958.
[2] S. H. Caldwell, *Switching Circuits and Logical Design*, Wiley, New York, 1958.
[3] W. H. Ware, *Digital Computer Technology and Design*, Wiley, New York, 1963.
[4] H. J. Gray, *Digital Computer Engineering*, Prentice-Hall, Englewood Cliffs, N.J., 1963.
[5] H. W. Bode, *Network Analysis and Feedback Amplifier Design*, Van Nostrand, Princeton, N.J., 1945.
[6] M. E. Van Valkenberg, *Introduction to Modern Network Synthesis*, Wiley, New York, 1960.
[7] R. K. Richards, *Digital Computer Components and Circuits*, Van Nostrand, Princeton, N.J., 1957.
[8] J. Millman and H. Taub, *Pulse and Digital Circuits*, McGraw-Hill, New York, 1956.
[9] R. B. Hurley, *Transistor Logic Circuits*, Wiley, New York, 1961.
[10] A. I. Pressman, *Design of Transistorized Circuits for Digital Computers*, Rider, New York, 1959.
[11] A. W. Lo, "Some Thoughts on Digital Components and Circuit Techniques," *IRE Trans. Electron. Computers*, **EC-10**, 3, 416–425 (September 1961).
[12] C. A. Renton and B. Rabinovici, "Composite Characteristics of Negative Resistance Devices and Their Application in Digital Circuits," *Proc. IRE*, **50**, 7, 1648–1656 (July 1962).
[13] J. L. Haynes, "Logic Circuits Using Square-Loop Magnetic Devices: A Survey," *IRE Trans. Electron Computers*, **EC-10**, 2, 191–203 (June 1961).
[14] W. H. Pierce, "Predicting Signal Deterioration and Gate Compatibility in Logic Circuits," *IEEE Trans Electron. Computers*, **EC-12**, 3, 277–281 (June 1963).

EXERCISES

1.1 Which functions of one input variable can be mechanized by a single decision element consisting of a resistor current-summing mapping operation and a negative-gain decision operation amplifier if input variables only (not their complements) are available? If both input variables and their complements are available?

1.2 Same as Problem 1.1 for two input variables.

1.3 Same as Problem 1.1 for three input variables.

1.4 If the mapping operation of Problem 1.1 consists of a diode current-summing network (Equations 1.3 and 1.4), what functions of one variable can be mechanized? Of two? Of three?

1.5 Suppose a mapping operation is of the algebraic form,

$$f_1 = a_0 + (a_{11} + a_{12}x_1)(a_{21} + a_{22}x_2)(a_{31} + a_{32}x_3)\ldots.$$

What functions of one input variable can be mechanized if the decision operation is a perfect (infinite-gain) negative amplifier? Of two input variables? Of three?

1.6 If both input variables and their complements are available and the decision operation provides both the function output and its complement, which classes of three variable

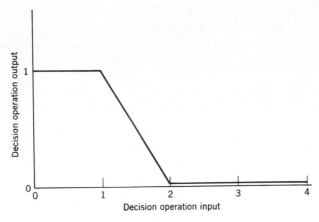

Figure P.1.1

functions (Figure 1.12) can be mechanized by a single linear-input decision element of the type described in Problem 1.1?

1.7 Same as Problem 1.6 by two such decision elements in cascade, if the output of one can be an input to the other?

1.8 If the decision element of Problem 1.6 is a diode-input type as described in Problem 1.4, which classes of three-variable function can be mechanized by a single decision element? By two such decision elements? By three?

1.9 Assume a decision-operation amplifier characteristic as in Figure P.1.1; input variables and their complements are available. Can we mechanize the OR function of two input variables with a linear-mapping operation? The AND function?

1.10 Same as Problem 1.9 for three input variables.

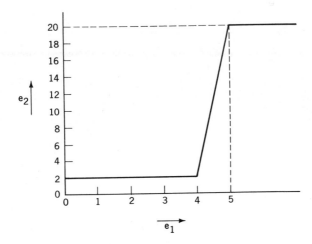

Figure P.1.2

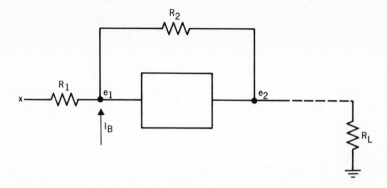

Figure P.1.3

1.11 If the mapping operation of Problems 1.9 and 1.10 is composed of ideal diodes rather than linear, can the functions be mechanized with the same decision amplifier (Figure P.1.1)?

1.12 If input variables and their complements are available, what is the minimum value of decision-amplifier gain f_2' in terms of the number of input variables, N, in order to mechanize the worst-case function if linear mapping is employed? If ideal diode mapping is employed?

1.13 Given the positive-gain amplifier characteristic of Figure P.1.2, what is the minimum value of feedback coefficient β necessary to cause the circuit to display bistable behavior?

1.14 Using the same amplifier characteristic, find a feedback network of the form illustrated in Figure P.1.3 and having the following properties: with the input at some quiescent

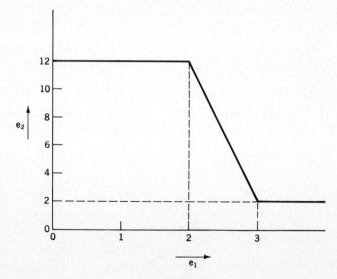

Figure P.1.4

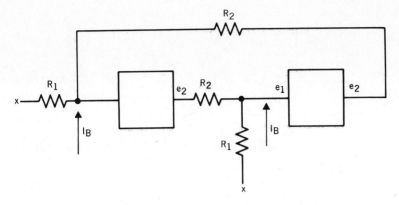

Figure P.1.5

value x_0 the device is bistable; a change in input of $\Delta x = 1$ forces the amplifier into its 1 state (high output); a change in input of $\Delta x = -1$ forces the amplifier into its 0 state (low output). Assume $x_0 = +11$ and $I_B = \pm 1$, as required, and infinite amplifier input impedance. R_L is not connected.

1.15 Suppose the characteristic of Figure P.1.2 were measured under open-circuit conditions, but it is known that the amplifier possesses an output impedance $R_g = 5$. How are the resistor values in the feedback network of Figure P.1.3 affected (R_L not connected)?

1.16 If R_L represents the load presented by the inputs to n similar devices in parallel, each of an impedance equal to R_1, write the equations which constrain the maximum value of the fan-out, n (other quantities given, as in Problem 1.14).

1.17 Given the negative-gain amplifier characteristic of Figure P.1.4, what is the minimum value of feedback coefficient β to cause two such amplifiers in cascade to exhibit bistable behavior? Assume identical coupling networks between each amplifier.

1.18 Using the amplifier characteristic of Figure P.1.4, find a feedback network of the form illustrated in Figure P.1.5 and having the following properties: a quiescent value of x is $x_0 = 7$; a change in either (but not both) x of ± 2 at the appropriate input will drive the circuit into its opposite state; $I_B = \pm 1$, as required. Amplifier input impedance is assumed infinite, and there is no external load.

Chapter 2

Diode and Transistor Equivalent Circuits

It is unusual for a single equivalent circuit of the nonlinear components with which digital designers must deal to serve satisfactorily all design purposes. The common practice is to employ a variety of equivalent circuits, each valid in a different operating region and each suited to a particular design question. This procedure tends to fragment the problem, however, and often fails to afford any unifying insight into the basic mechanisms of device behavior. We shall therefore attempt to build up a universal equivalent circuit for the most common component, the semiconductor transistor, employing conventional circuit elements to represent specific physical phenomena. From these, we shall see what simplifying assumptions can be made, and, correspondingly, which elements of the equivalent circuit can be safely removed, to reduce design complexity in special cases or in particular regions of operation.

The *p-n* Junction

The derivation of semiconductor junction equations rests upon three phenomena. One is *drift* of charge carriers under the influence of an electric field strength. This is simply Ohm's law, in which current and voltage obey the relationships

$$i_p = q\mu_p pE,$$
$$i_n = q\mu_n nE, \tag{2.1}$$

in which

$\qquad i$ = current, in amperes per unit area,
$\qquad q$ = electronic charge, in coulombs,

μ = mobility of charge carriers, in square centimeters per volt-second,

E = field strength, in volts per centimeter,

p, n = charge carrier density per unit volume.

Appropriate subscripts will be used throughout to indicate positive, p, or negative, n, carriers.

The second effect is that of *diffusion*, obeying the one-dimensional relationships

$$i_p = -qD_p \frac{dp}{dx},$$

$$i_n = qD_n \frac{dn}{dx}, \quad (2.2)$$

in which

D = diffusion constant, in square centimeters per second,

x = distance, in centimeters.

Combining (2.1) and (2.2), we see that the total current due to individual charge carriers is

$$i_p = q\mu_p pE - qD_p \frac{dp}{dx},$$

$$i_n = q\mu_n nE + qD_n \frac{dn}{dx}. \quad (2.3)$$

Charge density distributions in the vicinity of a junction have been sketched by Middlebrook [3]; a simplified version appears in Figure 2.1. In order to obtain boundary conditions for (2.3) in terms of the charge density $p(x_n, v)$ with an applied voltage V, it can be shown that i_p and i_n are negligible relative to the other terms in (2.3). This simply means that net current is the small difference between two large effects operating upon the charge carriers in the region of the junction. Setting, say, $i_p = 0$ in (2.3) and integrating the field E across the junction to obtain the applied voltage V, after some manipulation the boundary conditions become

$$p(x_n) = p_n e^{V\mu_p/D_p},$$

$$p(\infty) = p_n, \quad (2.4)$$

in which p_n is the thermal equilibrium $(V = 0)$ value of hole density in the n-type region.

The third phenomenon is the *continuity relationship*, which may be described as specifying the conservation of charge; it means that two phenomena can affect the total charge contained within a small region of material. The first phenomenon is a differential current flow across the region; that is, the current densities entering the region and leaving the region are not equal. Thus, if $i(x) \neq i(x + \Delta x)$, a net charge buildup is occurring within the region. Its magnitude, in number of charges per unit time, must therefore be $(1/q)$ (di/dx). The second phenomenon occurring is the recombination of charge carriers. If the quantity of charge is not at its thermal equilibrium value, the charges will recombine at a rate proportional to the excess. Thus the change in the number of charge carriers per unit time is $(p_n - p)/\tau_p$, in which p_n represents thermal equilibrium charge density, τ_p is the recombination time constant for holes, and the negative sign indicates a loss of holes for $p > p_n$. Actually, the quantity $(p_n - p)/\tau_p$ corresponds to two processes proceeding simultaneously, namely, thermal generation of holes at a rate p_n/τ_p and recombination at a rate p/τ_p. Because the sum of these effects must represent the net change in charge density per unit time,

$$\frac{dp}{dt} = \frac{p_n - p}{\tau_p} - \frac{1}{q}\frac{di_p}{dx} \tag{2.5}$$

for a one-dimensional model. Diffusion effects (Equation 2.2) dominate the current flow outside the junction region. Differentiating (2.2) with respect to x and substituting in (2.5), we have

$$\frac{dp}{dt} = \frac{p_n - p}{\tau_p} + D_p\frac{dp^2}{dx^2}. \tag{2.6}$$

Finally, note that under dc conditions charge cannot accumulate indefinitely. Therefore dp/dt must be zero, or

$$\frac{d^2p}{dx^2} + \frac{p_n - p}{D_p\tau_p} = 0. \tag{2.7}$$

This is the basic equation for charge-density distribution and can readily be solved subject to the boundary conditions (2.4). Defining an effective *diffusion length* by $L_p \equiv \sqrt{D_p\tau_p}$, we obtain the result

$$(p - p_n) = p_n(e^{V\mu_p/D_p} - 1)\, e^{(x_n - x)/L_p}. \tag{2.8}$$

This gives the hole-density distribution from x_n to $+\infty$. An entirely analogous result can be obtained for electron densities in the p-type region from $-x_p$ to $-\infty$.

Knowing the charge densities, we can simply obtain the currents crossing into the junction region at x_n by use of the diffusion relationship (2.2).

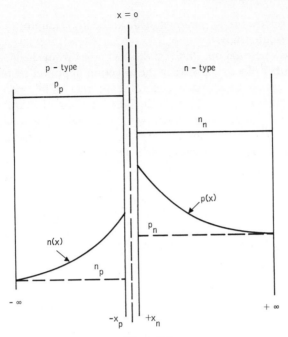

Figure 2.1

Another relationship, which we shall not attempt to show here, is $\mu/D = q/kT$. Taking the sum of $i_p(x_n)$ and $i_n(-x_p)$ and substituting for μ/D, we have

$$i = i_n + i_p = q\left(\frac{D_n n_p}{L_n} + \frac{D_p p_n}{L_p}\right)(e^{qV/kT} - 1). \qquad (2.9)$$

This is the *ideal diode* equation. For convenience, we define the constant term as I_s, the saturation or limiting current when V is a large negative quantity. Also, we define a "threshold voltage" E_T by the relationship $E_T \equiv kT/q$. (The value of E_T is 0.026 V at 25°C.) Equation 2.9 may therefore be written

$$i = I_s(e^{V/E_T} - 1). \qquad (2.10)$$

Because ideal diodes will often be assumed in subsequent discussions, the specific characteristic of (2.10) will be given the conventional diode symbol shown in Figure 2.2*a*. Its characteristic is plotted in Figure 2.2*b*. The same symbol will, of course, be used for real diodes not possessing the characteristic of (2.10), particularly in later circuit discussions. Context will make the distinction clear. We shall also make a distinction, on occasion, between (2.10) and a perfectly square *V*-*i* characteristic, as illustrated in Figure 2.2*c*.

This will be referred to as a *perfect diode* and will be employed on occasion to simplify circuit analysis.

Moll [8] has pointed out that real diodes depart from the ideal characteristic in a number of ways, always in the direction of poorer rectification. The departures can be categorized according to region of operation. These regions are illustrated in Figure 2.3. Included in the real characteristic is an avalanche breakdown (Zener breakdown). Although the avalanche mode of operation is, in fact, a most important phenomenon, it will not be considered in developing an equivalent circuit at this time.

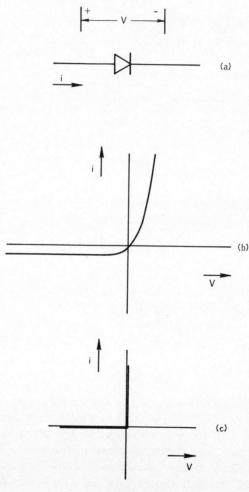

Figure 2.2

For moderate values of reverse bias, the actual current is larger than that predicted by (2.10). This is partly a result of surface leakage and partly a result of carrier generation by defects in the barrier region. This current may dominate the intrinsic carrier generation rate inherent in (2.10) when the intrinsic carrier density is small, as in silicon at room temperature and germanium at low temperatures. Because the current so generated is proportional to the volume in which generation occurs and because effective barrier width increases with reverse voltage, no reverse saturation of current occurs. These effects may be approximated over a wide range by assuming a linear resistor in parallel with the ideal diode.

At low forward voltages, the above effects may also predominate, leading to higher forward currents than predicted by (2.10). The regions marked *a* in Figure 2.3 indicate the effect. For intermediate forward bias, (2.10) is closely approximated, as indicated in region *b*. For higher forward bias, the assumption that minority carrier density is negligible compared to majority

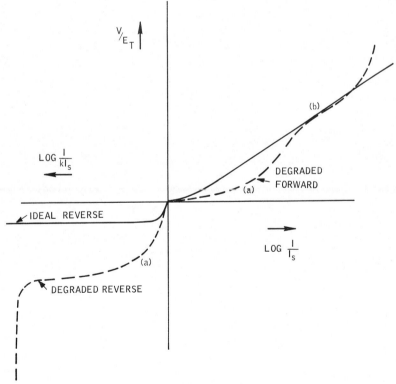

Figure 2.3

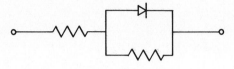

Figure 2.4

carrier density is no longer true. In addition, the assumption of zero electric field strength in the neutral region is no longer justified. Exact solutions for this mode of operation are difficult to obtain. However, the effect is to decrease current from the value predicted by (2.10).

The net result of the above considerations is that a reasonable dc equivalent circuit of the junction is given by an ideal junction, together with a series resistor and a parallel resistor, as shown in Figure 2.4. This circuit closely approximates the dc behavior of most diodes over a large range.

A good large-signal ac equivalent circuit for the diode must account for observed diode behavior under large-signal switching conditions. Two switching transients are of interest. The first, called *forward recovery*, has to do with the forward current that flows in a diode suddenly forward-biased after a long period of reverse bias. This transient is generally quite short, and will not be considered further. A more serious effect is *reverse recovery*, having to do with the behavior of a diode whose current is suddenly switched from the forward to the reverse direction. Figure 2.5 illustrates typical responses. The figure assumes that the diode is previously forward-biased by a constant current I_f that is the same in all cases. At $t = 0$, the current is suddenly reversed and instantaneously takes on the particular value I_r. The value of I_r is determined by the source voltage E and by some external resistance R. The initial drop across the diode is small compared to E, and

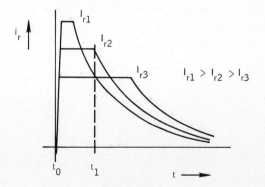

Figure 2.5

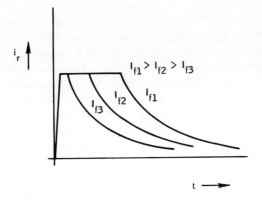

Figure 2.6

the source can therefore be considered a constant current I_r. After some period of time, during which little or no drop appears across the diode and $i_r = I_r$ remains constant, a voltage begins to build up at the diode. The diode drop is no longer negligible compared to E, and i_r falls toward an ideal final value of $-I_s$ at $t \to \infty$.

Waveforms of reverse current under different conditions are sketched in Figure 2.6, in which the forward current prior to $t = 0$ is varied while the reverse voltage and series resistance are maintained constant to produce the same initial value of reverse current. Note that i_r remains at the initial value $I_r \cong E/R$ for different durations of time, depending upon the value of I_f, before falling toward its quiescent state. The duration of the period of approximately constant reverse current increases as I_f increases.

This behavior can be explained if we focus attention on the minority carrier distribution in the diode and employ the same concepts used in deriving dc characteristics. A simplified model that makes the process easy to visualize is obtained if we imagine that carriers are confined to a region of width W very near the junction. This amounts to linearizing (2.8) by making $x \ll L_p$. Letting $e^y \cong 1 + y$ for $y \ll 1$ and taking starting coordinates at $x = x_n$, we see that the charge density above equilibrium prior to $t = 0$ varies with x as

$$p - p_n \cong p_n f(V) \left(1 - \frac{x}{W}\right). \tag{2.11}$$

Therefore the hole contribution to forward current is

$$i_p = -qD_p \frac{dp}{dx} = \frac{qD_p p_n}{W} f(V). \tag{2.12}$$

The carrier density distribution and forward current direction are sketched in Figure 2.7. At $t = 0$ the current is reversed. Because no charge can re-combine instantaneously, the distribution must rearrange itself to support this current in accordance with the diffusion equation (2.2). Because I_r is negative, dp/dx is proportional to I_r/qD_p. The case in which $|I_r| < |I_f|$ is illustrated in Figure 2.7. As time proceeds, recombination of stored charges occurs, but so long as $p(0) > 0$, I_r continues to flow because the charge-density gradient can be maintained, as indicated by dotted distributions in Figure 2.7. At $t = t_1$, $p(0) = 0$, and further recombination forces the charge-density gradient to decrease, thereby decreasing i_r, until equilibrium is

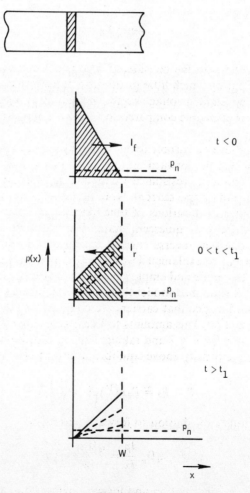

Figure 2.7

reached. For $t > t_1$, the reverse current is proportional to the total remaining charge, because for a linear distribution,

$$\frac{dp(t)}{dx} = \frac{p(w, t)}{W}$$

and

$$Q(t) = -\frac{qp(w, t)W}{2}. \tag{2.13}$$

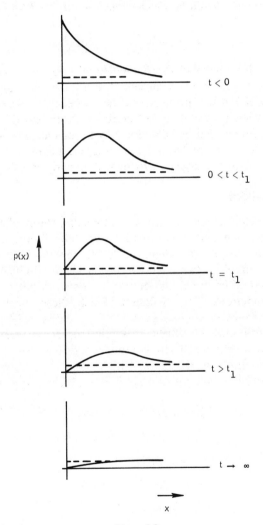

Figure 2.8

Therefore

$$i_r(t) = -qD_p \frac{dp}{dx} = \frac{2D_p}{W^2} Q(t), \qquad t > t_1. \qquad (2.14)$$

Note that (2.13) can also be used to find $Q(0)$ at $t < 0$, because total initial charge and I_f are similarly related. With the charge recombining at a rate proportional to its magnitude $Q(t)$ decays exponentially toward zero with a time constant τ_R, the recombination rate. With $Q(0)$ determined by I_f the time for the reverse current to reach some fraction $k < 1$ of its initial value is

$$T_R = \tau_R \ln \frac{I_f}{kI_r}, \qquad T_R > t_1. \qquad (2.15)$$

This highly simplified and approximate model does not, of course, describe all the details of the reverse transient. A more complete analysis has been developed by Ko [16], who sketches the sequence of charge distributions during the reverse transient, as in Figure 2.8. Note that the previous linear model approximates charge behavior only in regions close to the junction; that is, at $x = 0$ in Figure 2.8, charge-density gradient goes from negative to positive at $t = 0$, and charge density drops to nearly zero (p_n) at $t = t_1$.

The Transistor

Use of the ideal rectifier equations developed for the single junction allows the calculation of an ideal transistor equivalent circuit to proceed in a straightforward manner. We begin the development by noting that, because the transistor involves two junctions, each similar in principle to those previously considered, it will not be too surprising to find that the resulting currents are functions of two voltages, namely those appearing at the two junctions. The generalized p-n-p geometry considered by Ebers and Moll [5] is illustrated in Figure 2.9. In this case minority carrier density is made up of two positive and two negative components of the form of (2.9). First are the densities of holes in the base region (n-type) due, respectively, to the emitter and collector junctions. They are of the form (for an assumed planar junction),

$$p_{e1} = p_n(e^{Ve/E_T} - 1)f(x),$$

$$f(x) = 0 \text{ on } J_2 \qquad \text{(see Figure 2.9)},$$

$$= 1 \text{ on } J_1,$$

and

$$p_{e2} = p_n(e^{Vc/E_T} - 1)g(x),$$

$$g(x) = 1 \text{ on } J_2,$$

$$= 0 \text{ on } J_1. \qquad (2.16)$$

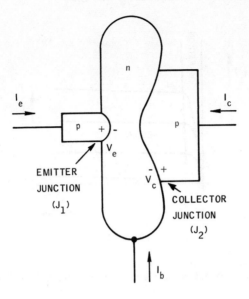

Figure 2.9

in which

V_e = voltage across emitter junction,
V_c = voltage across collector junction,
p_e = excess hole density due to emitter (1) and collector (2) junctions,
p_n = equilibrium density of holes in the base.

With both V_e and V_c applied the total hole density is $p_e = p_{e1} + p_{e2}$. This is also true of the electron densities in the two p-type regions.

The functions $f(x)$ and $g(x)$ are again determined by the continuity condition. Carrying out a development similar to that for the diode results in the *ideal-transistor* equations,

$$I_e = a_{11}(e^{V_e/E_T} - 1) + a_{12}(e^{V_c/E_T} - 1),$$

$$I_c = a_{21}(e^{V_e/E_T} - 1) + a_{22}(e^{V_c/E_T} - 1), \qquad (2.17)$$

in which the constants a_{ij} represent lumped material properties, and $a_{12} = a_{21}$.

In accordance with the Ebers-Moll treatment we shall consider three regions of operation, as illustrated in Figure 2.10. They are as follows:

1. Both emitter and collector junctions reverse-biased (cutoff region), with $e^{V/E_T} \ll 1$ for both junctions.

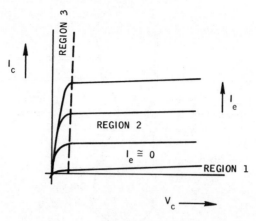

Figure 2.10

2. Emitter forward-biased, collector reverse-biased (active region), with $e^{V_c/E_T} \ll 1$.

3. Both junctions forward-biased (saturation region).

For region 2, (2.17) becomes approximately

$$I_e = a_{11}(e^{V_e/E_T} - 1) - a_{12},$$
$$I_c = a_{21}(e^{V_e/E_T} - 1) - a_{22}. \tag{2.18}$$

Eliminating the exponential term between the above two equations, we have

$$I_c = \frac{a_{21}}{a_{11}} I_e + \left(\frac{a_{12}a_{21}}{a_{11}} - a_{22} \right). \tag{2.19}$$

The constants above may be redefined by the equation

$$I_c = -\alpha_N I_e - I_{c0} \tag{2.20}$$

The quantity $\alpha_N \equiv -a_{21}/a_{11}$ is known as the *normal short-circuit current gain* and can also be defined by the measurable relationship $\alpha_N = -dI_c/dI_e$. The negative sign enters only because of the conventions for current flow chosen in Figure 2.9. Because conventions are arbitrary and in particular may be chosen differently for an *n-p-n* device, we shall commonly speak of α_N as a magnitude $\alpha_N \equiv |dI_c/dI_e|$. The quantity I_{c0} in (2.20) also corresponds to one of the measurable parameters of the transistor. By open-circuiting the emitter, we make $I_e = 0$ and $I_c \equiv -I_{cb0} = -I_{c0}$. The subscripts refer to the fact that this is the collector-to-base current measured under the condition of an open-circuit emitter. We can also obtain the value of collector current under the condition of an open-circuit base lead, $I_b = 0$, in which case $I_e = -I_c$. Substituting in (2.20) and solving, we have

$$I_c = -I_{ce0} = -\frac{I_{c0}}{1 - \alpha_N}. \tag{2.21}$$

Again, subscripts refer to the fact that this is collector-to-emitter current measured under the conditions of an open-circuit base.

When the transistor is reversed (emitter reverse-biased, collector forward-biased), the exponential term in V_e drops out of (2.17). The remaining terms in V_c can be eliminated between the two equations, as in (2.18) and (2.19). The result is

$$I_e = \frac{a_{12}}{a_{22}}I_c + \left(\frac{a_{12}a_{21}}{a_{22}} - a_{11}\right). \tag{2.22}$$

New constants may be similarly defined by writing the above as

$$I_e = -\alpha_I I_c - I_{e0}. \tag{2.23}$$

Again we may note that α_I, the inverse short-circuit current gain, is given by the measurable definition $\alpha_I \equiv |dI_e/dI_c|$. Care must be taken that the two current gains, normal and inverse, are measured under different operating conditions (collector or emitter reverse-biased, respectively). Emitter currents under various conditions corresponding to I_{ce0} and I_{cb0} could also be defined, but the results are quantities not frequently employed. Using (2.20) and (2.23), we find that the coefficients in (2.17) may be evaluated in terms of the more readily measured parameters α_N, α_I, I_{c0}, and I_{e0}. When this is done and the results are substituted in (2.17), the set of ideal-transistor equations becomes

$$I_e = \frac{I_{e0}}{1 - \alpha_N\alpha_I}(e^{V_e/E_T} - 1) - \frac{\alpha_I I_{c0}}{1 - \alpha_N\alpha_I}(e^{V_c/E_T} - 1),$$

$$I_c = -\frac{\alpha_N I_{e0}}{1 - \alpha_N\alpha_I}(e^{V_e/E_T} - 1) + \frac{I_{c0}}{1 - \alpha_N\alpha_I}(e^{V_c/E_T} - 1). \tag{2.24}$$

An additional useful result that can be derived from the definitions of the parameters α_N, α_I, I_{c0}, and I_{e0}, and from the relationship $a_{12} = a_{21}$ is that

$$\alpha_I I_{c0} = \alpha_N I_{e0}. \tag{2.25}$$

Note that (2.24) consists of two equations involving two exponential quantities. We can therefore solve for the exponential quantities in terms of the currents I_e and I_c. Alternatively, we can express the solution in the form of an implicit relationship that sheds a great deal of light on equivalent circuit behavior. Eliminating first the terms in V_c and then the terms in V_e in (2.24),

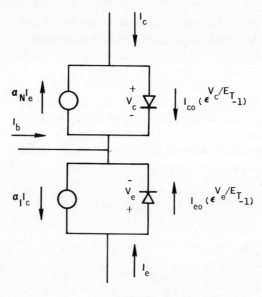

Figure 2.11

one can obtain

$$I_e + \alpha_I I_c - I_{e0}(e^{V_e/E_T} - 1) = 0,$$
$$I_c + \alpha_N I_e - I_{c0}(e^{V_c/E_T} - 1) = 0. \qquad (2.26)$$

These equations represent a single transistor equivalent circuit which is equally valid in all three regions of operation; that is, the first equation can

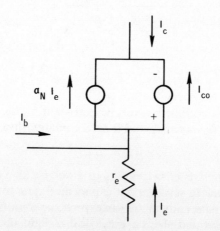

Figure 2.12

be taken to represent current summing at the emitter terminal, with I_e the external input, $\alpha_I I_c$ an equivalent current generator feeding into the terminal, and $I_{e0}(e^{V_e/E_T} - 1)$ representing the current leaving the terminal through an ideal diode. Similar observations apply to the collector terminal. Finally, it is necessary for charge conservation that $I_c + I_e + I_b = 0$. The resulting over-all equivalent circuit is shown in Figure 2.11. A similar model, with diodes and current generators reversed, can be obtained for the n-p-n transistor.

A variety of simplified equivalent circuits that are valid in the different modes of operation can be readily derived from (2.26) and Figure 2.11. For example, in region 2, the active region, the emitter is forward-biased, and the collector is reverse-biased. In this case

$$I_c \cong -\alpha_N I_e - I_{c0},$$
$$I_e = I_{e0}(e^{V_e/E_T} - 1) - \alpha_I I_c. \tag{2.27}$$

We define $r_e \equiv dV_e/dI_e$, the dynamic emitter resistance. To evaluate r_e, the quantity I_c can be eliminated from (2.27) and the result differentiated with respect to I_e. An equivalent circuit corresponding to (2.27) is given by Figure 2.12.

In many cases, input currents are derived from sources whose impedances are much larger than r_e. This means that, provided the emitter diode of Figure 2.11 is forward-biased, only small voltages may appear across it for reasonable current flows. Thus r_e (or V_e) may be neglected, the inputs being entirely determined by external sources I_b and I_e. Note further that, for moderate values of I_c in the active region, $|I_c| \gg I_{c0}$. Therefore, it is also reasonable to neglect the shunting diode across V_c. When these approximations are taken into account, the circuit of Figure 2.12 reduces to a simple current-generator equivalent.

The junction in the transistor is, of course, no different in principle from the junction in a diode and therefore suffers from the same defects that tend to degrade its ideal performance. Referring to the previous discussion of diodes and to Figure 2.4, we can see that the complete dc equivalent circuit of the transistor must be represented by something like Figure 2.13. The resistors r_{eL} and r_{cL} represent leakage resistances, whereas other values represent finite semiconductor bulk resistances. Unfortunately, even these parameters are not constants. The base resistance r_b in region 3, for example, may be a factor of 10 smaller than its value in the active region. This is a result of changes in the position of the effective point at which carriers are emitted from the emitter and of the presence of a source of base current at the collector. Although these effects are important in any precise analysis of circuit limitations, they are frequently of secondary importance in comparison

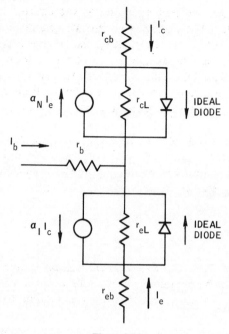

Figure 2.13

with more serious constraints that arise in switching behavior. Two ways of handling this problem are therefore common. For determining switching times it is usually sufficiently accurate to estimate maximum values of such quantities as base-emitter voltage or collector saturation voltage. When treating circuit behavior under dc conditions, it is usually more convenient to use published or measured curves of such quantities. Because we are primarily concerned with developing a general framework for circuit design, these effects will be treated as second-order corrections to the basic circuit constraints.

In completing the model of dc transistor behavior, it is worth noting the usefulness of an additional parameter. This is the forward-current transfer ratio h_{FE}, the transfer ratio between collector and base under the conditions of a forward-biased emitter. Thus $h_{FE} \equiv |dI_c/dI_b|$. This same quantity is also referred to as β or β_0, the dc common-emitter current gain. For convenience we shall retain the latter notation throughout the remainder of this text. The value of β follows from the ideal-transistor active region equations (2.27). Noting that $I_c + I_e + I_b = 0$, one obtains

$$I_c = \frac{\alpha_N}{1 - \alpha_N} I_b - \frac{I_{c0}}{(1 - \alpha_N)} = \beta I_b - I_{ce0}. \tag{2.28}$$

Because generally $\beta I_b \gg I_{ce0}$, a first approximation to the transistor in the active region is given by a current generator of value βI_b. Equivalent expressions can also be derived for inverse operation of the transistor. If a distinction needs to be made, normal and inverse parameters will be indicated by appropriate subscript, as β_N and β_I. If no subscript appears, β_N is implied.

It is unfortunate, in a way, that the convention has arisen of labeling the equivalent current generator in the collector junction $\alpha_N I_e$ and that in the emitter junction $\alpha_I I_c$. Note that the quantities α_N and α_I were defined arbitrarily in order to write (2.19) and (2.22) in the simpler forms of (2.20) and (2.23), respectively. A difficulty with this choice of nomenclature arises, however, when treating circuit behavior in region 3, in which both junctions are forward-biased. In this case it is more convenient to return to the original ideal-transistor equations (2.17) and identify quantities as follows:

$$I_{ef} \equiv a_{11}(e^{V_e/E_T} - 1) \equiv I'_{e0}(e^{V_e/E_T} - 1),$$

$$I_{cf} \equiv a_{22}(e^{V_c/E_T} - 1) \equiv I'_{c0}(e^{V_c/E_T} - 1),$$

$$a_{12} \equiv -\alpha_I I'_{c0},$$

$$a_{21} \equiv -\alpha_N I'_{e0}. \tag{2.29}$$

Substituting these quantities in the ideal-transistor equations, we may write the collector and emitter currents in terms of voltage functions, as

$$I_e = I'_{e0}(e^{V_e/E_T} - 1) - \alpha_I I'_{c0}(e^{V_c/E_T} - 1),$$

$$I_c = I'_{c0}(e^{V_c/E_T} - 1) - \alpha_N I'_{e0}(e^{V_e/E_T} - 1). \tag{2.30}$$

These equations may also be written in terms of currents only, as

$$I_c = I_{cf} - \alpha_N I_{ef},$$

$$I_e = I_{ef} - \alpha_I I_{cf}. \tag{2.31}$$

Physically, I_{cf} and I_{ef} can be interpreted as the forward currents emitted across the collector and emitter junctions, respectively, in accordance with the ideal-diode equation. The quantities α_N and α_I are the fractions of the forward currents collected at the collector and emitter, respectively. Note that we have not changed the basic definition of α_N and α_I in terms of the physical properties of the transistor. It is still true that $\alpha_N = -(a_{21}/a_{11})$, and $\alpha_I = -(a_{12}/a_{22})$. It is also still true that in the active region $\alpha_N \cong |dI_c/dI_e|$, although the route for arriving at this approximation is somewhat different. It is this difference in viewpoint that we shall find useful later on in studying the transient behavior of I_{cf}, the quantity of interest in the saturation region.

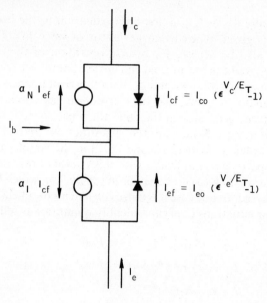

Figure 2.14

The resulting equivalent circuit amounts to a relabeling of the former model, as shown in Figure 2.14. Note that this model is also equally valid in all three regions of operation.

In determining transient behavior in the saturation region the quantity of interest is I_{cf}, because its direction determines whether the collector is forward- or reverse-biased. Solving (2.31) for this quantity, we have

$$I_{cf} = \frac{I_c + \alpha_N I_e}{1 - \alpha_N \alpha_I}. \tag{2.32}$$

This relationship can be used to obtain expressions for switching behavior in the saturation region.

Transistor ac Equivalent Circuit

There exist at least three main routes for obtaining useful ac equivalent circuits for transistors. One route assumes a single-pole frequency-dependent characteristic for the equivalent circuit current generators α_N and α_I. Derivation of switching transients then proceeds strictly in accordance with standard network analysis. The second route remains somewhat closer to the underlying physical phenomena by focusing attention on the behavior of charges in the base region of the transistor. It represents a viewpoint that is particularly useful under certain drive conditions. The third route is based on a lumped spatial approximation to the actual continuously variable behavior of minority carriers in the base region. It offers the potential advantage of being limited in accuracy only by the number of lumped elements employed in the approximation.

All three routes lead to identical first-order results, but they appeal to somewhat different backgrounds and physical intuition on the part of the reader. We shall, therefore, pursue all three approaches, attempting along the way to show their basic similarities.

The Ebers–Moll Model

In this section a particular form for the frequency dependence of $\alpha_N = \alpha_N(s)$ is assumed, in accordance with the classic treatment by Ebers and Moll [5]. In particular, let

$$\alpha_N(s) = \frac{\alpha_N}{s/\omega_N + 1},$$

$$\alpha_I(s) = \frac{\alpha_I}{s/\omega_I + 1}. \tag{2.33}$$

These are, of course, only approximations whose accuracy drops off as ω approaches ω_N or ω_I. Because of the diffusion mechanism, the form of the frequency response is, in reality, more like that of a transmission line. However, experience has shown that the approximations of (2.33) are satisfactory for good first-order prediction of switching behavior in the usual modes of operation, and that the added difficulty of more complex, though more precise, representations is not usually warranted.

In accordance with Moll's development and the above frequency dependence of α_N the complete equivalent circuit that can be considered for transistor active-region representation is given in Figure 2.15. Here junction capacitances are included in the emitter and collector impedances, which are

$$Z_c = \frac{r_c}{r_c C_c s + 1},$$

$$Z_e = \frac{r_e}{r_e C_e s + 1}, \tag{2.34}$$

in which r_c may be interpreted as the reverse junction leakage resistance r_{cL}, and r_e represents the current-dependent equivalent emitter resistance. Diode reverse currents are neglected. In order to simplify the switching-time equations, however, it is useful to assume that short-circuit operating

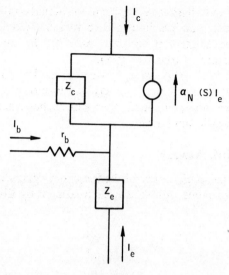

Figure 2.15

conditions hold. These conditions, with load resistance R_L, may be summarized for the common base as follows:

$$\frac{R_L + r_b}{r_c} \ll 1,$$

and

$$r_c C_c \omega_N \ll 1.$$

The active-region equivalent circuit therefore degenerates to that of Figure 2.16.

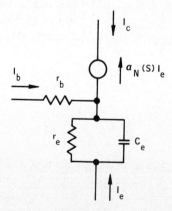

Figure 2.16

Based on this circuit, the transfer relationships with various inputs can be readily written. For simplicity of notation it is convenient to define $\tau_N \equiv 1/\omega_N$. Then the transfer function from emitter to collector is

$$\frac{I_c(s)}{I_e(s)} = -\alpha_N \frac{1}{\tau_N s + 1}. \tag{2.35}$$

In response to a step input $1/s$ of I_e,

$$i_c(t) = -\alpha_N I_e(1 - e^{-t/\tau_N}). \tag{2.36}$$

For the common-emitter connection, short-circuit operating conditions hold when

$$\frac{R_L}{r_c} \ll 1 - \alpha_N$$

and

$$\omega_N C_c r_c \ll 1.$$

Under these conditions

$$\frac{I_c(s)}{I_b(s)} = \frac{\alpha_N}{1 - \alpha_N + s/\omega_N}. \tag{2.37}$$

For a step input of I_b

$$i_c(t) = \frac{\alpha_N}{1 - \alpha_N} I_b(1 - e^{-\omega_N(1 - \alpha_N)t}). \tag{2.38}$$

To simplify notation again, we define

$$\tau_a \equiv \frac{1}{\omega_N(1 - \alpha_N)}.$$

The time constant τ_a will be referred to as the *active-region time-constant* (common-emitter connection implied). Note that by definition

$$\tau_a = \tau_N \frac{1}{1 - \alpha_N} \cong \beta \tau_N \tag{2.39}$$

for α_N near unity or $\beta \gg 1$. Using this definition, we can write (2.38) as

$$i_c(t) = \beta I_b(1 - e^{-t/\tau_a}). \tag{2.40}$$

The base-emitter voltage as a function of base-current drive is given by

$$\frac{V_{be}(s)}{I_b(s)} = r_b + \frac{r_e(\tau_N s + 1)}{\tau_N s + 1 - \alpha_N}. \tag{2.41}$$

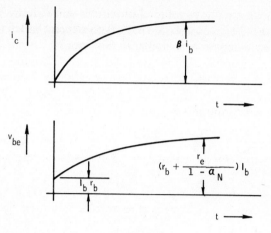

Figure 2.17

For a step input of base current this becomes

$$V_{be}(t) = I_b r_b + I_b \frac{r_e}{1 - \alpha_N}(1 - \alpha_N\, e^{-t/\tau_a}) \tag{2.42}$$

Waveforms for (2.40) and (2.42) are shown in Figure 2.17.

The same short-circuit conditions as before hold for the common collector. In this case

$$\frac{I_e(s)}{I_b(s)} = \frac{\tau_N s + 1}{\tau_N s + 1 - \alpha_N}. \tag{2.43}$$

For a step input

$$i_e(t) = I_b \frac{1}{1 - \alpha_N}(1 - \alpha_N\, e^{-\omega_N(1 - \alpha_N)t}). \tag{2.44}$$

The foregoing results for transient behavior have been confined to the active region; that is, the conditions imposed essentially imply that the transistor can be adequately represented by the equivalent circuit of Figure 2.16. In the saturation region, however, we must take account of the fact that the collector junction is forward-biased. Under this condition the collector is also injecting charge carriers into the base, and the equivalent current generator α_I is no longer inactive.

The simplest way of attacking this problem is to employ the modified equivalent circuit of Figure 2.14. From this representation it was possible to derive (2.32), which describes I_{cf} in terms of the currents into the transistor.

However, I_{cf} is precisely the quantity of interest in determining saturation-region switching times, because the sign of I_{cf} determines whether the transistor is in the saturation region or not; that is, the sign of I_{cf} is the same as the sign of the voltage V_c across the junction. To obtain a complex frequency equation for I_{cf} it is merely necessary to substitute complex frequency values for α_N and α_I from (2.33) and to take advantage of the fact that $I_c + I_e + I_b = 0$. Performing these substitutions, we obtain the equation

$$\frac{I_{cf}(s)}{I_b(s)} = \left\{ \frac{1 - \alpha_N/(1 + s/\omega_N)}{1 - \left[\dfrac{\alpha_N}{(1 + s/\omega_N)}\right]\left[\dfrac{\alpha_I}{(1 + s/\omega_I)}\right]} \right\} \frac{I_s}{I_b(s)}$$

$$- \left\{ \frac{\alpha_N/(1 + s/\omega_N)}{1 - \left[\dfrac{\alpha_N}{(1 + s/\omega_N)}\right]\left[\dfrac{\alpha_I}{(1 + s/\omega_I)}\right]} \right\}. \qquad (2.45)$$

It can be assumed that the external collector current, $I_c = I_s$, is a constant for the duration of the transient, because the transistor impedance is low compared to the external circuitry, and collector current is therefore determined by the supply voltage and load resistor. The characteristic equation from the denominator of (2.45) is

$$s^2 + (\omega_N + \omega_I)s + \omega_N\omega_I(1 - \alpha_N\alpha_I) = 0. \qquad (2.46)$$

Therefore, the roots may be written as

$$s = -\frac{\omega_N + \omega_I}{2}\left\{1 \pm \left[1 - \frac{4\omega_N\omega_I(1 - \alpha_N\alpha_I)}{(\omega_N + \omega_I)^2}\right]^{\frac{1}{2}}\right\}. \qquad (2.47)$$

The quantity in brackets in (2.47) has a maximum value when $\omega_N = \omega_I = \omega$. For α_N nearly unity the maximum value is $1 - \alpha_I$. Generally α_I is sufficiently large that $1 - \alpha_I \ll 1$. Therefore we can use the approximation that $\sqrt{1 + x} \cong 1 + x/2$. With this approximation the two roots of (2.46) become

$$S_1 \cong -(\omega_N + \omega_I)$$

and

$$S_2 \cong -\frac{\omega_N\omega_I(1 - \alpha_N\alpha_I)}{\omega_N + \omega_I}. \qquad (2.48)$$

Note that $S_2 \ll S_1$—their ratio is a maximum of approximately $(1 - \alpha_I)/4$ for $\omega_N = \omega_I$—and S_2 will therefore dominate the response. For reasons of simplicity and for reasons to be brought out later in connection with the charge-control model, we can thus treat the saturation-region transient as predominantly single-pole. The form of the transient is thereby determined, and it is only necessary to insert initial conditions to obtain a reasonable expression for the time behavior of $i_{cf}(t)$. For a step input of base current having the parameters

$$I_b = I_{b1}, \qquad t < 0,$$

$$I_b = I_{b2}, \qquad t > 0,$$

the single-pole form of (2.45) can be written in the time domain as

$$i_{cf} = (I_s - \beta I_{b1}) - \beta(I_{b2} - I_{b1})(1 - e^{-t/\tau_s}), \qquad (2.49)$$

in which

$$\tau_s \equiv \frac{\omega_N + \omega_I}{\omega_N \omega_I (1 - \alpha_N \alpha_I)}.$$

Note that for the *p-n-p* transistor, $I_c < 0, I_{b1} < 0, I_{b2} > 0$, and for any transistor in saturation,

$$|\beta I_{b1}| > |I_c|.$$

During the portion of the transient of interest, the transistor is effectively traversing the path from point 1 to point 2 in Figure 2.18.

Switching-Time Expressions

We are now in a position to evaluate all of the switching times associated with the ideal transistor. For the saturated case the transient ends when

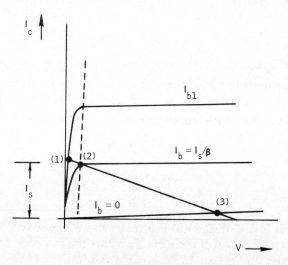

Figure 2.18

$i_{cf} = 0$. Solving (2.49) and writing currents in terms of *magnitudes*, we have

$$T_s = \tau_s \ln \frac{I_{b2} + I_{b1}}{I_{b2} + I_s/\beta}. \tag{2.50}$$

In the active region the governing equation is (2.40) for either fall or rise times. In the case of fall time the initial condition is $i_c = I_s$, and the transient is complete when $i_c = 0$. Inserting the initial condition, setting $i_c(t) = 0$, and solving again in terms of current *magnitudes*, we obtain

$$T_f = \tau_a \ln \frac{\beta I_{b2} + I_s}{\beta I_{b2}}. \tag{2.51}$$

For the rise time the initial condition is $i_c = i_b = 0$. The transient is complete when $i_c = I_s$, the value of collector current being determined by the external circuitry. Solving (2.40) for this case with current *magnitudes* gives

$$T_r = \tau_a \ln \frac{I_b}{I_b - I_s/\beta} \tag{2.52}$$

The waveforms associated with these switching times are illustrated in Figure 2.19.

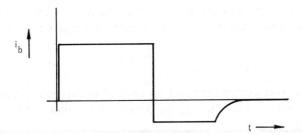

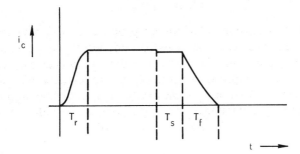

Figure 2.19

The Charge-Control Model

The basis of the charge-control model is the assumption that the continuity equation (2.5) holds true not only at each elemental volume within the base region, but also over the total volume of the base at all times. Thus the change in total base charge per unit time is made up of a recombination quantity proportional to the total charge present plus a quantity corresponding to the net current inflow or outflow. Writing (2.5) in terms of excess minority carrier density, we have

$$\frac{dp(x)}{dt} = -\frac{p(x)}{\tau} - \frac{1}{q}\frac{di(x)}{dx}. \tag{2.53}$$

Now integration over the volume of the base, from $x = 0$ at the emitter to $x = W$ at the collector, gives

$$\frac{d}{dt}\int_{x=0}^{x=W} p(x) = -\frac{1}{\tau}\int_{x=0}^{x=W} p(x) - \frac{1}{q}\int_{x=0}^{x=W} di(x).$$

The first and second integrals simply represent the total number of charge carriers in the base. The number of charge carriers is, in turn, the total charge contained in the base, Q, divided by the unit charge q. The last integral is the difference between the current across the emitter junction and the current across the collector junction. However, this is the base current I_b. Therefore, with the choice of current convention from Figure 2.9, we obtain

$$\frac{dQ}{dt} = -\frac{Q}{\tau} + I_b. \tag{2.54}$$

Although (2.53) is written for holes, an equivalent expression holds true for the electron contribution to charge density. Rewriting (2.54) in complex frequency form gives

$$\frac{Q(s)}{I_b(s)} = \frac{\tau}{\tau s + 1}. \tag{2.55}$$

It remains to determine a relationship between total base charge and collector or emitter currents. The basis for a simple form of such a relationship lies in the assumption that the effective recombination current, represented by $-Q/\tau$, is small compared to either emitter or collector currents. This is equivalent to saying that in the steady state, $I_b \ll I_c$, or that dc current gain $\beta \gg 1$. Neglecting recombination and solving (2.7) with the boundary conditions $p = p_0$ at $x = 0$, and $p = 0$ at $x = W$, gives

$$p = p_0\left(1 - \frac{x}{W}\right). \tag{2.56}$$

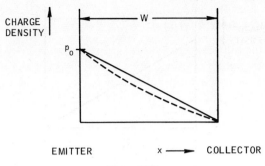

Figure 2.20

A more exact expression can be obtained if it is desired to include the effects of recombination. Solving (2.7) with the same boundary conditions (p now defined as the excess carrier density, $p - p_n$) yields

$$p = p_0\left(\frac{e^{x/L}}{1 - e^{2W/L}} + \frac{e^{-x/L}}{1 - e^{-2W/L}}\right),\tag{2.57}$$

in which L is the diffusion length. In the usual case $W \ll L$, and (2.57) closely approaches (2.56). Figure 2.20 shows the charge-density distributions for these two cases.

Using the approximation (2.56), it is clear from the geometry that the total charge in the base region is

$$Q = q\frac{p_0 W}{2}.\tag{2.58}$$

The diffusion current which flows as the result of the charge-density gradient is proportional to dp/dx, whose magnitude from (2.56) is p_0/W. Solving (2.58) for p_0 and substituting in the diffusion equation, we find that the magnitude of the diffusion current flowing through the base and therefore reaching the collector is

$$I_{cd} = \frac{2D}{W^2}Q \equiv \omega_N Q,\tag{2.59}$$

in which D can be taken to be a composite diffusion constant made up of contributions from both holes and electrons. The quantity ω_N may be regarded for the moment as an arbitrary constant of proportionality.

Whenever the collector voltage rises to a point at which the collector junction is no longer reverse-biased, the collector diffusion current ceases to increase. Additional charge can still be inserted into the base, however. This excess charge serves no useful purpose and simply disappears at a

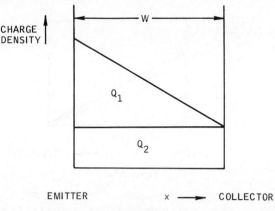

Figure 2.21

rate determined by recombination. The situation can be represented as in Figure 2.21, in which Q_2 is the excess charge present when the transistor is in saturation, and Q_1 is that portion of the total charge corresponding to (2.59), which constitutes the collector current.

Assume for the moment that (2.59) is obeyed instantaneously; that is $I_{cd}(s) = \omega_N Q_1(s)$. However, total charge is given in terms of base current by (2.55). In the active region, $Q_2 = 0$, and Q_1 is the total charge. Therefore, because $I_c = I_{cd}$ in magnitude,

$$I_c(s) = \omega_N Q(s) = \omega_N \tau I_b(s)\frac{1}{\tau s + 1}. \tag{2.60}$$

For a step input of base current,

$$i_c(t) = \omega_N \tau I_b(1 - e^{-t/\tau}). \tag{2.61}$$

Compare (2.61) and (2.40). It is apparent that we can identify the recombination time constant τ with the active-region time constant τ_a previously assumed to result from a rather arbitrary frequency-dependent form for α_N. Note that the dc current gain can also be identified as

$$\beta = \omega_N \tau = \frac{2D}{W^2}\tau. \tag{2.62}$$

It has already been observed that $\beta \cong \tau_a/\tau_N$. Therefore, ω_N can also be identified as $1/\tau_N$, the inverse of the common-base time constant, or the α cutoff frequency. The limiting frequency response of a transistor is commonly specified by f_T. the frequency at which the current gain drops to unity.

Because

$$\beta(j\omega) = \frac{\beta}{j\omega\tau_a + 1},$$

the frequency at which current gain drops to unity must occur approximately when $2\pi f_T \tau_a = \beta$. Thus

$$f_T \cong \frac{\beta}{2\pi\tau_a} = \frac{1}{2\pi\tau_N} = \frac{\omega_N}{2\pi}. \tag{2.63}$$

Because (2.61) is identical to (2.40), the rise and fall times previously calculated are unchanged. The storage time, however, can be derived on a much simpler basis than before. To determine the storage time, it is only necessary to consider the excess charge Q_2 (Figure 2.21), which must be present if the transistor is in saturation. The question then becomes: When does Q_2 reach zero under the influence of a turn-OFF base current? The form of the charge transient behavior is given by (2.55). Therefore we can immediately write, in general, for a step change in base current ΔI_b,

$$Q(t) = Q(0) + \tau_a \Delta I_b (1 - e^{-t/\tau_a}). \tag{2.64}$$

Let the value of base current be I_{b1} prior to time zero and I_{b2} after time zero. The initial value of charge is assumed to be the quiescent value τI_b. The transient ends when $Q = Q_1$, the charge required to support the saturation current I_s. However, this, by (2.59), is $Q_1 = \tau_N I_s = \tau_a I_s / \beta$. The excess charge can thus be written

$$Q_2(t) = Q(t) - Q_1 = \frac{\tau_a}{\beta}[(\beta I_{b1} - I_s) + \beta(I_{b2} - I_{b1})(1 - e^{-t/\tau_a})]. \tag{2.65}$$

This is identical in form to (2.49) with signs reversed in accordance with the convention implicit in (2.60). In both (2.65) and (2.49) the storage time ends when the excess charge Q_2 or excess current I_{cf} goes to zero. Therefore the storage time previously calculated is also unchanged if τ_a is used in place of τ_s. Such a substitution appeals to intuition because it implies that the same physical mechanism is responsible for transient behavior, regardless of region of operation.

The unifying concept that the recombination rate is the primary determining factor in transient response allows us to treat saturation and active-region behavior as a single unit; that is, rather than regarding storage and fall times as somehow different, we can simply ask how long it takes to turn the transistor OFF ($I_c = 0$), starting from any initial point. This is equivalent to asking when the total base charge reaches zero. The latter is given by (2.64). With the same initial conditions, the total turn-OFF time $T_F = T_s + T_f$

becomes, in terms of current *magnitudes*,

$$T_F = \tau_a \ln \frac{I_{b2} + I_{b1}}{I_{b2}}. \qquad (2.66)$$

Although the foregoing results provide us with a set of useful tools for predicting transistor transient behavior, it must be remembered that they are based on approximations to the real situation. For example, integration of the continuity equation (2.5) causes loss of detailed information about the actual transient distribution of charges in the base region. Another way of stating this is to observe that (2.60) holds true only if charges can rearrange themselves instantaneously to conform to the model of Figure 2.20. Bader [16] has pointed out that, in fact, the instantaneous injection of a charge increment is followed some characteristic delay time τ_D later by a corresponding step of collector current. The relationship between collector current and base charge should therefore have the form

$$I_c(s) = \omega_N Q(s) e^{-\tau_D s}, \qquad (2.67)$$

corresponding to a delay-line response. It is awkward to deal with this form, however, and a suitable approximation is to represent the delay line by its first-order term. The form of the collector current then becomes

$$I_c(s) = \omega_N Q(s) \frac{1}{\tau_D s + 1}. \qquad (2.68)$$

For a step input of base current,

$$I_c(s) = \beta I_b \left[\frac{1}{s(\tau_D s + 1)(\tau_a s + 1)} \right], \qquad (2.69)$$

the characteristic time response of which is sketched in Figure 2.22.

The Lumped Model

A more formal recognition of the delay-line concept is contained in the lumped-model approach. In this model the base is broken into a finite number of separate regions, and the diffusion and continuity equations for each region are written in finite-difference form. For example, between some elemental region at point 1 and its neighbor at point 2, the charge-density gradient is

$$\frac{dp}{dx} \simeq \frac{p_1 - p_2}{\Delta x}. \qquad (2.70)$$

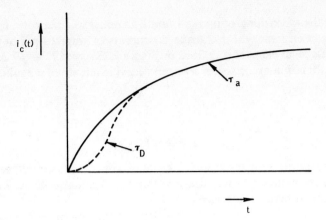

Figure 2.22

The current flow between the two points due to diffusion is

$$i = qD\frac{dp}{dx} = \frac{qD}{\Delta x}(p_1 - p_2),$$ (2.71)

or

$$I(s) = F[P_1(s) - P_2(s)].$$

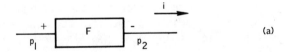

(a)

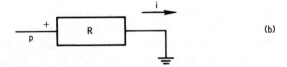

(b)

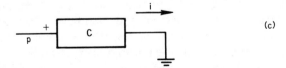

(c)

Figure 2.23

We may therefore think of points 1 and 2 as nodes in a network, connected by a circuit element whose transfer function is in amperes per unit charge density. Such an element is defined in Figure 2.23a, with $F \equiv qD/\Delta x$.

Recombination accounts for a current leaving a node of magnitude

$$i = \frac{Q}{\tau} = \frac{q\,\Delta x}{2\tau}\,p,$$

or

$$I(s) = RP(s). \tag{2.72}$$

A corresponding element is indicated by Figure 2.23b, with R defined by (2.72). Thermal generation (corresponding to p_n) can also be accounted for but is omitted here for simplicity.

Finally, we remove the restriction that charge cannot accumulate in an elemental volume and let the net inflow of current be

$$i = \frac{dQ}{dt} = \frac{q\,\Delta x}{2}\frac{dp}{dt}$$

or

$$I(s) = \frac{q\,\Delta x}{2}\,sP(s) \equiv CsP(s). \tag{2.73}$$

Such an element is indicated in Figure 2.23c. The equivalent circuit of a given node may now be drawn as a lumped-element network. It appears in Figure 2.24a, in which the node p_2 is connected to its neighbors through F components and to a source (or sink) through R and C components. Although no external current source is shown, it may be readily incorporated as appropriate.

The beauty of the lumped-model representation of the transistor is that accuracy as high as required may be obtained by increasing the number of nodes employed to represent the base. In addition, the values of the components F, R, and C, may be chosen to be different at different points to represent changes in properties of the material. The foregoing development was based upon per-unit-area currents, whereas the actual model must incorporate device area. Base cross section generally varies through the base, and because the equivalent component values depend directly upon this area, they will differ systematically through the base.

In order to relate this model to those previously developed the same first-order representation can be employed; that is, let us represent the base region by a single π section with two nodes, one at the emitter and one at the collector junction. Representing the junction by its ideal-diode equation, the equivalent circuit reduces to that of Figure 2.24b. This model is valid in all three regions of operation, just as are the previous models. Relationships

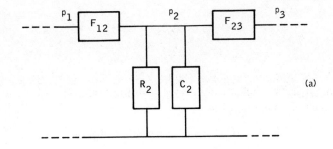

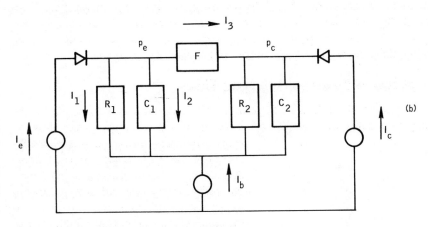

Figure 2.24

pertaining to particular regions can be readily obtained. For example, in the active region with the emitter forward-biased and the collector reverse-biased, the charge density at the collector node is zero. With $p_c = 0$, the circuit elements R_2 and C_2 carry no current, and $I_c = -I_3$. We can also write $I_e = I_1 + I_2 + I_3$. The ratio of collector to emitter current is then

$$-\frac{I_c(s)}{I_e(s)} = \frac{I_3(s)}{I_1(s) + I_2(s) + I_3(s)} = \left(\frac{1}{1 + R_1/F}\right)\left\{\frac{1}{[(C_1/F)/(1 + R_1/F)]s + 1}\right\}.$$

$$(2.74)$$

As expected, (2.74) has precisely the same form as $\alpha_N(s)$ in (2.33). Because $\Delta x = W$ in the two-node approximation,

$$\frac{R_1}{F} = \frac{W^2}{2D\tau} = \frac{W^2}{2L^2} = \frac{\tau_N}{\tau_a} = \frac{1}{\beta}$$

and

$$\frac{C_1}{F} = \frac{W^2}{2D} = \tau_N. \tag{2.75}$$

Because the lumped model of Figure 2.24b is valid in all three regions of operation, transient behavior in the saturation region can be obtained from the pair of node equations

$$I_e(s) = (C_1 s + R_1 + F)P_e(s) - FP_c(s),$$

$$I_c(s) = -FP_e(s) + (C_2 s + R_2 + F)P_c(s). \tag{2.76}$$

It can be readily shown that (2.76) possesses the same characteristic equation as (2.46). Note that this model gives some plausibility to the notion that $\tau_s \neq \tau_a$, because the physical quantities involved are not necessarily the same in the neighborhoods of the emitter and collector junctions.

Effects of Loads on Switching Times

The effect of loading on transistor switching times is manifested via the changing collector voltage. We have implicitly assumed so far that the

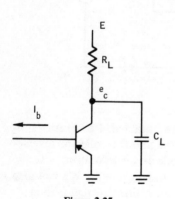

Figure 2.25

transistor enters saturation at a particular current value I_s given by a supply voltage divided by a series resistor, E_1/R. When the collector current and voltage are no longer so simply related, we must revert to the fundamental definition of the saturation region, namely the condition in which both junctions are forward-biased. It is therefore necessary to examine the collector voltage transient to determine when the collector junction changes from reverse to forward bias. Because the forward-biased emitter junction represents only a small voltage drop and because supply voltages are generally relatively large, we shall for simplicity approximate the condition that the collector junction reaches zero volts by the condition that the collector voltage reaches zero volts.

One case of interest is that of capacitive loading, as in Figure 2.25. For initial simplicity, assume that $\tau_L \equiv R_L C_L \gg \tau_a$. Relative to the collector voltage, therefore, collector current will rise virtually instantaneously to its final value βI_b. The time response under these conditions is

$$\Delta e_c(t) = \beta \Delta I_b R_L (1 - e^{-t/\tau_L}). \tag{2.77}$$

Rise-time equations can be readily obtained, as before. For fall time a more realistic circuit must be considered, because turning OFF the collector current in Figure 2.25 simply allows e_c to decay exponentially toward zero, for an infinite fall time. In practice, the collector is almost always clamped to some signal voltage V by a circuit such as that of Figure 2.26. If $I_c < I_1 = (E - V)/R_L$, then $I_d > 0$, the clamp diode is forward-biased, and $e_c = V$. If $I_c > I_1$, the clamp diode is reverse-biased. The change in base drive is that required to change the collector current from $I_c = I_s = E/R_L$ to $I_c = 0$.

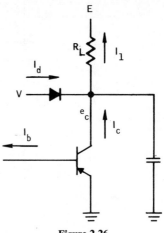

Figure 2.26

Placing a clamp in the collector circuit affects the rise time when τ_a is not negligible. Under such a condition the rise time must be broken into two periods; thus $T_r = T_{r1} + T_{r2}$. During the first period $i_c < I_1$, $I_d > 0$, and the collector voltage remains clamped. The transient is therefore entirely determined by τ_a. As soon as $I_d = 0$, the collector voltage begins to rise, but at a rate determined by both τ_a and τ_L. The sequence is therefore as shown in Figure 2.27. Note that i_c may temporarily exceed I_s because collector current continues to rise at a rate determined by τ_a, but collector voltage may be delayed by τ_L.

Changes in collector voltage can also affect switching time via the collector junction capacitance. This capacitance is effectively an internal feedback path between the collector and base, as indicated in Figure 2.28. It has the effect of subtracting a current $C_c(de_c/dt)$ from the external base drive current I_b. The effective ac equivalent circuit is represented in Figure 2.29, from which the governing equations are

$$E_c = -I_c Z_L,$$

$$I_x = C_c s E_c,$$

$$I_e + I_c + I_b = 0,$$

$$I_c + \alpha_N I_e - I_x = 0. \tag{2.78}$$

Combining Equations (2.78) results in

$$\frac{E_c}{I_b} = \frac{\beta R}{\tau_a(\tau_L + \tau_c)} \left\{ \frac{1}{s^2 + [(\tau_a + \tau_L + \beta\tau_c)/\tau_a(\tau_L + \tau_c)]s + 1/\tau_a(\tau_L + \tau_c)} \right\}. \tag{2.79}$$

It can be verified that the roots of this expression cannot be oscillatory.

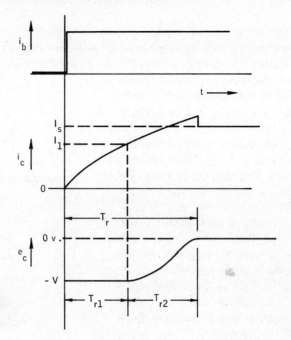

Figure 2.27

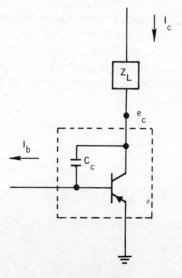

Figure 2.28

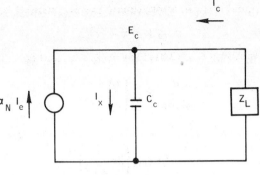

Figure 2.29

In order to obtain a solution in a readily identifiable form, let us make two approximations. The first is $\tau_L \gg \tau_c$, or $C_L \gg C_c$. This is generally justifiable because the collector junction capacitance is usually small compared to stray wiring capacitance. The second approximation is $\sqrt{1-x} \cong 1 - x/2$. This is a reasonably good approximation for x not near 1. For example, it is only in error by 6 percent for $x = \frac{1}{2}$; by 16 percent for $x = \frac{2}{3}$; etc. Using these approximations, we can reduce (2.79) to

$$\frac{E_c}{I_b} \cong \beta R \left[\frac{1}{(k\tau_T s + 1)(\tau_T s + 1)} \right], \tag{2.80}$$

in which

$$\tau_T = \tau_a + \tau_L + \beta \tau_c,$$

$$k = \frac{1}{(1 + \tau_L/\tau_a + \beta \tau_c/\tau_a)(1 + \tau_a/\tau_L + \beta \tau_c/\tau_L)}.$$

Even with $\tau_c \rightarrow 0$, the maximum value of k is $\frac{1}{4}$. In common situations $k \ll 1$.

The collector voltage transient may therefore be described as made up of two exponential terms with, in general, one having a time constant much shorter than the other. A reasonable estimate of transient behavior may be obtained from τ_T alone if $k \ll 1$. In this case for $t \gg k\tau_T$, (2.80) reduces to a single-pole expression from which, for step input, al' of the previous switching-time results follow with τ_T replacing τ_a.

For those cases in which the above linear approx.mations are not sufficiently accurate, higher-order treatment is required. The time function corresponding to (2.80) for a step input is

$$\frac{e_c}{\beta I_b R} = 1 + \frac{k}{1-k} e^{-t/k\tau_T} - \frac{1}{1-k} e^{-t/\tau_T} \tag{2.81}$$

Using an additional term to approximate the exponential, as

$$\varepsilon^x \cong 1 + x + x^2/2,$$

we obtain a reduced form of (2.81) for $t = T_r \ll \tau$; that is,

$$\frac{V}{\beta I_b R_L} \cong \frac{1}{2k}\left(\frac{T_r}{\tau_T}\right)^2,$$

or

$$T_r \cong \tau_T\left(2k\frac{V}{\beta I_b R}\right)^{\frac{1}{2}}. \tag{2.82}$$

This simply means that T_r no longer varies directly as the ratio of saturation to drive current, but rather as the square root of that ratio. Additional drive may therefore be required to achieve a given switching time.

In addition to the above effects upon rise time, we have noted in (2.67) that the collector current experiences a characteristic delay as a result of the finite time required to establish the charge-density distribution required for current flow. This delay time is frequently treated as a separate and relatively fixed quantity, T_D, which simply adds to the total time taken to turn a transistor ON. A further delay occurs due to the time required to discharge the capacitances of the emitter and collector junctions.

If the emitter junction is previously reverse-biased by a turn-OFF voltage ΔV, some time must be allowed for discharging ΔV to zero before the transistor can enter the active region. We define $C = C_c + C_e$ and let the input impedance of the transistor be regarded as effectively infinite until ΔV reaches zero. The charge stored on the input capacitance is $\Delta Q = C\Delta V$. A good first estimate for the discharge time is thus

$$T_0 \cong \frac{Q}{I} = \frac{\Delta Q}{I_b}. \tag{2.83}$$

The total turn-ON time for the transistor, T_N, is now $T_N = T_0 + T_D + T_r$. Because we are usually interested in the exponential value of T_N/τ_a, we can write

$$e^{T_N/\tau_a} = e^{\Delta Q/\tau_a I_b} e^{T_D/\tau_a} \frac{1}{1 - I_s/\beta I_b}. \tag{2.84}$$

Because the delay due to junction capacitance charging is usually small compared to the recombination rate τ_a, (2.84) becomes approximately

$$e^{(T_N - T_D)/\tau_a} \cong \frac{1 + \Delta Q/\tau_a I_b}{1 - I_s/\beta I_b}. \tag{2.85}$$

We see that adding the correction term due to junction capacitance charging increases the estimated turn-ON time, but that the general relationship between increased drive I_b and decreased switching time prevails.

REFERENCES

[1] W. Shockley, "The Theory of *p-n* Junctions in Semiconductors and *p-n* Junction Transistors," *Bell Sys. Tech. J.*, **28**, 435–489 (July 1949).

[2] W. Shockley, *Electrons and Holes in Semiconductors*, Van Nostrand, Princeton, N.J., 1950.

[3] R. D. Middlebrook, *An Introduction to Junction Transistor Theory*, Wiley, New York, 1957.

[4] D. Dewitt and A. L. Rossoff, *Transistor Electronics*, McGraw-Hill, New York, 1957.

[5] J. J. Ebers and J. L. Moll, "Large-Signal Behavior of Junction Transistors," *Proc. IRE*, **42**, 1761–1772 (December 1954).

[6] J. L. Moll, "Large Signal Transient Response of Junction Transistors," *Proc. IRE*, **42**, 1773–1784 (December 1954).

[7] J. L. Moll, "Junction Transistor Electronics," *Proc. IRE*, **43**, 1807–1819 (December 1955).

[8] J. L. Moll, "The Evolution of the Theory for the Voltage-Current Characteristic of *p-n* Junctions," *Proc. IRE*, **46**, 1076–1082 (June 1958).

[9] A. Uhlir, Jr., "The Potential of Semiconductor Diodes in High-Frequency Communication," *Proc. IRE*, **46**, 1099–1115 (June 1958).

[10] J. H. Forster and P. Zuk, "Millimicrosecond Diffused Silicon Computer Diodes," *1958 IRE Wescon Conv. Rec.*, part 3, 122–130.

[11] G. Wolff, "Back-Transient Diode Logic," *AIEE Commun. Electron.*, **47**, 4–9 (March 1960).

[12] R. H. Beeson, "A Complete Transistor Equivalent Circuit," *Proc. IRE*, **49**, 825–6 (April 1961).

[13] R. S. C. Cobbold, "On the Application of the Base Charge Concept to the Design of Transistor Switching Circuits," *IRE Trans. Circuit Theory*, **CT-7**, 1, 12–18 (March 1960).

[14] J. J. Sparkes, "A Study of Charge Control Parameters in Transistors," *Proc. IRE*, **48**, 1696–1705 (October 1960).

[15] W. H. Ko, "The Reverse Transient Behavior of Semiconductor Junction Diodes," *IRE Trans. Electron. Devices* **ED-8**, 2, 123–131 (March 1961).

[16] C. J. Bader, "Charge-Step-Derived Transfer Functions for the Junction Transistor," *IEEE Trans. Commun. Electron.*, 66, 179–185 (May 1963).

[17] J. A. Ekiss, "Applications of the Charge-Control Theory," *IRE Trans. Electron. Comp.*, **EC-11**, 3, 374–381 (June 1962).

[18] J. G. Linvill, "Lumped Models of Transistors and Diodes," *Proc. IRE*, **46**, 1141–1152 (June 1958).

[19] J. G. Linvill and J. F. Gibbons, *Transistors and Active Circuits*, McGraw-Hill, New York, 1961.

[20] D. J. Hamilton, F. A. Lindholm, and J. A. Narud, "Comparison of Large Signal Models of Junction Transistors," *Proc. IEEE*, **52**, 3, 239–248 (March 1964).

[21] C. L. Hegedus, "Charge Model of Fast Transistors and the Measurement of Charge Parameters by High Resolution Electronic Integrator," *Solid State Design*, **5**, 8, 23–36 (August 1964).

[22] K. G. Ashar, "The Method of Estimating Delay in Switching Circuits and the Figure of Merit of a Switching Transistor," *IEEE Trans. Electron. Devices*, **ED-11**, 11, 497–506 (November 1964).

EXERCISES

2.1 What is the output voltage of the circuit of Figure P.2.1 for each of the following combinations of input voltages:

e_1 (volts)	e_2 (volts)
0	0
-4	0
-4	-4

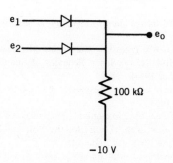

Figure P.2.1

2.2 if diodes are ideal, with $I_s = 10\ \mu A$, and (a) $T = 25°C$, (b) $T = -50°C$, (c) $T = 150°C$. Same as Problem 2.1, with diode leakage resistance of 1 MΩ and forward ohmic resistance of 1 kΩ.

2.3 After being OFF for a long time, the current generator I of Figure P.2.2 is suddenly turned ON to $I = 10$ mA. The effective diffusion length and diffusion constant of the diode are such that $L^2/2D = 1\ \mu s$, and the diode material recombination rate is also 1 μs. Sketch the output waveform, $e_0(t)$. At what time t does e_0 reach (a) $+1$ V and (b) $+4$ V?

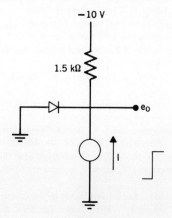

Figure P.2.2

2.4 A capacitor $C = 10$ pF is connected from the output e_0 of the gate of Figure P.2.1 to ground. After having been at 0 V for a long time, both e_1 and e_2 change suddenly to -4 V. Assuming the same diode characteristics as in Problem 2.3, sketch the output waveform. At what time t does e_0 reach -2 V?

2.5 A capacitor $C = 100$ pF is connected from the output e_0 of the clamp circuit of Figure P.2.2 to ground. Repeat Problem 2.3 with this component taken into account.

2.6 An unloaded *DCTL* bistable circuit is shown in Figure P.2.3. Find the collector voltages of e_{c1} and e_{c2} if both transistors are characterized by the parameters $\alpha_n = 0.98$, $I_{co} = 10\ \mu A$, $I_{eo} = 8\ \mu A$, and $r_b = 0$.

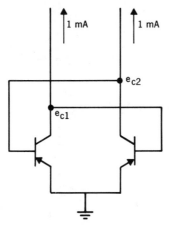

Figure P.2.3

2.7 Repeat Problem 2.6 if $r_b = 1$ kΩ.

2.8 Find V_{be} and V_{ce} for the transistor in Figure P.2.4 for $e_1 = -4$ V (a) if the parameters of Problem 2.6 apply, (b) if $r_b = 1$ kΩ.

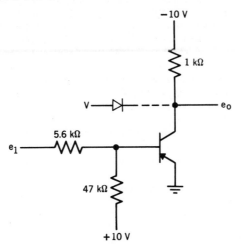

Figure P.2.4

2.9 Plot the dc transfer characteristic, e_0 versus e_1, for the circuit of Figure P.2.4 with a transistor having the parameters of (a) Problem 2.6 (b) Problem 2.7.

2.10 Plot the dc transfer characteristics, e_{01} and e_{02} versus e_1, for the circuit of Figure P.2.5 with a transistor having the parameters of Problem 2.6 and (a) $R = 0$, (b) $R = 1\ \text{k}\Omega$.

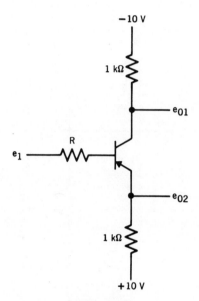

Figure P.2.5

2.11 Find T_r, T_s, and T_f for the circuit of Figure P.2.4 if e_1 is a step input voltage from 0 to -4 V for ON and from -4 to 0 V for OFF, $\tau_a = \tau_s = 1.0\ \mu\text{s}$, and $\beta = 50$, (a) assuming V_{be} is negligible, (b) assuming V_{be} is determined by the parameters of Problem 2.6.

2.12 Repeat Problem 2.11 if the base junction capacitances are known to be $C_c = C_e = 10\ \text{pF}$.

2.13 A perfect-diode clamp is added to the collector circuit of Figure P.2.4 (dotted) with $V = -4$ V. Find T_r, T_s and T_f for the same conditions as in (a) Problem 2.11, (b) Problem 2.12. Sketch the output voltage e_0 waveform. What is the fall time of e_0?

2.14 The collector circuit e_0 of Figure P.2.4 is required to drive a load consisting of a $1\ \text{k}\Omega$ resistor in parallel with a 1,000-pF capacitor to ground. Find T_r, T_s, and T_f for the conditions of Problems 2.11 and 2.12.

2.15 Derive expressions for rise, storage, and fall times for the circuit of Figure P.2.6 based upon the assumption that for a step input all of the capacitor charge $\Delta Q = CV$ is delivered to (or removed from) the base at the initial instant. Assume $V/R_g = I_g > I_b = I_s/\beta$, $I_s = E/R_1$.

2.16 Repeat Problem 2.15 with the assumption that the time constant $\tau = r_b C$ is small compared with either τ_a or $R_g C$, and that base junction capacitances $C_c + C_e$ must be charged by the voltage V_{be} before base charge storage begins.

2.17 Show that the two-node lumped equivalent circuit of the transistor leads to a two-pole complex-frequency transfer function when the transistor is in the saturation region. Find ω_N and ω_I in terms of C_1, R_1, C_2, R_2, and F.

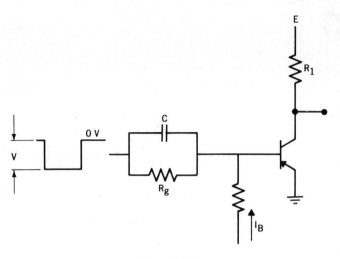

Figure P.2.6

2.18 Show that a three-node lumped equivalent circuit of the transistor leads to a two-pole complex-frequency transfer function when the transistor is in the active region.

2.19 Find T_r, T_s, and T_f for the circuit of Figure P.2.6 with the following parameters and circuit values:

$$\tau_a = \tau_s = 1.0\ \mu s$$
$$\beta = 50$$
$$E = -10\ V$$
$$R_1 = 1\ k\Omega$$
$$I_B = 0.2\ mA$$
$$R_g = 7.5\ k\Omega$$
$$V = 4\ V$$

when (a) $C = 25\ pF$, (b) $C = 100\ pF$. Assume V_{be} is negligible.

2.20 Repeat Problem 2.19 with $r_b = 1\ k\Omega$ and $C_c = C_e = 10\ pF$.

2.21 Plot the time response of the collector (e_{01}) and emitter (e_{02}) voltages of the circuit of Figure P.2.5 for a step input e_1 from $+2$ to $0\ V$ with $R = 1\ k\Omega$.

2.22 Find the value of R_1 for the circuit of Figure P.2.6 to yield a rise time, $T_r = 0.75\ \mu s$ for the same circuit values as in Problem 2.19, with $C = 0$, and $C_c = 10\ pF$.

2.23 Using the ideal-transistor large-signal dc model, show that for a transistor in saturation, the voltage across the collector junction (V_{cb}) is equivalent to that appearing across a diode carrying a forward current of magnitude $\beta I_b - I_s$ and having a reverse saturation current of magnitude $I_{co}/(1 - \alpha_N)$.

Chapter 3

Magnetic-Core
Equivalent Circuits

Although circuits containing ferromagnetic components have been employed extensively in digital computers for a number of years, the development of a single equivalent circuit model simple enough to be useful yet applicable to a broad class of situations has proved curiously stubborn. This is at least partly a result of the fact that the switching behavior of practical ferromagnetic configurations is still incompletely understood or can be described only in terms of rather cumbersome mathematical models, each applicable to limited ranges of property variations. In addition, the effects of a large number of factors such as material composition, anneal, stress, and impurities upon magnetic properties are far from being exhaustively cataloged or predictable in many cases. The result is that a great deal of practical circuit design is performed in the laboratory, and predictions based upon equivalent circuit models are treated as rough first-order approximations.

With these limitations in mind, however, magnetic-circuit design work is considerably eased with the aid of one or more simple equivalent circuit models that yield some insight into the effects of parameters that are the most significant and offer some point of departure for experimentation and refinement.

Single Closed-Magnetic-Path Model

When a step function of voltage is applied to a winding on a closed magnetic path, in the simplest case the observed current flowing into the winding will rise linearly with time. The circuit may then be represented by the conventional linear inductor, for which $e = L(di/dt)$. If the time integral of e

is plotted versus i, the result is simply a straight line. The variables $\int e \, dt$ and i may be linearly scaled by the geometry of the magnetic circuit and relabeled to give the familiar B-H plot; that is, $B = k_1 \int e \, dt$, and $H = k_2 i$ in appropriate units. The scaling factors can be ignored, however, until we are interested in relating the elements of the equivalent electrical circuit to the parameters of a particular magnetic-core configuration.

If a second winding is placed on the magnetic core, the device is now a four-terminal network. Any load resistor present may, of course, be transformed back to the input winding by the square of the turns ratio, giving a two-terminal equivalent circuit.

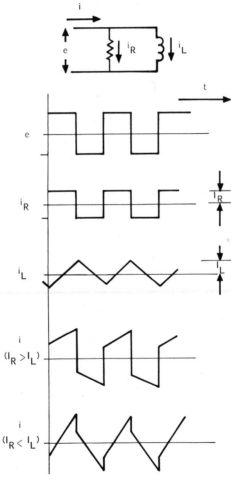

Figure 3.1

If an alternating square wave of voltage is applied to such a circuit, the resulting waveforms are as shown in Figure 3.1. When these waveforms are plotted on the equivalent *B-H* plane, the result is as shown in Figure 3.2 for two cases of relative magnitude of the maximum branch currents I_R and I_L.

Suppose that a square-wave voltage is applied to an unknown component, and the current waveform illustrated in Figure 3.3 is observed. Suppose the current is virtually zero until some time *T*, when it rises to a value limited only by the impedance external to the component in question, namely, *E/R*. Clearly, for $0 < t < T$, the impedance of the component must approach infinity. For $t > T$, it must approach zero. The equivalent circuit is therefore simply a normally open switch which closes at time *T*.

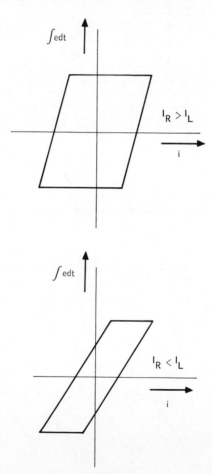

Figure 3.2

In order to characterize the properties of this ideal switch we can further observe that the time T is related to the magnitude of applied voltage such that the product ET is always a constant. Under this constraint the equivalent *B-H* plot is as shown in Figure 3.4. This representation is, indeed, a fair approximation of the characteristics of some magnetic-core materials, particularly in low-frequency applications in which small magnetizing currents can be neglected. If more careful measurements are made, it will be

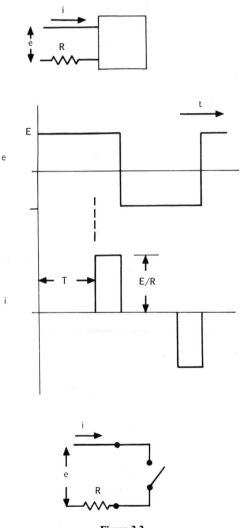

Figure 3.3

observed that the current during $0 < t < T$ is not zero but rather a constant, which will be called I_0.

Further tests would reveal that if the voltage driving source is replaced by a constant-current generator, for drive currents $I < I_0$, no output voltage

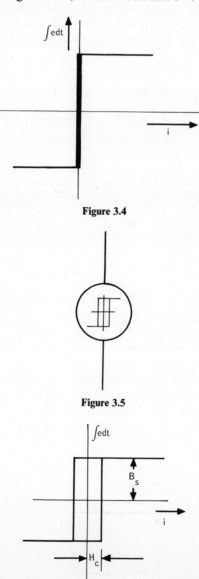

Figure 3.4

Figure 3.5

Figure 3.6

is observed; that is, $\int e\, dt$ does not change. The simple switch is not sufficient, therefore, to represent observed behavior, and we must postulate some kind of current-actuated switch. In order to agree with observation, the requirements are that the circuit must have a zero impedance for $|I| < I_0$, a constant-current source for $|I| = I_0$ and $\int e\, dt$ between its limiting values, and a zero impedance again for $|\int e\, dt|$ equal to or greater than its limit. A rather artificial component is necessary in order to represent these properties. Such a component will be referred to as an *ideal square-loop component* and will be represented by the symbol of Figure 3.5. The ideal square-loop circuit is characterized by two parameters: (a) H_c, corresponding to the constant-current generator I_0 and (b) B_s, corresponding to the limiting value of $\int e\, dt$, as represented in Figure 3.6.

A typical observation on real magnetic circuits would reveal that the value of I_0 is not independent of E but rather increases approximately according to the relationship $I = I_0 + KE$. Therefore we can regard I as made up of two components, one due to the ideal square-loop component, the other due to a parallel resistor ($R_c = 1/K$). The symbol R_c will be used to represent this effective internal core resistance. Although we shall see that the presence of a fixed, ideal resistor cannot be interpreted too literally, in general, R_c represents the power losses in switching a core, over and above those losses inherent in traversing the B-H loop at dc. Its effect shows up as power dissipation (i.e., heat) that must of course be supplied by external driving sources. Clearly, with R_c present, the equivalent B-H loop will expand under increased drive voltage, as shown in Figure 3.7.

The equivalent circuit now permits some meaning to be given to the case of a constant-current driving source. The voltage appearing across the

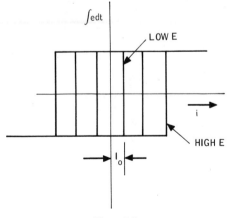

Figure 3.7

equivalent circuit must be $E = (I - I_0)R_c$, and the time required to switch from bottom to top of the loop can be obtained from

$$2 \int_{max} e \, dt = ET = (I - I_0)R_c T. \tag{3.1}$$

It may also be observed that i is not a constant during $0 < t < T$ but in fact changes value, rising approximately linearly. The waveforms and corresponding equivalent circuit are shown in Figure 3.8. The equivalent inductance L_m is commonly termed the magnetizing inductance. It represents

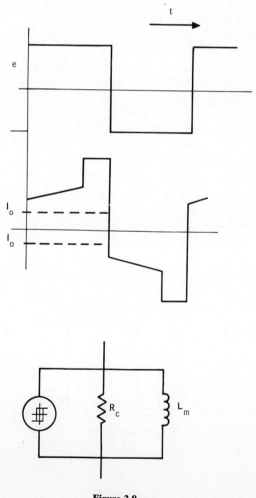

Figure 3.8

the additional drive current (and consequent increased magnetic field strength) required to subject increasingly stubborn local regions of the magnetic core to forces sufficient to insure switching (and consequent flux change). Because $L = \int e\,dt/i$ and $\int e\,dt$ is proportional to flux density, L_m is simply proportional to $\Delta B/\Delta H$, the change in flux density per unit change in

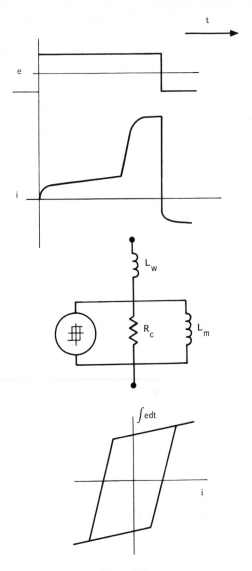

Figure 3.9

driving field strength. The ratio B/H is commonly termed the *permeability* of a medium, and L_m is therefore a direct measure of core permeability.

Finally, a refinement of the equivalent circuit can be made by noting that i does not rise instantaneously to $I_0 + E/R_c$ at $t = 0$, nor to E/R at $t = T$, but rather exhibits an exponential time constant usually short compared to other parameters. This behavior may be related to the saturated winding inductance (and resistance, if significant). Saturation refers to that condition of the core in which it is no longer possible to switch additional flux irreversibly. In the ideal circuit model it corresponds to the condition that the switch inherent in the constant current generator I_0 is closed. In the equivalent circuit, this has the effect of shorting out R_c and L_m. Thus the characteristics of the magnetic material that are of interest during switching are no longer present, and the circuit now looks much like a simple inductive winding on a passive (linear) medium. It can be represented by an impedance (usually a small inductance) in series with the parallel equivalent circuit, as shown in Figure 3.9. The winding inductance L_w and magnetizing inductance L_m are separately designated. Note that this refinement results in putting a slope on the top and bottom of the equivalent B-H plot, because $\int e\,dt$ can still increase with i after saturation (closing of ideal square-loop component switch). Additional effects may also be represented by adding to the equivalent circuit of Figure 3.9. The most important of these is commonly the capacitance between turns on the winding. It is usually satisfactory for first-order effects to regard such capacitance as a single lumped element C shunting whichever portion of the circuit is of interest at the time.

Parameters of Linear Magnetic Circuits

Ampere's law, relating current and magnetic field strength H, can be stated in at least two equivalent ways. They are

$$dH = \frac{I\,dp\,\sin\theta}{10r^2}$$

and

$$F = \oint \bar{H}\,d\bar{p} = \frac{4\pi I}{10}, \tag{3.3}$$

in which

$$I = \text{current, in amperes,}$$
$$r, p = \text{length, in centimeters,}$$
$$H = \text{field strength, oersteds,}$$
$$\theta = \text{angle between the current-carrying element } dp$$
$$\text{and the point at which } H \text{ is measured,}$$
$$F = \text{magnetomotive force, in gilberts.}$$

The first form of Ampere's law relates the field strength at a point in space at a distance r from a current-carrying element of incremental length dp and current I. The second form relates the value of the integral of H around a closed path of integration to the net current enclosed by the path. Appropriate constants are included in order to relate quantities in terms of the units commonly specified.

Either form of Ampere's law may be used, depending on convenience in evaluating the particular geometrical configuration. In most magnetic work a given path of integration will enclose more than one turn of the same coil. Therefore, because I in (3.3) is the total current enclosed by the path it may be rewritten as

$$\oint H \, dp = \frac{4\pi NI}{10},\tag{3.4}$$

where I is redefined to mean simply the current in the wire.

Faraday's law provides a second basic relationship in magnetic circuits and, together with Ampere's law, makes it possible to relate the physical variables, voltage and current, usually measured. It can be stated as

$$e = \frac{d\phi}{dt} \times 10^{-8},\tag{3.5}$$

where

e = voltage induced at the terminals of a path (usually a wire) which encloses flux, in volts,

ϕ = total flux enclosed, in maxwells,

t = time, in seconds.

Because each turn of a multiple-loop coil normally encloses (almost) the same flux, this relationship may be written as the familiar

$$e = N\frac{d\phi}{dt} \times 10^{-8},\tag{3.6}$$

with e being simply the voltage across the terminals of the winding.

Other quantities need definition. One of these is magnetic induction B, more generally referred to as *flux density*. It is commonly measured in gauss and is defined by

$$B = \frac{\phi}{A} = \text{flux per unit area, in maxwells per square centimeter.}$$

Permeability μ is a measure of the ease with which a field may be established in a medium and in the absence of internal induction is defined by $B = \mu H$.

It equals unity in vacuum and closely that in air but ranges from 10^3 to 10^6 in ferromagnetic materials. Permeability is commonly nonlinear. The high relative permeability of magnetic materials, however, allows some considerable simplifications when treating circuits containing such materials. Thus in many cases only the flux appearing in the magnetic material need be considered, and alternative paths in the low-permeability surrounding medium (usually air) can be neglected. This is exactly analagous to practices in the analysis of electrical circuits, in which, for example, it is sufficiently accurate to consider current confined to the low-resistance path (the wire or resistor) and neglect any effects in the high-resistance surrounding medium (wire insulation or air). Unfortunately, the relative magnitudes of magnetic permeabilities only range over some six orders of magnitude, whereas the resistance of materials ranges over more than 20 to 30 orders of magnitude. The result is that geometry plays a much more important role in magnetic circuit analysis than it does in electrical circuit analysis.

Consider a closed magnetic path enclosing a current-carrying wire. If the permeability of the magnetic material is at all high, practically all flux will remain confined within the material throughout the path. If the path is of uniform cross section and permeability, B must likewise be uniform, and consequently H is uniform along the path. Therefore the integral (3.4) may be written in the form

$$ H = \frac{4\pi NI}{10p}, \qquad (3.7) $$

where p is the total length of the path, independent of geometry.

Only two assumptions have been required to arrive at (3.7), namely, (a) that the great majority of flux is confined to the magnetic path, and (b) that the cross-sectional area and permeability are constant around the path.

This introduces the concept of flux tubes or shells of magnetization; that is, considerations of symmetry will usually permit us to make the approximation that flux is confined to tubes or filaments of incremental cross section ΔA, within which flux density is uniform. All flux entering one end leaves the other end (no leakage).

With μ uniform and B constant, H must be constant along the tube. By following such a tube around a path enclosing current, we can determine the total magnetomotive force Hp, corresponding to a given drive $0.4\pi NI$. If the geometry is uniform along one coordinate, the tube may become a shell. Two such cases are illustrated in Figure 3.10. In both cases the magnetic path length appropriate for any given region within the material may be readily determined. Therefore, H as a function of location may be written as $H = H(I, p)$. For example, in the case of a path symmetrically surrounding

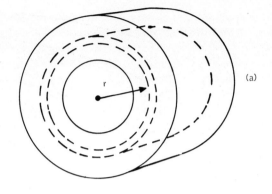

(a)

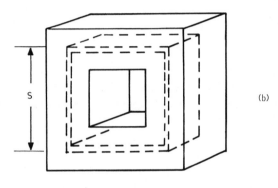

(b)

Figure 3.10

a circular central aperture, $p = 2\pi r$, and H becomes simply

$$H = \frac{NI}{5r}.$$ (3.8)

For a path symmetrically surrounding a central square aperture, $p = 4s$, and H becomes $\pi NI/10s$. It may appear unduly simplified to ignore the effects of actual flux distribution, especially around corners and other geometric irregularities. Experimental verification of flux distributions closely resembling the right-angle paths of Figure 3.10b have been obtained, however, by microscopic examination of surface magnetic domain arrangements.

The concept of the flux tube may be usefully extended to a number of cases in which cross section and permeability are not uniform around the

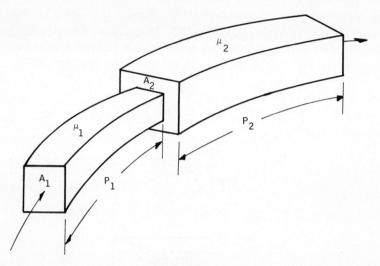

Figure 3.11

path, provided it is still reasonable to retain the assumption of flux confined to the path. The magnetomotive force integral may be written as a summation,

$$F = 0.4\pi NI = \oint H\, dp = H_1 p_1 + H_2 p_2 + \cdots, \qquad (3.9)$$

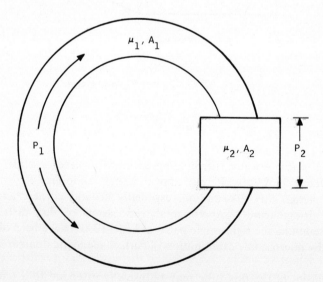

Figure 3.12

the general case illustrated by Figure 3.11. Assuming μ and A are constant in each segment, $B_i = \mu_i H_i$ and $B_i = \phi/A_i$. In this case, the only assumption made is that $\phi_1 = \phi_2 = \cdots = \phi_i = \phi$, that is, that the flux remains confined as it goes from one path segment to another. We can therefore write

$$\sum \frac{B_i p_i}{\mu_i} = \phi \sum \frac{p_i}{\mu_i A_i} = 0.4\pi NI. \tag{3.10}$$

Because flux is confined,

$$B_i = B_j \frac{A_j}{A_i},$$

and with $B_i = \mu_i H_i$

$$H_i = H_j \frac{\mu_j A_j}{\mu_i A_i}. \tag{3.11}$$

Equation (3.9) may therefore always be written in terms of a single variable; for example,

$$H_1 \left(p_1 + p_2 \frac{\mu_1 A_1}{\mu_2 A_2} + \cdots \right) = 0.4\pi NI. \tag{3.12}$$

For geometries consisting of two main path segments, as in Figure 3.12, it is sometimes useful to depict the relationships graphically. In this case,

$$H_1 p_1 + H_2 p_2 = 0.4\pi NI.$$

Substituting for H_2 by (3.11) and for H_1 by $B_1 = \mu_1 H_1$ and rearranging, we obtain

$$B_1 = \frac{0.4\pi N \mu_1}{p_1} I - \frac{\mu_1^2 A_1 p_2}{\mu_2 A_2 p_1} H_1. \tag{3.13}$$

The solution for B_1 may therefore be graphed as the intersection of two straight lines having the form $B_1 = \mu_1 H_1$ and, from (3.13), $B_1 = aI - bH_1$, as shown in Figure 3.13. The graphical approach is particularly useful when dealing with materials in which one magnetic path segment does *not* possess a linear characteristic; that is, if $B_1 = f(H_1)$, where f may be quite general, solutions may nevertheless be obtained graphically. Let $B_1 = f(H_1)$, some nonlinear characteristic. For a two-segment path, $B_1 A_1 = B_2 A_2 = \phi$, and $B_2 = (A_1/A_2)B_1$. However, $B_2 = \mu_2 H_2$ if the second path segment is linear, so $H_2 = (1/\mu_2)(A_1/A_2)B_1$. Now $H_1 p_1 + H_2 p_2 = 0.4\pi NI$. Substituting and rearranging, we obtain (3.13), which therefore does not depend on the linearity of the first segment. Solutions must consequently exist at the intersections of the curves $B_1 = f(H_1)$ and $B = aI - bH_1$, as given by (3.13).

The result is illustrated in Figure 3.14. In the particular case in which the segment designated by μ_2, p_2, and A_2 is an air gap, the slope of the line corresponding to the equation $B_1 = aI - bH_1$ is related to a quantity termed the *demagnetizing factor*. Specifically, the demagnetizing factor is generally taken to be 1/magnitude of slope, so that a large demagnetizing factor corresponds to slopes near zero (H axis), and a small demagnetizing factor corresponds to slopes near infinity (B axis). For small air gaps it is

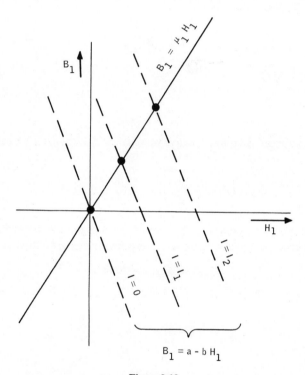

Figure 3.13

usually sufficiently accurate to neglect fringing flux or to correct for it by some factor related to the geometry.

For large air gaps determination of the demagnetizing factor is more difficult, because the basic assumption of uniform flux tubes is violated. Thus the distribution of flux in space surrounding a magnetic material is dependent on the geometry of the material as well as on any constraints imposed upon the surrounding space. This problem is of importance in the design of magnetic recording systems and will be considered in connection with them in Chapter 9.

Nonlinear Magnetic-Circuit Parameters

The preceding discussion has been generally confined to materials posses-sing a constant permeability μ. The nonlinearity of μ is, of course, precisely the property that makes ferromagnetic materials of great interest for switching circuits. In the remainder of this discussion attention will be focused on the static and dynamic characteristics of such materials and the derivation of the equivalent circuit values from measurable core parameters.

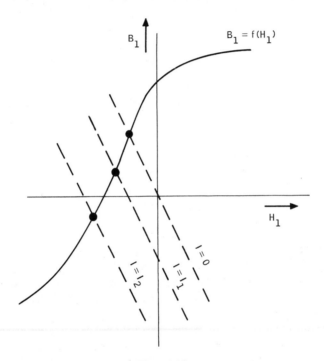

Figure 3.14

A generalized *B-H* relationship of the type of interest to us is shown in Figure 3.15. Points on this curve are defined as follows:

B_s = saturation flux density, the value of B at which no further change in magnetization of the material can occur with increase in H.

B_r = residual flux density (remanence); the value of B to which material will return on removal of magnetomotive force ($H = 0$) following saturation.

H_c = coercive force; or the value of applied field required to reduce residual flux density to zero following saturation.

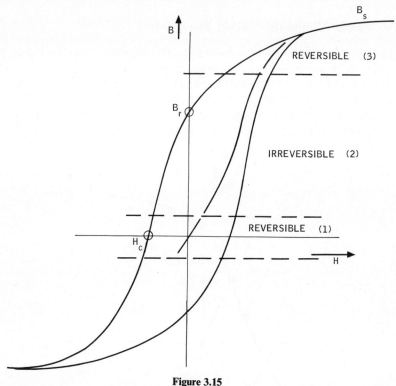

Figure 3.15

The ratio B_r/B_s is commonly defined as the *squareness ratio*.

Ferromagnetism in general and the shape of the curve of Figure 3.15 in particular can be briefly explained by means of domain theory as follows. Electron spins of individual atoms account for magnetization, because atoms thereby possess magnetic moments. In ferromagnetic materials there are five electrons in the outer shell of the atoms. Four are arranged in pairs, with one electron spinning in one direction and one in the other. The fifth, or uncompensated, electron spin causes each atom to appear as a saturated permanent magnet. In the course of crystal formation during cooking, atoms align themselves together in groups called *domains*. Within these domains nearly all atoms are aligned in parallel, making the domain a small permanent magnet. Domains have the property of acting as individual semielastic bodies. Accordingly, the regions of Figure 3.15 may be explained as follows:

1. Reversible domain boundary displacement. Domains tend to distort elastically to align with the field but will return to original condition upon its removal.

2. Irreversible domain growth or switching. Boundaries of domains nearly in line with the field grow and absorb nearby domains. Other domains, initially aligned in opposition to the field, will switch into alignment when the field becomes sufficiently strong. Both domain growth and switching are retained when the field is removed, giving the material its property of memory.

3. Reversible domain rotation. In the presence of high fields, domains tend to rotate in order to line up more precisely with the field. They will return to their original condition upon removal of the field.

The separate stages in this sequence of domain arrangements are, of course, not entirely distinct throughout the volume of the ferromagnetic body and therefore tend to merge into one another when the summation of individual events is viewed at the terminals of a winding.

TABLE 3.1

Hysteresis loop	Material	Typical B_s (gauss)	Typical H_c (oersteds)	Typical B_r/B_s	Application	Desirable characteristics
Figure 3.16a	Ferrite	1,600–2,200	0.5–1.2	0.95	Coincident-current magnetic-matrix memories; switch transformers	Fixed H_c; high squareness; fast switching
Figure 3.16b	Metal tape (or laminations)	7,000–15,000	0.02–0.3	0.8–0.95	Magnetic amplifiers; transformers; read-record heads	Low H_c; high B_s (high μ); high squareness
Figure 3.16c	Ferrite	2,000–3,000	0.1	Low	Read-record heads; linear transformers	High μ; low squareness (low loss)

The hysteresis loops of some typical useful magnetic materials are given in Figure 3.16, and the corresponding typical associated values are given in Table 3.1. The ferrite materials of Figure 3.16a are particularly desirable for coincident-current memory applications. They provide good discrimination between an applied H slightly less than H_c and one slightly greater, good information "retention" (high squareness), and low cost. They are poor for continuous high-frequency or transformer use because of high loss.

The tape materials of Figure 3.16b are commonly high-nickel-content irons, rolled to a range of from $\frac{1}{8}$ to 4 mils and wrapped upon a supporting nonferrous bobbin. They provide very high flux densities and low coercive force, thus requiring low drive currents.

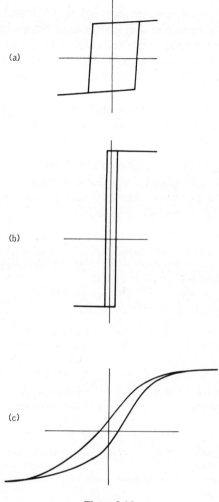

Figure 3.16

The ferrite materials of Figure 3.16c are commonly molded in a variety of shapes for assembly in a circuit with some air gap. Advantage can be taken of their reasonably high μ in linear transformers and in read-record heads. Other than the shape of the hysteresis curve, these ferrites are quite similar in mechanical and electrical properties to the square-loop type.

DC characteristics. Nearly all cores suitable for switching-circuit applications exhibit high permeability during low-speed flux reversal in the major hysteresis loop. The permeability, on the order of 10^3 to 10^4, is in fact sufficiently large so that it may be regarded as infinite relative to the effects due to geometry. It is therefore sufficiently accurate for early design purposes to consider the *material* properties to be specified by three points, namely H_c, B_r, and B_s, corresponding to the simplified B-H characteristic of Figure 3.9. This permits approximate determination of the values of two circuit elements in the magnetic-core equivalent circuit, namely, I_0 and L_w. Consider a thin circular band. Then from (3.8)

$$I_0 = \frac{5H_c r}{N}. \tag{3.14}$$

Because we shall want to evaluate I_0 for cores that are not infinitely thin, a question arises as to the value of r to be used. Note that the standard definition of H_c defines it as that field strength that just reduces flux density to zero. However, a finite core has an outer radius r_0 and an inner radius r_i, and I_0, the current required to initiate switching, is a function of the path length r; that is, $I_0 = I_0(r)$. Because we are usually interested in the value of current just required to begin switching the core at the inner radius, we shall take the value of I_0 to equal $I_0(r_i)$.

The second circuit parameter of interest is L_w. It can be related to core geometry and parameters by noting that

$$e = L\frac{di}{dt} = N\frac{d\phi}{dt} \times 10^{-8},$$

and therefore,

$$L = N\frac{d\phi}{di} \times 10^{-8}. \tag{3.15}$$

Using the characteristic of Figure 3.9, we find that the total flux switched when a core is driven from $H = 0$ to $H = H_c$ is $(B_s - B_r)A$. Let the core be a thin band of toroidal shape with constant width W and thickness Δr. Then the cross sectional area A is $\Delta r\, W$. The change in drive current required to produce this flux change is $\Delta i = I_0(r) = 5H_c r/N$. Substituting into (3.15), we obtain

$$L_w(N) = N^2\frac{B_s - B_r}{H_c}\frac{\Delta r}{r}\frac{W}{5} \times 10^{-8} \equiv N^2 L_w, \tag{3.16}$$

in which L_w is a function only of core material properties and geometry. The quantity $(B_s - B_r)/H_c$ will sometimes be referred to as the *saturation permeability* μ_s. It will enter into figures of merit for certain circuits.

The result (3.16) should not be taken too literally, because in fact ΔB may vary considerably over the actual hysteresis loop. In practice, the accuracy of (3.16) in predicting L_w depends on the sharpness of the knee of

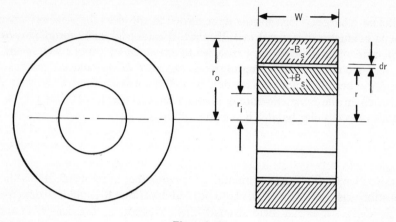

Figure 3.17

the major hysteresis loop and on the prior state of domain nucleation within the material. The value of L_w is affected not only by variations in μ_s, but also by the physical arrangement of windings around the magnetic core.

Geometry. The specific geometric form taken by a magnetic core enters the equivalent circuit primarily by means of the magnetizing inductance

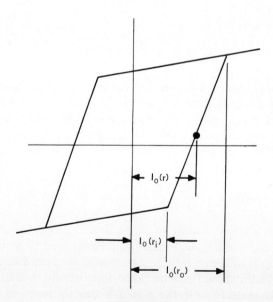

Figure 3.18

L_m. The effect can be described in terms of the domain wall-movement theory of magnetization reversal. This theory, experimentally verified in some switching modes, holds that magnetization reversal takes place along a roughly coherent "wall" representing the boundary between regions of opposite saturation within the core. The wall is considered to move, generally outward (toward regions of increasing mean magnetic path length), at a velocity such that the rate of change of magnetization just balances the applied voltage according to Faraday's law. Referring to Figure 3.17, consider a toroidal core initially at $-B_r$ being driven toward $+B_s$. During a small interval Δt a thin band of material Δr, located at radius r, switches from $-B_r$ to $+B_s$. The incremental flux switched is therefore $(B_s + B_r)W \Delta r$, and the outward velocity of the wall is obtained from

$$e = N\frac{\Delta\phi}{\Delta t} \times 10^{-8} = N(B_s + B_r)W \times 10^{-8}\frac{\Delta r}{\Delta t}. \qquad (3.17)$$

Material within r has already switched to $+B_s$ and therefore cannot contribute any further flux change, whereas material outside r has not yet been subjected to a field strength (H_c) sufficiently high to cause switching. For a toroidal core (3.14) gives the current just required for switching as a function of radius. Because the radius at which switching is taking place is progressively increasing, the effect upon the hysteresis loop is as shown in Figure 3.18. The rate of change of magnetizing current just required for switching is, in general, from (3.7),

$$\frac{dI_0}{dt} = \frac{H_c}{0.4\pi N}\frac{dp}{dt}. \qquad (3.18)$$

For a toroid (3.18) becomes

$$\frac{dI_0}{dt} = \frac{5H_c}{N}\frac{dr}{dt}. \qquad (3.19)$$

Substituting from (3.19) into (3.17), we obtain

$$e = N^2\frac{(B_s + B_r)W \times 10^{-8}}{5H_c}\frac{dI_0}{dt} = L_m\frac{dI_0}{dt}.$$

Therefore

$$L_m(N) = N^2\frac{(B_s + B_r)W \times 10^{-8}}{5H_c} \equiv N^2 L_m, \qquad (3.20)$$

with L_m again indicating an intrinsic property of a given core.

Transient Response. The third factor affecting the shape of the hysteresis loop involves the dynamics of magnetization reversal within the magnetic material. So far only ideal hysteresis loops have been considered; that is, it has been assumed that the loop completely defines voltage and current relationships by $e = NA \times 10^{-8} (dB/dt)$ and $i = Hp/0.4\pi N$, for all conditions of driving voltage or current. Discrepancies occur in a number of practical cases, however. For example, we have already noted that losses must be attributed to an equivalent core-shunt resistance, resulting in increased current flow with increases in switching rate. This observation may be explained on the basis of two effects. They are as follows:

1. Domain "inertia," which may be regarded as a reluctance of the domains to grow or rotate under the influence of an applied field. This is

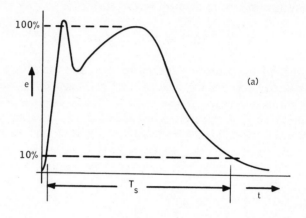

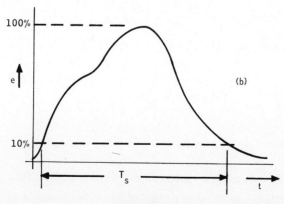

Figure 3.19

dominant in high-resistivity (ferrite) and ultrathin low-resistivity (metal tape) materials.

2. Eddy currents (induced in the thicker low-resistivity materials), which act to cancel the applied field as far as internal portions of the core are concerned.

These are very complex phenomena. We shall make an attempt to deal with simple forms of such processes later in this discussion. Meanwhile, it is fortunate that on a gross basis, at least, the effects can be lumped together and treated by using a rather simple relationship. In particular, it has been observed experimentally that the following relationship holds over a reasonably wide range:

$$S_w = (H_a - H_c)T_s, \tag{3.21}$$

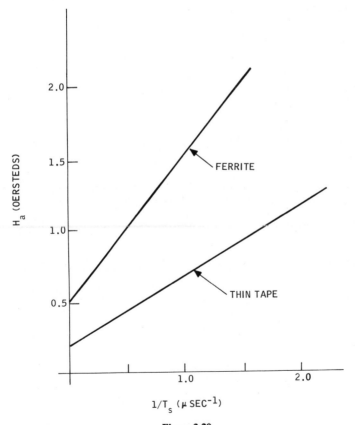

Figure 3.20

in which

    S_w = a constant (for a given core), in oersted-seconds,
    H_a = applied field, in oersteds,
    H_c = threshold field (at which the average domain wall velocity is zero), or dc coercive force,
    T_s = time required to reverse magnetization in the material (usually taken as the time between 10 percent points on the voltage output resulting from a step of applied current).

Figure 3.19 shows typical voltage transients in response to an applied current step for (a) metal tape cores and (b) ferrite cores. The initial voltage spike in the case of tape cores is generally attributed to the rapid formation of domains throughout the material at the instant the field is applied. Switching times versus applied field for typical materials are given in Figure 3.20.

Explanations for this observed phenomenon are of considerable interest in high-speed switching applications and yield insight into the mechanisms of magnetization reversal. Unfortunately, detailed switching models usually result in expressions that are not readily applied in the early stages of practical circuit design. It is initially more useful, therefore, to approximate switching behavior based on the idealization of (3.21) or on the characteristics of Figure 3.20, and on an assumption that the response to a step current drive can be approximated by a step voltage, as illustrated in Figure 3.21. This assumption simply means that because the area under the voltage-time curves is proportional to total flux, this area can be arranged in the form of a step response of voltage to approximate the shape of the actual curve, while retaining the total area. Approximately, therefore, $d\phi/dt$ can be assumed to be constant during time of switching, T_s, and of such a magnitude as to cause the core to switch from $-B_r$ to $+B_s$ within that time. E and I are consequently related by (3.21), resulting in an equivalent resistance R_c that represents core loss in excess of that experienced in traversing the dc hysteresis loop.

The value of R_c may be derived as follows. Rearranging (3.21), we have

$$H_a = H_c + \frac{S_w}{T_s} = H_c\left(1 + \frac{S_w/H_c}{T_s}\right) \equiv H_c\left(1 + \frac{\tau_m}{T_s}\right). \qquad (3.22)$$

The quantity, $\tau_m \equiv S_w/H_c$, an intrinsic property of the core material, will be referred to as the *magnetic time constant*. Note that by equating the switching time to a frequency $T_s \equiv 1/\omega$ and then replacing ω by the complex frequency s, we can write (3.22) as

$$G(s) \equiv \frac{H_c}{H_a(s)} = \frac{1}{\tau_m s + 1}. \qquad (3.23)$$

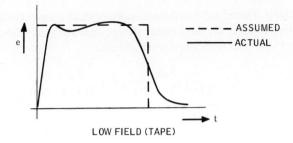

LOW FIELD (TAPE)

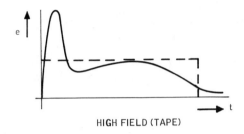

HIGH FIELD (TAPE)

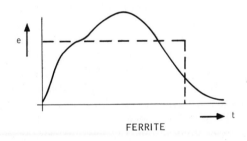

FERRITE

Figure 3.21

It will not be too surprising, therefore, to find that the behavior of magnetic logic and memory circuits is dominated by a single-pole form of frequency response, in much the same way that transistor circuits are to a first order limited by the form of $\alpha_N(s)$. For a constant (square-pulse) voltage E appearing across the core during switching, $ET_s = N\phi \times 10^{-8}$, with $\phi = (B_s + B_r)A$. Solving for the switching time gives $T_s = (N\phi/E) \times 10^{-8}$. We also know that H_a is related to the total current flowing in the core by $H_a = 0.4\pi NI/p$. This, of course, neglects the effects of the magnetizing inductance L_m. However, (3.21) holds approximately true for cores with very large magnetizing inductance, in which no appreciable change in driving current occurs during switching. Therefore the two equivalent circuit elements can be treated independently, at least for first-order design calculations. Substituting for H_a and T_s in (3.22) and rearranging, we obtain

$$I = \frac{S_w p}{0.4\pi N^2 \phi \times 10^{-8}} E + \frac{H_c p}{0.4\pi N}. \tag{3.24}$$

However, (3.24) is of the form

$$I = \frac{E}{R_c} + I_0.$$

Therefore

$$R_c(N) = N^2 \frac{0.4\pi\phi \times 10^{-8}}{S_w p} = N^2 R_c. \tag{3.25}$$

Again, R_c is only a function of core material and geometry. For $\phi = (B_r + B_s)A$ and a thin band of material such that $p = 2\pi r$, (3.25) becomes

$$R_c = \frac{(B_r + B_s)A \times 10^{-8}}{5S_w r}. \tag{3.26}$$

Again the question arises as to the appropriate value of r to use in (3.26). If the 10 percent points are representative of the time it requires to switch material even at the outer radius extremity, r_0 should be used. This point deserves some elaboration. If we again consider thin bands of material at various radii, note that (3.21) means, in effect, that the bands nearest r_i will switch more rapidly, because H_a is largest there. If the bands are taken to have equal cross-sectional area, however, rapid switching means that the inner bands will complete the traversal of the loop from $-B_r$ to $+B_s$ sooner than the outer bands. Thus the output waveform should be composed of large but brief contributions (high E, low T_s) from inner bands and small but longer contributions (low E, long T_s) from outer bands. This implies that the 10 percent points of Figure 3.19 are representative of outer regions corresponding to r_0.

Switching Models

The foregoing analysis, although approximating gross magnetic-core behavior on the basis of a linearized hysteresis loop, fails to account for many of the observable details of a switching waveform. In particular, initial high-velocity flux reversal and nonlinear turnover rates such as illustrated in Figure 3.21 can only be crudely approximated by a linear model. Some study of the internal details of core switching is therefore in order.

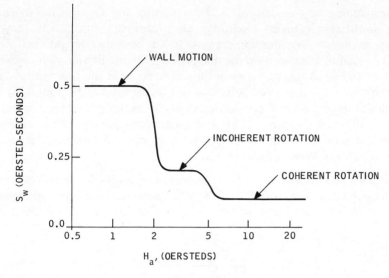

Figure 3.22

Models of magnetization reversal in both metal tape and ferrite cores can be placed into one of two broad categories: (a) reversal by domain wall motion and (b) reversal by domain rotation. The former category applies primarily to low-resistivity materials in which the switching mechanism is dominated by eddy currents and to high-resistivity materials subjected to low driving fields (up to the general order of twice the coercive force). The latter applies to high-resistivity materials under high-field drive (more than twice the coercive force). In both cases it is possible to demonstrate experimentally and theoretically that the net rate of magnetization reversal at a point within the material is related to the product of excess field strength and magnetization at that point by a coefficient β, which may be regarded as a viscous damping parameter. Equations of motion, therefore, exhibit the general form, in one dimension,

$$\beta v = (H_a - H_c)M_s,\tag{3.27}$$

where v is the velocity of magnetization reversal, H_a is the applied field, and M_s is the saturated domain magnetization and equals $B_s/4\pi$ in appropriate units. If a domain wall moves a distance d in time τ (or domain rotation produces an equivalent change in magnetization), $v = d/\tau$, and the reversal equation may be put in the form

$$(H_a - H_c) = \frac{\beta d}{M_s}\frac{1}{\tau},\tag{3.28}$$

corresponding to the form of (3.21). For both domain wall motion and domain rotation, equivalent velocity v and vector magnetization M_s may be functions in two or three dimensions. It has been determined that these functions, in turn, depend on the magnitude and distribution of the applied field as well as on the geometry of the magnetic material. The result is that appropriate factors can be derived to suit the form of (3.27) and (3.28) for particular materials and geometries, and the resulting switching constant S_w is valid over a limited range of applied fields. Shevell [1] has obtained switching constants experimentally for a number of ferrite materials; a typical curve is shown in Figure 3.22.

Domain wall motion. Consider a magnetic core, initially driven to saturation, to which is applied a field in the reverse direction. It has been experimentally observed that under very low applied fields, particularly in single

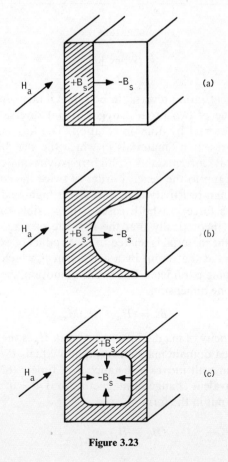

Figure 3.23

crystals, domain nucleation will take place along the inner boundary of the core and will travel outward with time. The state of the core at any instant thus corresponds to magnetized domains carrying flux around the path length in opposite directions. The boundary between oppositely magnetized domains is known as a *Block wall*, or 180° domain wall. In this simple case, the position of the wall is directly related to net magnetization of the core, that is, to the net flux level or position of the core on the vertical coordinate of the hysteresis loop. As the applied field is increased, the velocity of the domain wall increases according to the form of (3.27). At low fields the boundary may be regarded as a plane, as in Figure 3.23*a*. As the driving field is increased, generating larger eddy currents in conductive materials, the interior of the core tends to be shielded by the internal eddy currents. The result is an equilibrium akin to that of surface tension, giving the reversal distribution of Figure 3.23*b*. At high fields nucleation occurs at all surfaces, and the 180° walls move inward approximately symmetrically from all surfaces simultaneously, as shown in Figure 3.23*c*.

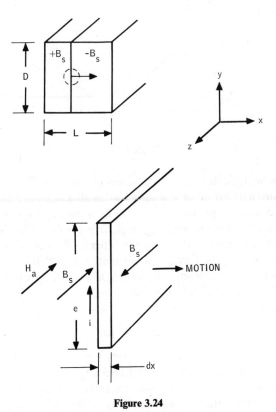

Figure 3.24

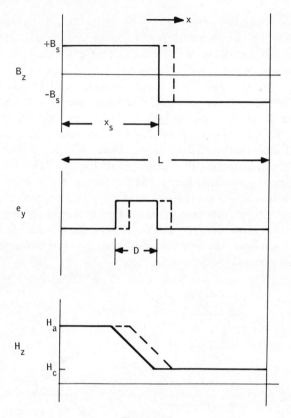

Figure 3.25

An example will illustrate how the model of Figure 3.23*a* can be analyzed. Consider a small portion of a plane 180° wall extending in the *y* and *z* directions and moving in the *x* direction, as shown in Figure 3.24. Initially it may be assumed that the wall has negligible thickness in the *x* direction. On one side, flux density is $+B_s$, on the other $-B_s$. Let the position of the wall be x_s. For $x > x_s$ the flux is not yet switched, and $B = -B_s$ uniformly. At $x = x_s$ flux is reversing from $-B_s$ to $+B_s$ at the rate (per unit *y* dimension) $2B_s(dx_s/dt)$. The electrical field intensity in volts per centimeter induced at the wall is therefore

$$e_y = 2B_s v \times 10^{-8}, \tag{3.29}$$

where *v* is the wall velocity in centimeters per second. For $x < x_s$ the induced field intensity and resulting current density will have some distribution that decreases in magnitude at distances remote from the wall. An exact solution of the distributions for this model of low-field switching has been made by Williams, Shockley, and Kittel [3]. The result shows that it is sufficiently accurate to assume that the field intensity and current density are uniform and are confined to an equivalent region D, the total width of the specimen in the y direction. The instantaneous relationships are shown as a function of x in Figure 3.25. The current density in the xz plane is

$$i_y = \frac{e_y}{\rho} = \frac{2B_s v}{\rho \times 10^8}, \tag{3.30}$$

where ρ is resistivity in ohm-centimeters. Because magnetomotive force possesses a component in the z direction only, the integral of H around a unit area in the xz plane will consist of $H_z(x) - H_z(x + \Delta x)$ and will enclose a current density i_y. Because $i_y = 0$ is assumed everywhere outside the region $x_s - D < x < x_s$, H is uniform for $x < x_s - D$ and $x > x_s$. However, at $x = 0$, $H = H_a$, and at $x = x_s$, H must equal H_c for switching to occur. Therefore integration of $H\,dp$ may be taken around the region D only. In this case, considering a unit distance in the z direction, we have

$$\oint H \cdot dp = H_a - H_c = 0.4\pi D i_y = \frac{0.8\pi B_s D}{\rho \times 10^8} v, \tag{3.31}$$

which is of the form of (3.27). The total switching time is given by $T_s = L/v$. Substituting for v from (3.31), we obtain

$$T_s = \frac{L}{v} = \frac{0.4\pi \phi_s}{\rho \times 10^8} \frac{1}{H_a - H_c}, \tag{3.32}$$

in which $\phi_s \simeq 2B_s A = B_s L D$. This is the same form as (3.21), with

$$S_w = \frac{0.4\pi \phi_s}{\rho \times 10^8}, \tag{3.33}$$

a function of material properties only. The voltage produced on a winding around the specimen is

$$e = 2NB_s D \frac{dx_s}{dt} \times 10^{-8}. \tag{3.34}$$

Substituting from (3.31) for v gives

$$e = \frac{N\rho}{0.4\pi}(H_a - H_c). \tag{3.35}$$

Assuming $H_a \gg H_c$ and finding H_a from Ampere's law, we obtain an equivalent core resistance,

$$R_c(N) = \frac{e}{i} = \frac{N^2 \rho}{p}. \tag{3.36}$$

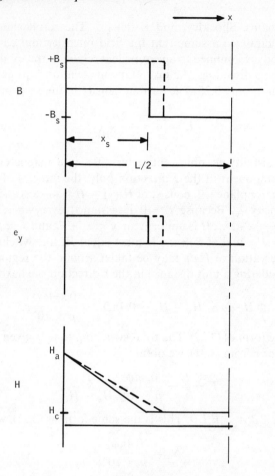

Figure 3.26

This is simply (3.25), which was derived on the basis of an experimental switching constant S_w, with (3.33) for S_w.

This simple model is, of course, inadequate to handle the waveforms observed experimentally. A considerably closer match is obtained if it is assumed that the dimensions of the material are such that field intensity and current density are not confined to a fixed region D, but rather to the region $0 < x < x_s$, in accordance with the work of Chen and Papoulis [4] and Ekstein [5]. The resulting field distributions are illustrated in Figure 3.26, in which wall movement is considered to be inward from both sides of the specimen, and the transient is considered complete when $x_s = L/2$. It can be seen qualitatively that as x_s increases, the total magnitude of eddy currents opposing switching likewise increases, and switching will therefore proceed progressively more slowly. The current density in the region $0 < x < x_s$ is still considered to be given by (3.30), but the magnetomotive force integral becomes (enclosing the current region)

$$\oint H \, dp = H_a - H_c = 0.4\pi i_y x_s = \frac{0.8\pi B_s}{\rho \times 10^8} x_s \frac{dx_s}{dt}. \tag{3.37}$$

For a constant applied field the solution to this differential equation is

$$x_s(t) = \left[\frac{\rho(H_a - H_c) \times 10^8}{0.4\pi B_s} t \right]^{\frac{1}{2}}. \tag{3.38}$$

The output voltage appearing at the terminals of a winding around this specimen must be

$$e = N \frac{d\phi}{dt} \times 10^{-8} = 2NB_s \frac{dA}{dt} \times 10^{-8} = 2NB_s D \frac{dx_s}{dt} \times 10^{-8}. \tag{3.39}$$

Differentiating (3.38) and substituting in the above gives

$$e = N \left[\frac{B_s \rho D^2 (H_a - H_c)}{(0.4\pi \times 10^8)} \right]^{\frac{1}{2}} t^{-\frac{1}{2}}. \tag{3.40}$$

Switching is complete when $x_s(T_s) = L/2$, which assumes the 180° walls are moving inward simultaneously from two sides, corresponding to Figure 3.23c. Solving (3.38), we obtain

$$T_s = \frac{\pi B_s L^2}{\rho \times 10^9} \frac{1}{H_a - H_c}. \tag{3.41}$$

This result simply means that the general switching relationship of (3.21) still holds true, although the output voltage waveform is of a different form.

At higher field strengths or in materials in which eddy currents may be neglected, neither of the above models is adequate. More general approaches to the description of switching by domain wall movement have been made by Menyuk and Goodenough [6] and by Chen and Papoulis [4]. In these analyses it is consistently assumed that wall motion is limited only by a viscous damping factor and is proportional to the excess applied field. For a constant applied field, local wall velocity is taken to be a constant. The instantaneous rate of magnetization reversal therefore depends upon the instantaneous geometrical configuration of domain walls; that is, the change in magnetization is dependent upon the total instantaneous domain wall area, which in turn is a function of the state of magnetization. The change of magnetization is therefore proportional to the wall velocity v and to the instantaneous wall area. Letting $f(B)$ be proportional to the domain wall area per unit volume, we have

$$\frac{dB}{dt} = 2B_s(H_a - H_c)f(B).\qquad(3.42)$$

For simple assumed geometries, the form of $f(B)$ can be determined. For example, if domains are regarded as uniform needles increasing in number, $f(B)$ is simply proportional to B. For spherical domains, B is proportional to r^3, and the wall area is proportional to r^2. For cylindrical domains, B is proportional to r^2, and the wall area is proportional to r. Thus

$$f(B) = n2\pi r, \qquad B = n\pi r^2.\qquad(3.43)$$

Therefore

$$f(B) = 2(n\pi)^{\frac{1}{2}}B^{\frac{1}{2}}.\qquad(3.44)$$

Experimental results indicate some agreement with the cylindrical model, at least during the period of increasing $f(B)$. The model can be visualized as in Figure 3.27. Using $f(B)$ from (3.44), we find that the solution of (3.42) is of the form $B = (k^2/4)t^2$, and the corresponding output voltage, proportional to dB/dt is

$$e \sim \frac{dB}{dt} = [8n\pi B_s^{\,2}(H_a - H_c)^2]t.\qquad(3.45)$$

This simply means that during the portion of switching that occurs before expanding reversal cylinders begin to collide, the output voltage rises linearly with time. The output of a typical core driven by a step current is shown in Figure 3.28 with portions of the waveform corresponding to the model indicated.

Rotational models. At sufficiently high field strengths it can be observed that at least part of the magnetization reversal is accomplished by the mechanism of domain rotation. During such a switching transient, domains rotate through directions in which they possess a component of magnetization orthogonal to the direction of the applied field. Rotation may be incoherent, in which case the direction of individual domain rotation is random, or coherent. In this instance all domains tend to go through the same angular sequence. Models have been proposed by Gyorgy [8] and by Smith [10] for describing switching behavior in these modes. They consist of equations of motion of the magnetization vector $\overline{M}$ of the form $\partial \overline{M}/\partial t = f(\overline{M}, \overline{H})$. The proposed functions consist of a term that is proportional to the torque $\overline{T}$ (or, in terms of the applied field, the product $\overline{M} \times \overline{H}$) acting on the magnetization vector, plus a loss term that has taken various forms. Some experimental

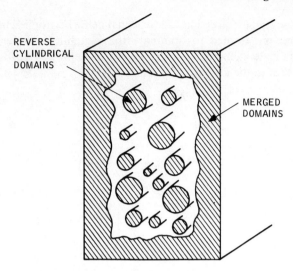

REVERSE CYLINDRICAL DOMAINS

MERGED DOMAINS

Figure 3.27

confirmation of calculated results has been obtained for cores switching in the several-nanosecond region.

A physical interpretation of Gyorgy's model can be readily obtained in simple cases, in which it is assumed that the magnetic core (toroid) is equivalent to an infinite cylinder extending in the z direction (see Figure 3.29a). The core is initially in a condition in which the saturated magnetization

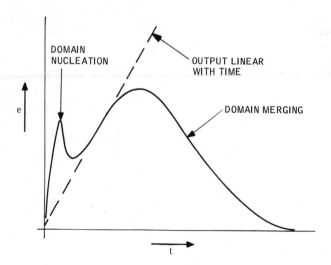

DOMAIN NUCLEATION

OUTPUT LINEAR WITH TIME

DOMAIN MERGING

e

t

Figure 3.28

vector $\overline{M}_s$ is in the $+z$ direction. The external field H is applied in the reverse direction and is assumed to dominate the switching transient (H much greater than the demagnetizing fields). The torque applied by H is, in general, $\overline{M} \times \overline{H}$, so that in the absence of losses

$$\frac{\partial \overline{M}_s}{\partial t} = \frac{\overline{M}_s \times \overline{H}}{S}, \tag{3.46}$$

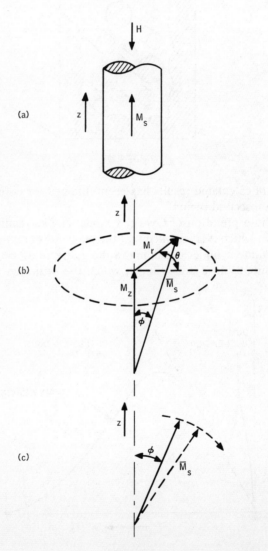

Figure 3.29

in which the constant of proportionality, S, has the dimensions of oersted-seconds. For $\overline{H}$ in the z direction and $\overline{M}_s$ in an arbitrary position, $\overline{M}_s = \overline{M}_z + \overline{M}_r$, the only component of $\overline{M}_s$ contributing to (3.46) is $\overline{M}_r$. The torque is therefore in the θ direction (Figure 3.29b), and the magnetization vector precesses about the z axis according to the relation

$$M_r \frac{d\theta}{dt} = \frac{M_r H}{S}. \tag{3.47}$$

The frequency of precession is thus simply H/S. The loss or dissipative term is assumed to have the general form

$$\frac{\partial \overline{M}_s}{\partial t} = \frac{\alpha}{M_s}\left(\overline{M}_s \times \frac{\partial \overline{M}_s}{\partial t}\right), \tag{3.48}$$

in which the constant of proportionality α is dimensionless. For $\alpha \ll 1$ we may think of the precession described by (3.47) as being virtually unaffected. However, (3.48) will cause $\overline{M}_s$ to move slowly in the direction of $\overline{H}$ at a rate determined by the precessional velocity. The relatively slow motion of $\overline{M}_s$ in the plane determined by $\overline{M}_s$ and the z axis is depicted in Figure 3.29c. The contribution to $\partial \overline{M}_s/\partial t$ due to $d\phi/dt$ can be neglected relative to the much larger velocity in the θ direction. Thus

$$\overline{M}_s \times \frac{\partial \overline{M}_s}{\partial t} \cong M_s M_r \frac{d\theta}{dt}. \tag{3.49}$$

Using (3.47) and (3.49), we obtain

$$\frac{d\phi}{dt} \cong \frac{\alpha H}{S}\frac{M_r}{M_s} = \frac{\alpha H}{S}\sin\phi. \tag{3.50}$$

Generally, we are only interested in the z component of flux and in the length of time for switching. However, $M_z = M_s \cos\phi$. Differentiating and using (3.50) gives

$$\frac{dM_z}{dt} = \frac{\alpha M_s H}{S}\left[1 - \left(\frac{M_z}{M_s}\right)^2\right]. \tag{3.51}$$

Note that (3.51) is of the form of (3.42) with $H_a \gg H_c$ and $f(B)$ given by the term in brackets. The time function corresponding to (3.51) appears as in Figure 3.30. It is also noteworthy that integration of (3.51) leads to a linear relationship between total switching time and reciprocal driving field of the same form as (3.21) for $H_a \gg H_c$.

Consideration of magnetization and field strength components in three dimensions yields some insight into the switching behavior of special geometries, particularly multipath arrangements and thin film, and is

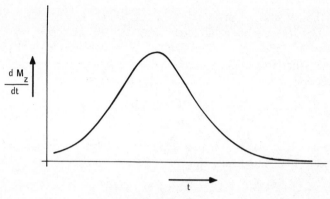

Figure 3.30

therefore useful for design purposes even at relatively low switching speeds. In accordance with the treatment by Smith [10], a small element of magnetic material is depicted in Figure 3.31, in which the field may have components in any direction, and the instantaneous magnetization is given by the direction of the saturated magnetization vector $\overline{M}_s$. In any single-crystal or polycrystalline material with regular crystal structure there exist certain directions of "easy" magnetization corresponding to the crystal axes. Alignment of domains in a direction other than along the crystal axes results

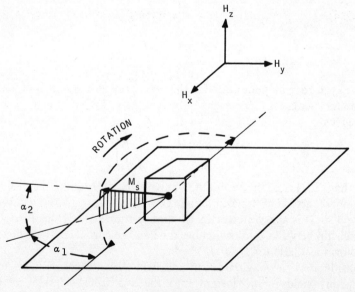

Figure 3.31

in elastic distortion or strain of the crystal structure. This elastic distortion results in energy being stored in the mechanical structure in a manner analogous to the storage of energy in a stretched spring. This stored energy is called the *anisotropic energy* (lack of uniform or isotropic energy distribution in all vector directions). It is proportional only to a crystal anisotropy constant and not to the magnitude of the magnetization vector. Assuming a single easy axis in the x direction, the anisotropic energy is generally taken to have the form

$$E_c = E_a \sin^2 \alpha_1, \tag{3.52}$$

which is 0 at $\alpha_1 = 0$ or π and some value E_a (which is a property of the material) at $\alpha_1 = \pi/2$. There is also energy due to the external field E_f. Because B (or M) is proportional to the voltage-time integral and H is proportional to current, the HM product represents such energy (corresponding to the BH product plots commonly drawn for permanent magnets). The total energy is therefore given by $E = E_c + E_f$, or in the rectangular coordinates of Figure 3.31,

$$E = E_a \sin^2 \alpha_1 + H_x M_s \cos \alpha_1 \cos \alpha_2 - H_y M_s \sin \alpha_1 \cos \alpha_2 + H_z M_s \sin \alpha_2. \tag{3.53}$$

Consider a special case in which a sample is subjected to external applied fields in the x and y directions, but only to a demagnetizing field in the z direction. H_z is therefore dependent upon the geometry of the sample. As we have seen previously, there is a linear relationship between B (or M) and H determined by the demagnetizing factor. Therefore the field in the z direction can be described by

$$H_z = -K_d B_z = -K_d M_s \sin \alpha_2. \tag{3.54}$$

The total free energy is therefore

$$E = E_a \sin^2 \alpha_1 + H_x M_s \cos \alpha_1 \cos \alpha_2 - H_y M_s \sin \alpha_1 \cos \alpha_2 - K_d M_s^2 \sin^2 \alpha_2. \tag{3.55}$$

Consider now the case of low-speed or static switching in which the magnetization vector rotates only in the xy plane. This is frequently appropriate for thin-film materials in which the demagnetizing factor in the z direction is sufficiently large relative to other fields to confine the magnetization to the plane of the film. Thus $\alpha_2 = 0$. In order to determine the hysteresis loop in the x direction it is necessary to calculate $M_s \cos \alpha$ versus H_x ($\alpha = \alpha_1$) for various fixed values of H_y. The state of magnetization will always adjust so that total energy is minimized. This can be verified by setting $\partial E/\partial \alpha = 0$. As pointed out by Smith, we also wish this minimum energy to be maximized, consistent with the equations, in order to determine the major hysteresis loop (discontinuity in the α-versus-H_x curve). Therefore set $\partial^2 E/\partial \alpha^2 = 0$. Carrying out these operations on (3.55) gives

$$\frac{\partial E}{\partial \alpha} = 2E_a \sin \alpha \cos \alpha - H_x M_s \sin \alpha - H_y M_s \cos \alpha = 0,$$

$$\frac{\partial^2 E}{\partial \alpha^2} = 2E_a(\cos^2 \alpha - \sin^2 \alpha) - H_x M_s \cos \alpha + H_y M_s \sin \alpha = 0. \tag{3.56}$$

Define field strengths normalized to the anisotropy energy as follows:

$$h_x \equiv \frac{H_x M_s}{2E_a},$$

$$h_y \equiv \frac{H_y M_s}{2E_a}. \tag{3.57}$$

We shall also regard the magnetization as normalized to the maximum M_s, according to $m \equiv M_s \cos \alpha/M_s = \cos \alpha$. Equations (3.56) then become

$$1 - \frac{h_x}{\cos \alpha} - \frac{h_y}{\sin \alpha} = 0,$$

$$1 - h_x \frac{\cos \alpha}{\cos^2 \alpha - \sin^2 \alpha} + h_y \frac{\sin \alpha}{\cos^2 \alpha - \sin^2 \alpha} = 0. \tag{3.58}$$

For the case $h_y = 0$, it can be readily verified that the solution is $h_x^2 = 1$, or $h_x = \pm 1$, and $\cos^2 \alpha = 1$, or $\alpha = 0°$, $180°$. This simply means that the minimum energy reaches a maximum at precisely that field strength at which the irreversible flux change of $2m$ occurs. The result is the ideal square hysteresis loop. For $h_y \neq 0$, the equations above can be solved by obtaining h_x and h_y in terms of α. For example, multiplying the first equation of (3.58) by $\sin^2 \alpha \cos \alpha$, the second by $\cos \alpha(\cos^2 \alpha - \sin^2 \alpha)$, and adding, eliminates h_y. By similar manipulation h_x is removed. The two results are

$$h_x = \cos^3 \alpha,$$

$$h_y = \sin^3 \alpha. \tag{3.59}$$

Taking the $\frac{2}{3}$ power of each of (3.59), and adding, gives

$$h_x^{2/3} + h_y^{2/3} = 1. \tag{3.60}$$

This is a widely used relationship, whose parameters (3.57) can be readily determined experimentally. It defines the regions in the $h_x h_y$ plane within which switching will and will not occur, as depicted in Figure 3.32. For h_x and h_y not on the curve of Figure 3.32, the shape of the hysteresis loop can

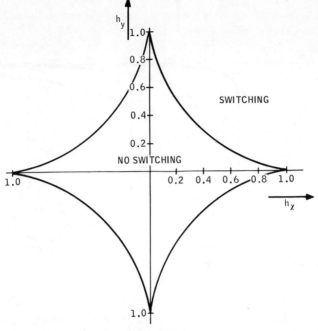

Figure 3.32

be determined by returning to the minimum-energy condition corresponding to the first equation of (3.58). Note that for h_x and h_y not satisfying (3.60), the second equation of (3.58), implying an extremum, is violated. Because $m = \cos \alpha$, the first equation of (3.58) can be written

$$h_x = m\left(1 - \frac{h_y}{\sqrt{1 - m^2}}\right). \tag{3.61}$$

This function is plotted for several values of h_y in Figure 3.33. The extrema, corresponding to the major hysteresis loop, occur at points satisfying (3.60).

If h_x is maintained at zero while h_y is increased from 0 to nearly 1.0 and returned to zero, the observed flux in the x-direction will reduce from m to zero and return to m. This is because the model assumes no crystal axes in the y direction. Thus any component of the magnetization vector in the y direction implies an energy that must be supplied by the external field. When this field is removed, the domains are compelled to return to the minimum-energy condition, which corresponds to alignment with the axis of anisotropy. The change in observed flux corresponds to the movement of the intersection of (3.61) with the $h_x = 0$ axis in Figure 3.33, as h_y takes on

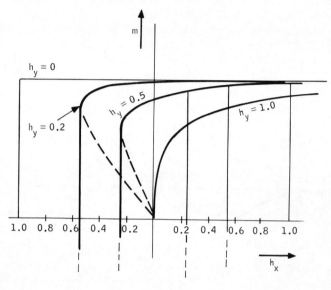

Figure 3.33

values other than zero. It is therefore possible to observe the stored flux in a core nondestructively by applying a field slightly less than that sufficient to align domains to 90° with the normal direction of magnetization, and subsequently removing the field, allowing domains to return to their original condition. If flux is originally stored in the 0° direction, the observed flux change is negative during rotation toward 90° alignment, and positive during return. If initial magnetization is in the 180° direction, the sign of the flux change sequence is reversed.

REFERENCES

[1] W. L. Shevell, "Millimicrosecond Switching Properties of Ferrite Computer Elements," *J. Appl. Phys.*, **30**, 4 (April 1959).

[2] C. P. Bean and D. S. Rodbell, "Kinetics of Magnetization in Some Square Loop Magnetic Tapes," *J. Appl. Phys.*, **26**, 1 (January 1955).

[3] H. J. Williams, W. Shockley, and C. Kittel, "Studies of the Propagation Velocity of a Ferromagnetic Domain Boundary," *Phys. Rev.*, **80**, 6 (December 1950).

[4] T. C. Chen and A. Papoulis, "Terminal Properties of Magnetic Cores," *Proc. IRE*, **46**, 5, 839–849 (May 1958).

[5] H. Ekstein, "Theory of Remagnetization of Thin Tapes," *J. Appl. Phys.*, **26**, 11, 1342–1343 (November 1955).

[6] N. Menyuk and J. B. Goodenough, "Magnetic Materials for Digital Computer Components. I. A Theory of Flux Reversal in Polycrystalline Ferromagnetics," *J. Appl. Phys.*, **26**, 1, 8–18 (January 1955).

[7] N. Menyuk, "Magnetic Materials for Digital Computer Components. II. Magnetic Characteristics of Ultra-thin Molybdenum-Permalloy Cores," *J. Appl. Phys.*, **26**, 6, 692–697 (June 1955).

[8] E. M. Gyorgy, "Rotational Model of Flux Reversal in Square Loop Ferrites," *J. Appl. Phys.*, **28**, 1011–1015 (September 1957).

[9] E. M. Gyorgy, "Flux Reversal in Soft Ferromagnetics," *J. Appl. Phys.*, **315**, 5, 1105–1175 (May 1960).

[10] D. O. Smith, "Magnetization Reversal and Thin Films," *J. Appl. Phys.*, **29**, 3, 264–273 (March 1958).

[11] R. W. Roberts and R. I. Van Nice, "Influence of ID-OD Ratio on Magnetic Properties of Toroidal Cores," *Electrical Eng.*, **74**, 10, 910–914 (September 1955).

[12] R. W. Roberts and R. I. Van Nice, "Proceedings of the Conference on Magnetism and Magnetic Materials," *J. Appl. Phys.*, **29**, 3 (March 1958).

[13] O. J. Van Sant, "Magnetic Viscosity in 4-79 Molybdenum Permalloy," *J. Appl. Phys.*, **28**, 4, 486–494 (April 1957).

[14] S. H. Chow, "On a Nonlinear Diffusion Equation Applied to the Magnetization of Saturable Reactors," *J. Appl. Phys.*, **25**, 3, 377–381 (March 1954).

[15] D. S. Rodbell and C. P. Bean, "Influence of Pulsed Magnetic Fields on the Reversal of Magnetization in Square-loop Metallic Tapes," *J. Appl. Phys.*, **26**, 11, 1318–1323 (November 1955).

EXERCISES

3.1 The following measurements are taken on a tape-wound magnetic core having an outside magnetic path length of 2.5 cm and made up of 10 wraps of $\frac{1}{8}$-in tape, each wrap of 10^{-4} cm^2 cross-sectional area:

No. turns (N)	Current drive	Switching time
10	60 mA	5 μs
10	120 mA	1.25 μs

In a further experiment it is found that with $N = 100$, the core will switch in 3 μs with an applied drive of 5 V. Find B_s, H_c, and S_w, and the equivalent one-turn core parameters I_0, R_c, and L_m.

3.2 What are the equivalent circuit characteristics I_0, ϕ_s, R_c, L_m, and L_w of a memory core having the following properties:

OD: 30 mils,
ID: 20 mils,
Width: 10 mils,
B_r: 1,900 G,
B_s: 2,000 G,
H_c: 1.0 Oe,
S_w: 1.0 Oe-μs.

3.3 In order to switch the core of Problem 3.2 in 2 μs what current drive should be applied? Estimate the voltage output for this switching rate.

3.4 It is desired to produce a 5-μs, 10-V pulse at a load $R_L = 500\ \Omega$ using the transformer of Figure P.3.1. The source is a switch limited to $E_1 = 20$ V peak amplitude. Find minimum

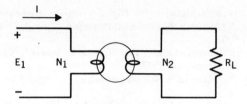

Figure P.3.1

N_2 and N_1, and maximum I (at the end of the pulse) if the core geometrical parameters are

$$OD = 0.375 \text{ in.},$$
$$ID = 0.260 \text{ in.},$$
$$\text{Width} = 0.125 \text{ in.},$$

and material parameters are the same as in Problem 3.2.

3.5 The same core as in Problem 3.4 is used in the circuit of Figure P.3.2*a*. The input and output waveforms are shown in Figure P.3.2*b*. It is desired to choose N such that $t_1 < 1\ \mu s$ and $t_2 > 6\ \mu s$ when the core is initially saturated. Is this possible? If so, what is the range of values of N?

3.6 Input and output waveforms for the loaded transformer of Figure P.3.1 when the input winding has a resistance of $10\ \Omega$ are shown in Figure P.3.3. Find E_0, ΔE_0, t_1, t_2, τ_1, and τ_2 if $N_1 = N_2 = 100$ turns, $R_L = 1\ k\Omega$, and if transformer core parameters are

$$H_c: 0.3 \text{ Oe},$$
$$B_s: 7,000 \text{ G},$$
$$L_w: 5\ \mu H/\text{turn}^2,$$
$$L_m: 50\mu H/\text{turn}^2,$$
$$R_c: 0.1\ \Omega/\text{turn}^2.$$

3.7 A blocking oscillator is to be constructed, using a square-loop core, to produce 20-V, 5-μs output pulses every 25 μs. The circuit is shown in Figure P.3.4, and is normally OFF until the transistor is triggered ON via a pulse through the capacitor. I_B is a constant-current source designed to return the core to its initial condition during the period between triggers, in preparation for the next output pulse. The transistor has a current gain $\beta = 50$, and for reliability is to be limited to a collector current $I_c = 100$ mA. Letting $N_1 = N_2 = N_3 = N_4 = N$, find N, I_B, maximum R, and minimum R_L. Neglect transistor switching times, and assume a perfect diode. Core parameters

$$I_0: 1.0 \text{ A/turn}$$
$$2\phi_s: 100 \text{ V-s/turn}$$
$$R_c: 0.1\ \Omega/\text{turn}^2$$
$$L_m: 10^{-6}\ \mu H/\text{turn}^2$$

3.8 Assuming that both modes of switching depicted in Figures 3.25 and 3.26 exist simultaneously, show from Equations (3.34) through (3.37) that

$$H_a - H_c = \frac{0.4\pi B_s}{10^8}\left(\frac{2}{\rho}x_s\frac{dx_s}{dt} + \frac{2D}{R_c p}\frac{dx_s}{dt}\right)$$

in which R_c is given by Equation (3.36) with $N = 1$ for internal eddy currents. Note that this corresponds to switching by the mode of Figure 3.26 while being loaded by an external

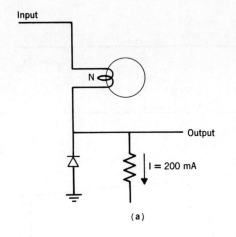

Input

N

Output

I = 200 mA

(a)

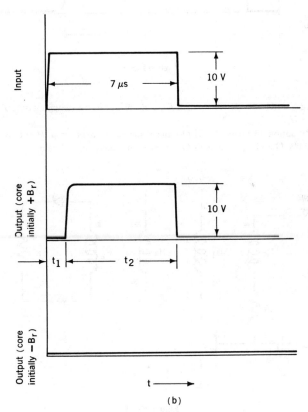

Input

7 μs

10 V

Output (core initially +Br)

10 V

t₁

t₂

Output (core initially −Br)

t

(b)

Figure P.3.2

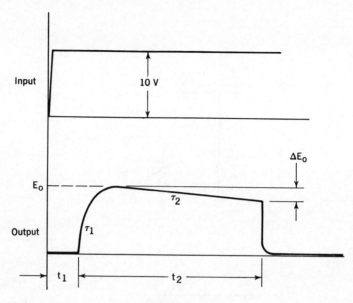

Figure P.3.3

winding N and load resistor R_L with $R_c = R_L/N^2$. Find the switching time T_s as a function of $H_a - H_c$.

3.9 Using Equations (3.57) and (3.61), plot the hysteresis loop (M versus H_x) of a material having $M_s = 2,000$ G and $E_a = 1,400$ G-Oe, if the orthogonal applied field is $H_y = 1.0$ Oe.

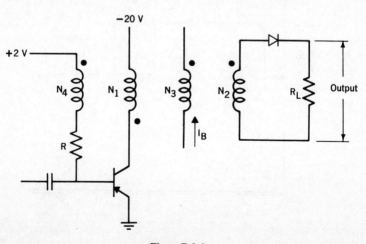

Figure P.3.4

Chapter 4

Coupling Networks and Mapping Functions

The purpose of the mapping or coupling operation in a network of decision elements is to convert the configurations that can be assumed by N binary input variables to corresponding values of a single variable, the mapping function f_m. The correspondence is determined by the form of the mapping or coupling network. Following the mapping process, a decision component performs the operation of "deciding" whether the value of f_m is to be classified as belonging to one or the other of two binary regions. The output of the decision component constitutes the output of the decision element, and the combination of mapping and decision operations is regarded as a complete decision element, in accordance with conventional logic terminology.

For reasons of simplicity and economy it is necessary to restrict the type of decision component under consideration. By far the most common type is one that ideally divides the signal input values into two regions. If the input is greater than some decision level d, the output assumes one value; if the input is below d, the output assumes a different value.

For similar reasons of simplicity and economy, it has been common to restrict the types of mapping networks employed. In general,

$$f_m = f_m(x_1, x_2, \ldots, x_N),$$

any multivariable function, for which an infinite variety of mathematical forms can be conceived. Few forms are convenient to mechanize from a circuit standpoint, however, and the following discussion is restricted to linear and exponential forms.

Linear-Mapping Operations

The linear-mapping operation may be described as the simple sum of the inputs, each multiplied by some coefficient. Thus

$$f_m = \sum_{i=1}^{N} c_i x_i, \qquad (4.1)$$

in which

$$x_i = \text{value of } i\text{th binary variable},$$

$$c_i = \text{constant coefficient}.$$

If we let $c_0 x_0 = - d$, in which x_0 may be regarded as an additional input variable whose value is constant (e.g., $+1$), we may write

$$\sum_{i=0}^{N} c_i x_i > 0, \qquad D = D_1,$$

$$\sum_{i=0}^{N} c_i x_i < 0, \qquad D = D_2. \qquad (4.2)$$

For simplicity, it is desirable to standardize the binary values that can be assumed by the input variables. One criterion that forms a useful basis for standardization is the following. It would simplify matters if we were free to use either input variables or their complements. In particular, it would be valuable if a logic function were left invariant under a transformation consisting of complementing an input variable *and* simultaneously reversing the sign of its coefficient. It would also be helpful if this transformation could be accomplished without adjusting other coefficients. Defining the two possible values of the input variables as v_1 and v_2, we see that the requirement is satisfied if $v_1 = - v_2$. Because the selection of a particular magnitude only amounts to a change in scale factor, we can choose $v_1 = - v_2 = + 1$. In the following discussion, therefore, we shall frequently consider binary input variables to possess the values ± 1, rather than the more conventional values 0 and 1. It is, of course, always possible to convert from these standard (and symmetrical) values to any other pair of values. Denoting the standard values by x_i and the transformed variable values by x_i' amounts to requiring the mapping operations to be linearly related by some scale factor, that is, it is necessary that

$$\sum_{i=0}^{N} c_i x_i = k \sum_{i=0}^{N} c_i' x_i'. \qquad (4.3)$$

For example, converting from ± 1 binary values to 0, 1 binary values, the coefficients are related by

$$c_i' = \frac{2}{k}c_i, \qquad i = 1, 2, \ldots, N,$$

and

$$c_0' = \frac{1}{k}\left(c_0 - \sum_{i=1}^{N} c_i\right). \tag{4.4}$$

This means that the transformation from ± 1 to $1, 0$ corresponds to a scale-factor change of 2 in signal range and a dc offset of one-half the signal range for each of the input variables. Other correspondences can similarly be derived for other choices of binary values.

The establishment of particular values for the coefficients in a linear-mapping operation depends upon combining the form of decision function (4.2) with the constraints of a particular logic function. We can conveniently identify the two decision levels with the values of a binary function. Thus, for binary levels 0 and 1, if $D = D_1, f = 1$, and if $D = D_2, f = 0$. For binary levels -1 and $+1$, if $D = D_1, f = +1$, and if $D = D_2, f = -1$. In the latter case, with a positive-going decision operation (4.2) may be written compactly as

$$f = \text{sgn} \sum_{i=0}^{N} c_i x_i. \tag{4.5}$$

Derivation of coefficients. It remains to determine the values of the coefficients c_i that will mechanize a given function. In general, 2^N function values are given, corresponding to the 2^N possible configurations of N input variables. These values represent 2^N constraints on the value of the mapping operation f_m, in accordance with (4.1). Thus it is required that

$$\sum_{i=0}^{N} c_i x_i \gtrless 0,$$

depending on whether $D = D_1$ or $D = D_2$, as in (4.2). Therefore 2^N simultaneous linear inequalities may be written and solved if a solution exists.

Fortunately, solutions for the inequalities representing classes of Boolean functions have already been found by several investigators: Minnick [1], Stafford [2], and Dertouzos [8]. Such solutions have been tabulated so that the only task for the circuit designer is one of table look-up.

Not all of the 2^{2^N} functions that can be specified for N variables can be obtained with a single decision element of this type (that is, some sets of inequalities will result in contradictions). A function that can be realized (for which a set of coefficients satisfies the inequalities) is termed a *linear-input logic* function.

A list prepared by Stafford of the coefficients of all *classes* of linear-input logic functions through $N = 5$ is reproduced in Table 4.1. The table can be used in the following manner. A linear-input function is located by means of its characteristic set of correlation coefficients. It can be shown that this set of numbers is unique to every function which is of the linear-input type.

TABLE 4.1   TABLE OF LINEAR-INPUT FUNCTION COEFFICIENTS

N	Class	$\rho = N_A - N_D$ (complete truth table)						Minimum integer coefficients						Σc_i
1	1					2	0					1	0	1
2	1				4	0	0				1	0	0	1
	2				2	2	2				1	1	1	3
3	1			8	0	0	0			1	0	0	0	1
	2			4	4	4	0			1	1	1	0	3
	3			6	2	2	2			2	1	1	1	5
4	1		16	0	0	0	0		1	0	0	0	0	1
	2		8	8	8	0	0		1	1	1	0	0	3
	3		12	4	4	4	0		2	1	1	1	0	5
	4		6	6	6	6	6		1	1	1	1	1	5
	5		8	8	4	4	4		2	2	1	1	1	7
	6		14	2	2	2	2		3	1	1	1	1	7
	7		10	6	6	2	2		3	2	2	1	1	9
5	1	32	0	0	0	0	0	1	0	0	0	0	0	1
	2	16	16	16	0	0	0	1	1	1	0	0	0	3
	3	24	8	8	8	0	0	2	1	1	1	0	0	5
	4	12	12	12	12	12	0	1	1	1	1	1	0	5
	5	16	16	8	8	8	0	2	2	1	1	1	0	7
	6	28	4	4	4	4	0	3	1	1	1	1	0	7
	7	20	12	12	4	4	0	3	2	2	1	1	0	9
	8	20	8	8	8	8	8	2	1	1	1	1	1	7
	9	14	14	14	6	6	6	2	2	2	1	1	1	9
	10	22	10	6	6	6	6	3	2	1	1	1	1	9
	11	18	10	10	10	6	6	3	2	2	2	1	1	11
	12	16	16	12	4	4	4	3	3	2	1	1	1	11
	13	14	14	10	10	10	2	3	3	2	2	2	1	13
	14	30	2	2	2	2	2	4	1	1	1	1	1	9
	15	24	8	8	4	4	4	4	2	2	1	1	1	11
	16	20	12	8	8	4	4	4	3	2	2	1	1	13
	17	18	14	14	2	2	2	4	3	3	1	1	1	13
	18	16	12	12	8	8	4	4	3	3	2	2	1	15
	19	26	6	6	6	2	2	5	2	2	2	1	1	13
	20	22	10	10	6	2	2	5	3	3	2	1	1	15
	21	18	14	10	6	6	2	5	4	3	2	2	1	17

The correlation coefficients are derived in a manner analogous to that for the conventional correlation function. For practical purposes, it is usually convenient to ignore the normalizing factors and work with sets of correlation numbers ρ_i. The values ρ_i can be obtained by simple inspection of the truth table, that is,

$$\rho_i = N_{A_i} - N_{D_i}, \tag{4.6}$$

in which

$N_{A_i} =$ number of times the value of the variable x_{ij} agrees with the function value f_j,

$N_{D_i} =$ number of times the value of the variable disagrees with the function value.

The quantity $N_A - N_D$ is listed in Table 4.1, together with the corresponding set of minimum integer coefficients for mechanizing the function.

We have already noted that the selection of values ± 1 leaves a function unchanged if both the variable and its coefficient are reversed in sign. Therefore, if the calculation of ρ results in some negative values, either the corresponding coefficient must also be negative or the complement of the input variable must be used in mechanizing the function. The correlation values ρ_i and corresponding minimum integer coefficients c_i are listed in decreasing order in Table 4.1. Clearly, the order of the input variables x_i and coefficients c_i must correspond. Therefore, any permutation of the ρ_i that agrees with that calculated for an unknown function must be satisfactory, and the c_i must be correspondingly permuted. Finally, Table 4.1 is based upon a decision characteristic given by (4.5). For a negative-gain amplifier, either the coefficients must be reversed in sign or the complements of the input variables must be used.

An example will illustrate the above rules. Consider the function $\bar{x}_3(\bar{x}_2 + \bar{x}_1)$. A count of $N_A - N_D$ will yield the following ρ_i (note that x_0 always equals $+1$): $\rho_3 = -6$, $\rho_2 = \rho_1 = \rho_0 = -2$. This corresponds to the set $(6, 2, 2, 2)$ appearing as class 3 under $N = 3$ in Table 4.1. Note that the ρ are all negative. If a negative-gain amplifier is used, however, they become positive, and the input variables (not their complements) are used. The appropriate circuit is therefore illustrated in Figure 4.1. The coefficients represent conductance (not resistance) values if the decision device is current-operated.

If a function is not of the type that can be mechanized in a linear-input circuit, it will be found to yield a set of correlation numbers ρ_i that do not appear in Table 4.1. Such functions can, of course, be mechanized by cascade linear-input circuits. Minnick [1] has prepared a tabulation including both single-stage and cascade mechanizations for all functions through $N = 4$.

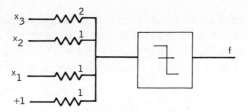

Figure 4.1

Most of the forms are minimum. Nonminimum mechanizations can, of course, always be demonstrated, for example, by standard AND and OR circuit forms.

Resistor mechanization. Circuits for mechanizing linear-mapping operations can take advantage of any physical phenomenon which relates two measurable quantities by a constant (but selectable) ratio. In particular, Ohm's law specifies a common way in which voltage and current are related, and the Kirchhoff adder provides a ready means of mechanizing functions of the foregoing type. The standard form for such a circuit is illustrated in Figure 4.2, in which e_i are the input voltages, g_i are the mapping-function coefficients, E_B and g_B are a bias voltage and conductance, respectively, and g_L is the load conductance. We are commonly interested in the voltage across the load conductance, e_L, or the current into the load conductance, $g_L e_L$. The former is given by

$$e_L = \frac{g_1 e_1 + g_2 e_2 + \cdots + g_N e_N + g_B E_B}{g_1 + g_2 + \cdots + g_N + g_B + g_L}. \tag{4.7}$$

Let us pick E_B and g_B such that some arbitrary conductance $g_0 = g_B + g_L$

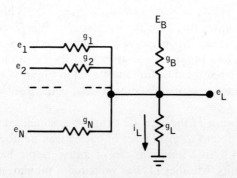

Figure 4.2

and some arbitrary voltage $e_0 = E_B g_B/(g_B + g_L)$. Then (4.7) becomes

$$e_L = \sum_{i=0}^{N} c_i e_i,$$

in which

$$c_i = \frac{g_i}{\displaystyle\sum_{i=0}^{N} g_i}. \tag{4.8}$$

Clearly, (4.8) is of the form of (4.1). Alternatively, we may let $g_L \to \infty$ and inquire as to the current i_L flowing to ground. With $g_L \to \infty$ and selecting $g_B = g_0$ and $E_B = e_0$, we obtain

$$i_L = \sum_{i=0}^{N} g_i e_i. \tag{4.9}$$

The discriminating ability of the decision component often must improve as the value of N increases, because the network of Figure 4.2 may suffer from considerable attenuation. The change in output for a change in one input variable, say e_i, must be capable of switching the decision component from one state to another. Assuming that g_i is the smallest conductance and taking its value as 1, we see that the change in output is

$$\frac{\partial e_L}{\partial e_i} = \frac{1}{\displaystyle\sum_{i=0}^{N} g_i}; \tag{4.10}$$

that is, the network attenuates by a factor corresponding to the sum of the coefficients. This attenuation factor is listed in Table 4.1.

Fan-in limits. In the general case, both conductances and input voltages will be subject to finite tolerances. In addition, the decision component is not perfect, that is, its gain is not infinite. For simplicity, let

$\pm r \equiv$ tolerance (fractional) on conductances,
$\pm v =$ tolerance (fractional) on input voltage levels,
$\pm d/2 =$ tolerance on decision component (see Figure 4.3).

In order for the decision component to operate correctly, the magnitude of the input sum must always be greater than $d/2$. Without loss of generality we can assume that the worst case occurs when the sum of all inputs except one equals zero, and that input, say x_N, equals ± 1. We can divide the summation into three components: (a) those conductances connected to positive inputs,

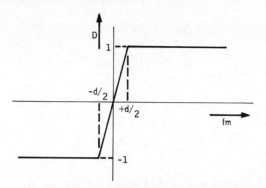

Figure 4.3

(b) those connected to negative inputs, and (c) c_N. Thus, nominally,

$$\sum_{i=0}^{j-1} c_i x_i + \sum_{i=j}^{N-1} c_i x_i = 0,$$

and

$$c_N x_N = \pm 1. \tag{4.11}$$

This assumes $|c_N| = 1$. Dividing by $|x_i|$, we have

$$\sum_{i=0}^{j-1} c_i - \sum_{i=j}^{N-1} c_i = 0. \tag{4.12}$$

Let the conductances associated with positive inputs decrease by a fraction r, and let those with negative inputs increase by an equal fraction. Then the new values of the coefficients c_i', $0 \le i \le j - 1$, are

$$c_i'(i < j) = \frac{g_i(1 - r)}{\displaystyle\sum_{i=0}^{j-1} g_i(1 - r) + \sum_{i=j}^{N-1} g_i(1 + r) + g_N(1 + r)}$$

$$= \frac{g_i(1 - r)}{\displaystyle\sum_{i=0}^{j-1} g_i + \sum_{i=j}^{N-1} g_i + g_N + r\left(-\sum_{i=0}^{j-1} g_i + \sum_{i=j}^{N-1} g_i + g_N\right)}. \tag{4.13}$$

Let

$$S = \sum_{i=0}^{N} g_i = \sum_{i=0}^{j-1} g_i + \sum_{i=j}^{N-1} g_i + g_N. \tag{4.14}$$

and note that

$$-\sum_{i=0}^{j-1} g_i + \sum_{i=j}^{N-1} g_i = 0,$$

from (4.12). Therefore (4.13) becomes

$$c_i' = \frac{g_i}{\sum\limits_{i=0}^{N} g_i} \frac{1-r}{1+r/(S/g_N)}. \tag{4.15}$$

Noting that $S/g_N \gg 1$ and $r \ll 1$, we can neglect the last term in the denominator and write

$$c_i' \cong c_i(1-r).$$

Similarly, for the coefficients connected to negative inputs,

$$c_i'(j \leq i \leq N-1) = c_i(1+r).$$

Note that this result also follows, more simply, if we are interested in current summation, as in (4.9). In this case c_i is directly proportional to g_i (there is no cross coupling, because the load is taken to be a short circuit, $g_L \to \infty$). Thus $c_i' = c_i(1 \pm r)$, as above.

Taking the complete summation, which must be at least as large as $d/2$, we obtain

$$\sum_{i=0}^{j-1} c_i(1-r)(1-v) - \sum_{i=j}^{N-1} c_i(1+r)(1+v) + c_N(1-r)(1-v) \geq \frac{d}{2} \tag{4.16}$$

Multiplying through, dropping second-order terms (rv products), and solving for the magnitude of the sum of the coefficients, we see that

$$\sum_{i=0}^{N} |c_i| \gtrsim \frac{1-d/2}{r+v}. \tag{4.17}$$

This result means that the limit on the sum of coefficients for a linear-input function is primarily established by component tolerances, that is, even if $d = 0$, Σc_i is limited by $1/(r+v)$. Only if $r+v \to 0$ is the limit removed. However, the foregoing analysis has ignored transient behavior. As we shall see in analyzing complete decision element circuits, imposing specifications on maximum switching times also has the effect of limiting Σc_i, even with perfect component tolerances.

It is worth noting that $\Sigma|c_i|$ is in some sense a direct measure of fan-in, that is, for every $c_i > 1$, we can pretend that the corresponding variable is split into two or more variables, equal in number to the numerical value of c_i. These imaginary variables can be given the property of acting in unison. When so regarded, every linear-input function can be treated as a form of "majority" function, that is, we can let $N' = \Sigma|c_i|$ and let m be the number of true inputs that are required to cause the circuit to change state.

It will be convenient to regard the circuit design problem from this stand-point in later chapters, because both fan-in and fan-out are dependent on N' rather than N.

Compensated networks. The time-dependent behavior of decision elements imposed by the properties of particular amplifying devices is considered in subsequent chapters. However, the circuit mechanizing the mapping opera-tion in a decision element may itself have a time-dependent response. In the case of linear-input elements mechanized with resistors, this situation fre-quently arises because of an unavoidable capacitance across the output of the summing circuit. The question therefore arises as to whether or not the sum-ming circuit can be compensated. The term *compensation* refers to an *RC* divider network, illustrated in Figure 4.4, in which the output capacitance

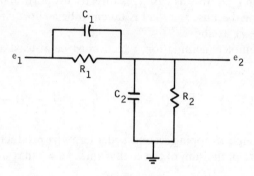

Figure 4.4

C_2 is usually fixed by the nature of the load. Compensation consists of finding a value of C_1 such that

$$\frac{e_2}{e_1} = \frac{R_2}{R_1 + R_2},\tag{4.18}$$

regardless of the driving waveform. The general expression for the transfer function can be put in the form

$$\frac{E_2(s)}{E_1(s)} = \frac{R_2}{R_2 + R_1[(R_2C_2s + 1)/(R_1C_1s + 1)]}.\tag{4.19}$$

Clearly (4.19) reduces to (4.18) when $R_1C_1 = R_2C_2$. The network is then referred to as being *compensated*. In the case of the linear-mapping operation the only quantities usually fixed by external considerations are the values of load C_L and g_L. We have already noted that an arbitrary bias source E_B and conductance g_B can be selected to yield the net effect of the zero-order conductance term g_0. The ac equivalent circuit is therefore as shown in Figure 4.5, in which $C_0 = C_L$ for notational simplicity. To compensate the circuit, capacitors C_j must be placed across each conductance g_j such that $R_jC_j = R_pC_p$, in which R_p is the parallel resistance of all resistors except R_j, and C_p is the corresponding parallel capacitance. However $R_p = 1/g_p$, so the above requirement can be written

$$C_j = g_j \frac{\sum_{i \neq j} C_i}{\sum_{i \neq j} g_i}. \tag{4.20}$$

Because C_0 and all conductances are fixed, (4.20) can be satisfied by letting $C_i = (C_0/g_0)g_i$, thus compensating the circuit.

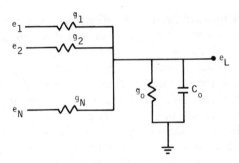

Figure 4.5

Unfortunately, the load conductance and capacitance corresponding to the input to a decision component are seldom constants, independent of signal level. Similarly, the output impedance of the device driving the mapping function is seldom zero, as required by the compensation criterion. Therefore the design is invariably a compromise made with consideration given to the particular decision component properties, desired switching speed, and power consumption.

Non-linear Mapping Functions

It is difficult to obtain an explicit expression for the mapping function performed by nonlinear components of the exponential type, even assuming the ideal-diode equation.

In view of these difficulties it has become common practice to linearize the ideal-diode characteristic by assuming that it consists of two regions of

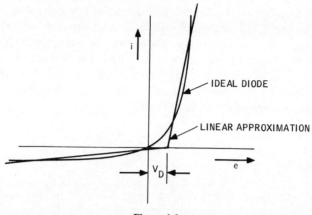

Figure 4.6

operation. When forward-biased, the diode is represented by a constant forward drop V_D or a constant dynamic forward conductance g_f, or both. When reverse-biased, the diode is represented either by a constant reverse current I_r or a constant dynamic reverse conductance g_r, or both. The relationships between these representations and an ideal-diode characteristic are illustrated in Figure 4.6. The linearizing assumptions are justified by several considerations: (a) simplicity of circuit analysis; (b) as we have seen, actual diodes differ from ideal diodes by equivalent resistors that tend toward linearizing the characteristic; and (c) as long as we are dealing with two-valued functions, the degree of correspondence between actual and assumed can be made quite close within the two restricted regions in which signals will occur.

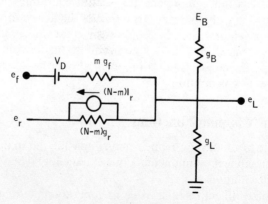

Figure 4.7

We may now write a complete expression for any diode gate with m diodes forward-biased and connected to an input e_f, and $N - m$ diodes reverse-biased and connected to an input e_r. The equivalent circuit is shown in Figure 4.7, from which

$$e_L = \frac{m(e_f - V_D) + (N - m)(g_r/g_f)e_r + (g_B/g_f)E_B - (N - m)(I_r/g_f)}{m + (N - m)(g_r/g_f) + g_B/g_f + g_L/g_f}. \quad (4.21)$$

Providing $g_f \gg g_B, g_L$, or g_r, the bias is of the proper polarity, and V_D is neglected, the output is $e_L \cong e_f$, if $m \geq 1$. If all inputs are at e_f, then $m = N$, and again $e_L \cong e_f$. If all inputs are at e_r, no diodes can be reverse-biased. Therefore $e_f = e_r$, $m = N$, and the output is $e_L \cong e_r$. Although the logic values of e_f and e_r are arbitrary, the polarities of e_f and e_r, and of E_B, are dependent on the polarity of diode connections. These statements simply formalize the rules:

1. Diode-mapping functions with common-p connections $(e_f < e_r)$ mechanize AND functions for positive signals and OR functions for negative signals.
2. Diode-mapping functions with common-n connections $(e_f > e_r)$ mechanize OR functions for positive signals and AND functions for negative signals.

An expression for the limit on the number of inputs under static conditions can be derived for the diode-mapping function in a manner similar to that for linear-mapping functions. Assuming e_f and e_r are symmetrical about some central value, say 0, (i.e., $e_f = -e_r = e$), and defining component and voltage tolerances r and v as before, we see that the output must be at least equal to $d/2$ when only one diode is forward-biased. Neglecting V_D, we obtain

$$e_L = \frac{d}{2} = \frac{\begin{aligned} g_f(1 - r)e(1 - v) - (N - 1)g_r(1 + r)e(1 + v) \\ + g_B(1 + r)E_B(1 + v) - (N - 1)I_r \end{aligned}}{g_f(1 - r) + (N - 1)g_r(1 + r) + (g_B + g_L)(1 + r)}. \quad (4.22)$$

Solving for N, we can put the results in the form

$$N \gtrless 1 + \frac{g_f}{g_r}(1 - 2r - 2v)\frac{1 - (d/2e)(F_1)}{1 + (d/2e)(F_2)}, \quad (4.23)$$

in which F_1 and F_2 are fixed functions of the bias voltage, tolerances, and conductance values, and in which second-order rv product terms have been neglected. In principle, e can be made as large as necessary relative to d. Therefore, to a first order, N is limited primarily by the forward-to-reverse conductance ratio g_f/g_r. This is an important result because the ratio g_f/g_r, can easily be on the order of 10^2 to 10^3, even over extreme temperature ranges.

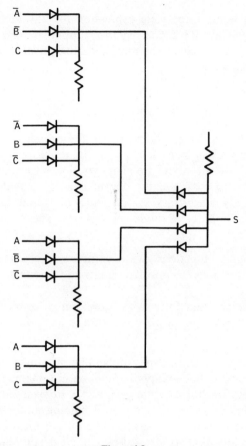

Figure 4.8

It is clear, therefore, that static conditions seldom limit diode-mapping functions. We shall see that more serious constraints are imposed in connection with transient behavior, cascade networks, and with the gain available in various decision components.

Cascade coupling logic. The mapping operation mechanization employing diodes possesses another important advantage over linear mapping, namely, that networks can be cascaded without appreciable interaction to produce complex over-all logic functions. Such cascade structures are useful in mechanizing AND-OR or OR-AND decision element structures corresponding to canonical sum-of-products or product-of-sum logic forms. These structures are typified by the full-adder network illustrated in Figure 4.8. The logic is $S = \bar{A}\bar{B}C + \bar{A}B\bar{C} + A\bar{B}\bar{C} + ABC$, and the truth table is

A	B	C	S
0	0	0	0
0	0	1	1
0	1	0	1
0	1	1	0
1	0	0	1
1	0	1	0
1	1	0	0
1	1	1	1

In Figure 4.8 the AND terms have $e_f > e_r$, and the OR terms $e_f < e_r$, and they are therefore satisfied by defining negative inputs as 1, positive inputs as 0. It is convenient as a first step in analyzing such structures to assume that the load conductance is zero and that the bias is supplied from a constant-current source. It is also convenient to assume that diodes reverse-biased by $|e_f - e_r|$ pass a constant current I_R, independent of small variations in the actual reverse voltage. Thus I_R is taken to include both constant-current leakage effects I_r, plus the voltage-dependent quantity $g_r|e_f - e_r|$. When all signals are at $e_r \equiv 1$ at the input to an active (true), AND element, the circuit appears in simplified form in Figure 4.9, with multiple-diode inputs replaced by single equivalent resistors. For proper operation, it is necessary that $I_f > 0$. Otherwise, $I_x > I_{B2}$, the input diodes will be reverse-biased,

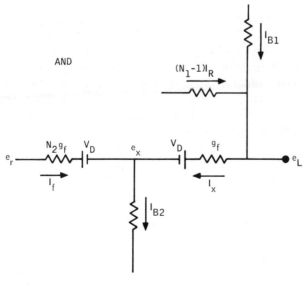

Figure 4.9

and e_L will no longer approximate e_r. With $I_f > 0$,

$$e_L = e_r - \frac{I_f}{N_2 g_f} - V_D + V_D + \frac{I_x}{g_f} = e_r + \frac{1}{g_f}\left(I_x - \frac{I_f}{N_2}\right). \qquad (4.24)$$

Note that the effects of V_D cancel. In general I_x and I_f are considerably different in magnitude. If g_f is large, however, the second term in (4.24) can be neglected. For $I_f > 0$, the current-summing requirement is that

$$I_{B2} \geq I_x = I_{B1} + (N_1 - 1)I_R. \qquad (4.25)$$

The subscripts used above refer to the logic level in the gating structure, counting backwards from the output. Thus N_1 and I_{B1} apply to the output, N_2 and I_{B2} apply to the preceding gate, and so on.

If the power supply and component tolerances affecting the bias current are lumped together as a variation Δ, the worst-case condition is

$$I_{B2}(1 - \Delta) \geq I_{B1}(1 + \Delta) + (N_1 - 1)I_R,$$

or

$$\frac{I_{B2}}{I_{B1}} \geq \left[1 + 2\Delta + (N_1 - 1)\frac{I_R}{I_{B1}}\right]. \qquad (4.26)$$

With tolerances of 5 percent on both supply and components, this factor is a minimum of 1.2. As will be seen later, higher ratios are required for fast response time, and factors of 2.0 or larger are common in practical computer systems.

When the AND-OR structure is in the 0 state, at least one input to each AND element is at e_f. It is important to know what current this input must supply to the gating structure under this condition, because active elements may be called on to supply this current to each of their n output loads. The worst case occurs when (a) only one input at the AND gate is at e_f, and (b) at least one other input to the OR gate is at e_r. The situation is illustrated in Figure 4.10. The total current flowing out of the AND gate summing junction, which must be supplied by the active driving source, is

$$I_A = I_{B2} + (N_2 - 1)I_R + I_R. \qquad (4.27)$$

Substituting for I_{B2} from (4.25), we obtain

$$I_A = I_{B1} + (N_1 + N_2 - 1)I_R. \qquad (4.28)$$

This simply indicates the (ideal) manner in which load current increases with additional logic levels. Component and supply tolerances, and nonideal bias current sources, of course, have an even greater effect upon loading.

If bias currents are in fact not supplied by constant-current generators but rather by voltage sources of magnitude E_{B1} and E_{B2}, in the 1 condition,

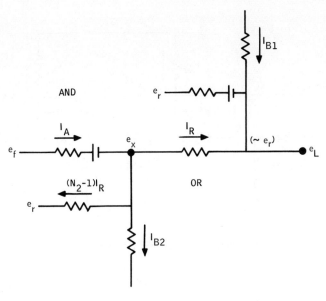

Figure 4.10

$I_{B1} = (E_{B1} + e_r)/R_{B1}$, and $I_{B2} = (E_{B2} - e_r)/R_{B2}$. For correct 1 condition operation,

$$\frac{E_{B2} - e_r}{R_{B2}} \geq \frac{E_{B1} + e_r}{R_{B1}} + (N_1 - 1)I_R. \qquad (4.29)$$

Letting $E_{B1} = E_{B2} = E_B$ and defining $I_{B1} \equiv E_B/R_{B1}$, we can put (4.29) in the form

$$\frac{R_{B2}}{R_{B1}} < \frac{1 - e_r/E_B}{1 + (e_r/E_B) + (N_1 - 1)(I_R/I_{B1})}. \qquad (4.30)$$

This ratio is always less than 1, or $g_{B2} > g_{B1}$. Because E_{B1} and E_{B2} can be symmetrical bias supplies about 0 V, we can consider signal levels symmetrical about 0 V as well, that is, let $e_f = -e_r = e$. In the 0 condition $I_{B2} = (E_B + e)/R_{B2}$, and the maximum current which must be supplied by an input is given by (4.27). Substituting in (4.30) and neglecting second-order terms, we obtain

$$I_A \cong \left(1 + \frac{3e}{E_B}\right)I_{B1} + (N_1 + N_2 - 1)I_R, \qquad (4.31)$$

representing an increase over (4.28).

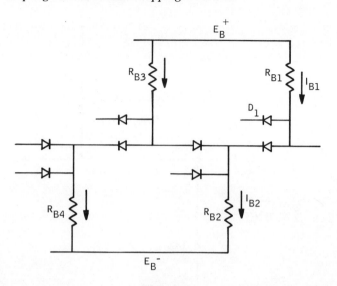

Figure 4.11

Power requirements. The AND-OR or OR-AND structure of Figure 4.8 can, of course, be readily extended to more logic levels. An OR-AND-OR-AND structure for positive signals representing 1 (or AND-OR-AND-OR for negative signals) is illustrated in Figure 4.11. Equations describing the ratio of AND gate resistor value for feeding an OR gate to the AND gate resistor being fed by an OR gate have been written by Yokelson and Ulrich [3] for both dc and pulse cases. In the dc case, however, simple verbal rules covering any type of gating structure are easy to derive. The worst case always occurs when the signals are such that the maximum current $I_{Bi}(\text{max})$ flows at some level i and the minimum current $I_{Bi+1}(\text{min})$ flows in the preceding level $i + 1$. This is the case, for example, in which e_L is at its most negative value in Figures 4.9 or 4.10. The maximum load current is augmented by the reverse current of nearly N reverse-biased diodes, such as D_1 in Figure 4.11. Let the maximum reverse current of a diode be represented by an equivalent conductance g_r connected in parallel with R_B; that is, let $g_r \equiv I_R/E_{B1(\text{max})}$, in which $E_{B1(\text{max})}$ is the worst-case voltage across R_{B1} and equals $E_B{}^+ - e_L$ when e_L is at its most negative value. The minimum current in the preceding stage is $I_{Bi+1(\text{min})} = E_{Bi+1(\text{min})}/R_{Bi+1}$, in which $E_{Bi+1(\text{min})}$ is the worst-case voltage across R_{Bi+1} and equals $e_L - E_B{}^-$. The requirement is that $I_{Bi+1} \geq I_{Bi} + NI_R$ (for $N \gg 1$), from which

$$g_{Bi+1(\text{min})} \geq (g_{Bi(\text{max})} + Ng_r)\frac{E_{Bi(\text{max})}}{E_{Bi+1(\text{min})}}. \qquad (4.32)$$

This equation simply means that the conductance loading the $(i + 1)$st logic level must be greater than the total effective conductance of the ith level, multiplied by the ratio of worst-case voltage differences across the two load levels. This rule can be extended to as many levels as desired, the practical limitation being the power that the driving source must supply. Because voltage levels are usually determined by other considerations (such as dynamic response) power becomes directly proportional to load current.

Although the foregoing establishes conductance (or current) ratios relative to the output stage, a specific value of conductance to use in the output stage must be determined. Because gates are extensively used throughout a computer system, a common basis for such a determination is the criterion of power dissipation. This can be used in the following way. Suppose the output stage is supplied from a bias source E_B and is driven by signals $\pm e$ around some average value E_A. Generally the circuit will be required to deliver some minimum current I_{min}. This may arise, for example, because of the necessity for charging a capacitance load in some specified time, or because an amplifier must be supplied with some minimum drive. The minimum current flows when the least voltage is across R, namely,

$$I_{min} = \frac{E_B - E_A - e}{R}.$$
(4.33)

The maximum power dissipation in R occurs, however, when the maximum voltage is across it, namely,

$$P_{max} = \frac{(E_B - E_A + e)^2}{R}.$$
(4.34)

Because the circuit may remain in the maximum-dissipation state for relatively long periods of time, we would like to minimize P_{max}. Solving (4.33) for $E_B - E_A$ and substituting in (34), we obtain

$$P_{max} = I_{min}^2 R + 4I_{min}e + \frac{4e^2}{R}.$$
(4.35)

Setting $dP_{max}/dR = 0$ and solving, we have $R = 2e/I_m$. Substituting in (4.33) gives $E_B - E_A = 3e$. This means, in effect, that for minimum power dissipation, E_B should be equal to the most positive (or negative) signal level plus (or minus) the total signal swing (in this case, $2e$). For $E_A = 0$, the value of P_{max} can be written in the form

$$\frac{P_{max}}{I_{min}e} = \frac{E_B}{e} \frac{(1 + e/E_B)^2}{(1 - e/E_B)}.$$
(4.36)

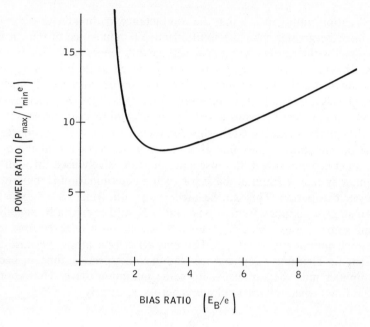

Figure 4.12

This function, plotted in Figure 4.12, has a minimum value of 8.0 at $E_B = 3e$, but the slope off the minimum is fairly shallow. For example, the value of (4.36) is less than 10 for E_B anywhere in the range $2e < E_B < 6e$. For various practical reasons, it is often desirable to use the higher values of E_B, despite slight deviations from minimum power. For example, it will be seen in later chapters that the fan-out or logic gain of circuitry is frequently proportional to the factor $1 - e/E_B$. Higher values of E_B are therefore justifiable in the interests of increasing the usefulness of a given circuit, particularly if some reduction in over-all component count can thereby be achieved.

Transient behavior. Depending on the various simplifying assumptions made, the transient behavior of nonlinear-mapping functions can be calculated with relative facility. The simplest case is to consider diodes as ideal switches (refer to Figure 4.13). At the instant the inputs go negative, the energy storage capacitor is charged to some voltage e_f. When the inputs go negative, the capacitor cannot discharge instantaneously, so the diodes become reverse-biased and may be considered to be out of the circuit. The capacitor must discharge through the load resistor R until a second voltage level e_r is reached. When the output equals e_r (or slightly less), the input diodes again conduct and the output remains at e_r, completing the switching opera-

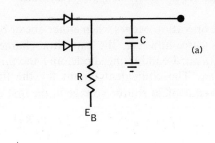

(a)

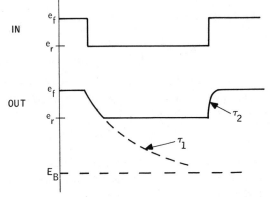

IN

OUT

Figure 4.13

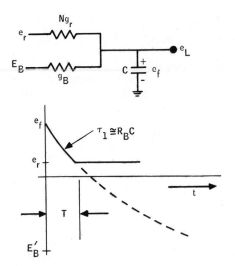

Figure 4.14

157

tion. The reverse operation occurs when either one or both of the inputs goes positive. In this case, however, the capacitor charges through the small resistance of a forward-conducting diode, and the time constant is correspondingly shorter. The equivalent circuit for the first case is shown in Figure 4.14. The equivalent source voltage in the first case is

$$E'_B = E_B \frac{1 + N(e_r/E_B)(g_r/g_B)}{1 + N(g_r/g_B)}, \tag{4.37}$$

and the equivalent resistance is

$$R'_B = R_B \frac{1}{1 + N(g_r/g_B)}. \tag{4.38}$$

In the usual case $g_r \ll g_B$ and $|E_B| \gg e_r$ so that, for small values of N, $E'_B \cong E_B$ and $R'_B \cong R_B$. The approximate time constant is therefore $\tau \cong R_B C$, and the equation of the output voltage is

$$e_L = e_f + (E_B - e_f)(1 - e^{-t/\tau}). \tag{4.39}$$

If the difference between e_f and e_r is relatively small compared to the difference between e_f and E_B, switching occurs on a nearly linear portion of the exponential, that is,

$$\frac{de_L}{dt} \cong \frac{de_L}{dt}\bigg|_{t=0} = \frac{1}{\tau}(E_B - e_f) \cong \frac{1}{\tau}E_B. \tag{4.40}$$

The total time required to switch from e_f to e_r is therefore

$$T \cong \frac{e_f - e_r}{de_L/dt} = \frac{\tau \, \Delta e}{|E_B - e_f|} \cong \frac{\tau \Delta e}{E_B}, \tag{4.41}$$

in which Δe is the voltage difference $e_f - e_r$. If E_B is sufficiently large, the bias source appears as a nearly constant-current device; that is $(E_B - e_f)/R_B \cong (E_B - e_r)/R_B \cong E_B/R_B = I_B$. Substitution for τ and E_B in (4.41) gives

$$T \cong \frac{R_B C \, \Delta e}{E_B} = \frac{C \, \Delta e}{I_B} = \frac{Q}{I_B}, \tag{4.42}$$

in which $Q = C \, \Delta e$ is the total change in charge on the capacitor. The switching operation may thus be regarded from the point of view of the time required for a constant-current source to supply (or remove) a fixed charge to the capacitor. Normally, C is determined by the input circuitry to the element being driven, as well as by packaging techniques that determine stray capacitance. Likewise, E_B is usually chosen on the basis of minimum power dissipation, or on the basis of existing supplies. T and Δe are determined by the desired operating frequency of the circuits and the input signal swing required to drive decision components reliably. Therefore R_B is completely determined by (4.42). I_{min} and P_{max} in (4.33) and (4.34) are thereby likewise determined.

Note that the value of T from (4.42) is exact, given the exponential approximation of (4.40), if E_B and R_B are replaced by E'_B and R'_B from (4.37) and (4.38), respectively. Performing the substitutions, we obtain

$$T = \frac{Q}{I_B} \frac{1}{1 + N(\Delta e/E_B)(g_r/g_B)}. \tag{4.43}$$

Thus the number of inputs to a diode-mapping function affects the response time by means of the factor N.

For practical purposes, it is useful to know how close various linear approximations are to actual switching behavior. Assuming that the effect of I_R has been taken into account, for example by (4.37) and (4.38), the question becomes one of approximating the one-time-constant behavior of (4.39). For this purpose, let the initial voltage across the capacitor be zero, and let the transient be complete when the voltage changes by the signal voltage $\Delta e \equiv V$. Solving (4.39) for the switching time T, we get

$$T = \tau \ln \frac{1}{1 - V/E_B}. \tag{4.44}$$

At least two approaches have been taken to approximating this function. One, the worst case, assumes that the current available for charging the capacitor is always at its minimum value. Thus $I_{B(min)} = (E_B - V)/R$, and using (4.42), we obtain

$$T \cong \tau \frac{V}{E_B} \frac{1}{1 - V/E_B}. \tag{4.45}$$

This approximation is very conservative, generally predicting a response time considerably greater than those actually present. It is a good estimate for $V/E_B \gtrsim 0.3$. Another approach to making a linear estimate of switching time is to assume that the charging current is approximately equal to its average value during the transient, $I_{B(ave)} = (E_B - V/2)/R$. Using (4.42) again, we obtain

$$T \cong \tau \frac{V}{E_B} \frac{1}{1 - V/2E_B}. \tag{4.46}$$

This function closely approximates the actual response (4.44) over a wide range but suffers from being slightly on the unconservative side (predicting shorter switching times than those actually present). The three functions

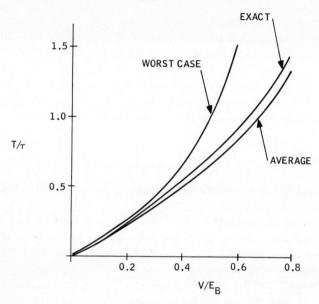

Figure 4.15

above are plotted for comparison in Figure 4.15. The figure suggests that an empirical value k can be found such that the function

$$T \cong \tau \frac{V}{E_B} \frac{1}{1 - (1/k)(V/E_B)} \tag{4.47}$$

will closely match the actual function. Indeed, with $k = 1.6$, the function (4.47) will match (4.44) over the range $0 < V/E_B < 0.8$ within 5 per cent, on the conservative side.

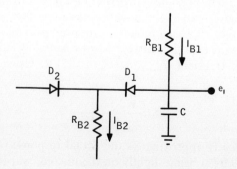

Figure 4.16

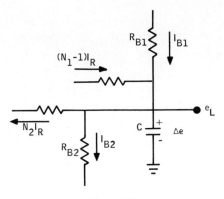

Figure 4.17

One of the most attractive features of diode-mapping operations is their ability to be cascaded with relatively little interaction. The transient behavior of such cascades is therefore of interest. Consider first the two-level cascade illustrated in Figure 4.16. If the input signal has just changed from negative to positive, diode D_2 is forward-conducting and diode D_1 is reverse-biased. The transient through D_2 will be short in comparison to the charging of C, and the output rise time is simply given by (4.43), with $N = N_1$ and $g_B = g_{B1}$, or

$$T \cong \frac{Q}{I_{B1} + N_1 I_R}. \tag{4.48}$$

Suppose that the input has just gone from positive to negative. Then diode D_2 is reverse-biased, and the equivalent circuit is shown in Figure 4.17, from which

$$T \cong \frac{Q}{I_{B2} - [I_{B1} + (N_1 - N_2 - 1)I_R]}. \tag{4.49}$$

If we desire a particular frequency response from the gating network, it may be necessary to obtain roughly equal rise and fall times. In this case equating (4.48) and (4.49) gives

$$I_{B2} = 2I_{B1}\left[1 + \left(N_1 - \frac{N_2 + 1}{2}\right)\frac{I_R}{I_{B1}}\right]. \tag{4.50}$$

A great variety of diode gate configurations have been developed to meet special system requirements. For example, if the gate structure is large (high fan-in) and reverse resistance becomes serious, AND functions can be cascaded as in Figure 4.18. Suppose an N-term AND function is required. In a

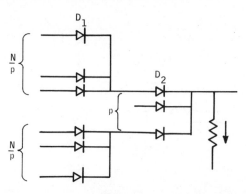

Figure 4.18

single-stage realization, the worst case is one in which one active input is loaded by the reverse resistance of $N - 1$ other diodes. However, let the function be divided into p approximately equal AND terms, and combine these terms in a second AND function. In the worst case, one active input, say D_1 (Figure 4.18), is loaded by $(N/p) - 1$ reverse resistances in the first stage and through the forward-conducting diode D_2, by the reverse resistances of $p - 1$ diodes in parallel, each in series with N/p diodes in parallel. Neglecting these latter diodes, the total load on a single input is now $N/p + p - 2$. The minimum value of the load, as a function of p, occurs when $p = \sqrt{N}$, at which value the load is $2(\sqrt{N} - 1)$.

Duality. We have confined all of the foregoing discussions to mapping functions in which voltage is taken to be the physical quantity representing the values of the binary variables. The principle of duality applies as well to diode-mapping functions, however, as it does to other types of circuits.

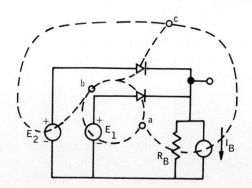

Figure 4.19

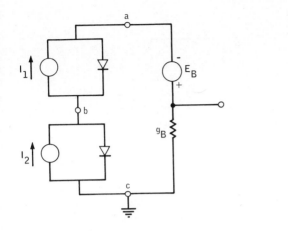

Figure 4.20

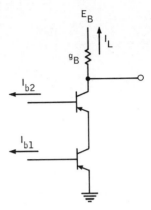

Figure 4.21

Reinecke [6] has pointed out, for example, that the dual of the voltage-operated gate of Figure 4.19 is given by the current-operated gate of Figure 4.20. The construction is as follows. From Figure 4.19 it can be seen that there are three nodes in the dual circuit. Node a connects to node b via a diode and a (dual) current source corresponding to the voltage source E_1. Node b connects to node c via a diode and (dual) current source (voltage source E_2). The parallel current source I_B and resistor R_B are replaced by a series voltage source and resistor between a and c. Consideration of Figure 4.20 shows that if both of the current sources I_1 and I_2 are in the direction indicated, both diodes are forward-biased, and the current through the load is $g_B E_B$. If either one or both of the source currents is reversed, at least one of the diodes is reverse-biased, and, letting $|I_1| = |I_2| = I$, the load current becomes $-I$. This principle may, of course, be extended to cascade diode networks.

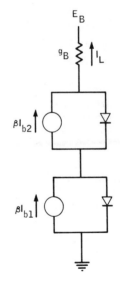

Figure 4.22

The above technique is sometimes called *current steering*, particularly when the current generator and diode input networks are conveniently mechanized by making use of the transistor collector equivalent circuit. Consider, for example, the circuit illustrated in Figure 4.21. If either one or both base currents I_b are negative, the corresponding transistor is in the inactive (cutoff) region, and the load current is virtually zero, or approximately I_{co}. If both base

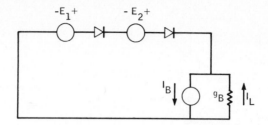

Figure 4.23

currents are positive and sufficiently large, both transistors are driven into saturation, and the load current is $g_B E_B$. Remembering that the equivalent circuit of the common-emitter transistor in the active region can be closely approximated by a collector current generator βI_b and a collector junction diode, we can derive a direct equivalent to the current-mode circuit of Figure 4.20 from Figure 4.21, as illustrated in Figure 4.22. The equivalence is theoretically exact if we observe the restriction that the transistor is not bilateral, so that $\beta \ll 1$ whenever $I_b < 0$, for the conventions of current direction chosen in the figure.

A somewhat artificial voltage-operated gate can be devised in order to construct the dual of the above circuit. Consider the series voltage-source network of Figure 4.23, in which $|E_1| = |E_2| = E$. Provided $E > I_B g_B$, it can be verified that the truth table of the network is

E_2	E_1	I_L
$-E$	$-E$	I_B
$-E$	0	I_B
0	$-E$	I_B
0	0	0

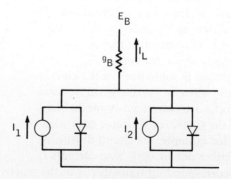

Figure 4.24

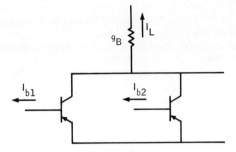

Figure 4.25

The current-mode dual of this circuit is shown in Figure 4.24. It can be seen that if both I_1 and I_2 are zero, $I_L = 0$. If either I_1 or I_2 are positive and greater than I_L, then $I_L = g_B E_B$. Realization of this circuit can be approximated by the transistor arrangement of Figure 4.25, in which $I_1 = \beta I_{b1}$, $I_2 = \beta I_{b2}$, and the collector junctions correspond to the ideal diodes of Figure 4.24.

The limiting static and dynamic conditions on the above transistor mechanizations of the mapping function can be readily derived. The worst load on a single element occurs when all but one of the transistors are in the inactive region. The current drawn by the cutoff transistors can be neglected, because they are not only in a high-impedance region, but also have only a small voltage appearing across them. Therefore, it is necessary that $\beta I_b \geq g_B E_B = I_L$. When all N input transistors are cut off, the load current becomes NI_{co}. At what level this becomes serious depends upon the decision component being driven. In circuits of this type the decision component is commonly another transistor, which must in turn drive a worst-case load of I_L, and so on. We shall consider this situation in connection with transistor amplifiers that mechanize the decision function.

We see that by realizing the current-mode gate with active components, we have introduced something more than a new type of mapping component. Gain has been introduced into the coupling network, and the previously clear-cut distinction between a mapping component and a decision component disappears.

REFERENCES

[1] R. C. Minnick, "Linear-Input Logic," *IRE Trans. Electron. Computers*, **EC-10**, 1, 6–16 (March 1961).

[2] R. A. Stafford, "A List of Weights for Six Variable Threshold Functions," in Annual Summary Report, "A Magnetic Integrator for the Perceptron Program", U-1405, *CFSTI* #AD264227 (September 1961).

[3] B. J. Yokelson and W. Ulrich, "Engineering Multistage Diode Logic Circuits," *AIEE Commun. Electron.* 20, 466–475 (September 1955).

[4] D. R. Brown and N. Rochester, "Rectifier Network for Multiposition Switching," *Proc. IRE*, 139–147 (February 1949).

[5] C. L. Wanlass, "Transistor Circuitry for Digital Computers," *IRE Trans. Electron. Computers*, **EC-4**, 1, 11–16 (March 1955).

[6] H. Reinecke, Jr., "Current-operated Diode Logic Gates," *AIEE Commun. Electron.* 52, 762–772 (January 1961).

[7] S. H. Caldwell, *Switching Circuits and Logical Design*, Wiley, New York, 1960.

[8] M. L. Dertouzos, "An Approach to Single-threshold-element Synthesis," *IRE Trans. Electron. Computers*, **EC-13**, 5, 519–528 (October 1964).

EXERCISES

4.1 It is desired to mechanize the function, $f = \bar{x}_4 x_3 + \bar{x}_4 \bar{x}_2 + \bar{x}_4 x_1 + x_3 \bar{x}_2$, using the circuit of Figure P.4.1. Inputs are to be 0 and -3 V and $E_B = \pm 12$ V. 0.5 mA forward base current is required to turn the transistor ON, and 0.05 mA reverse base current to be sure it is OFF.
(a) Find values of R_0 through R_4 if the transistor input impedance is regarded as negligible.
(b) Repeat if the tolerance on resistors is ± 5 percent.

4.2 What logic functions does the circuit of Figure P.4.2 generate with $\beta = 50$ if binary signal levels are defined as -4 V $= 1$ and 0 V $= 0$. Answer by filling in a table such as the following:

x_2	x_1	f_1	f_2
0	0		
0	1		
1	0		
1	1		

4.3 We have discussed logic functions which can be mechanized with linear combinations of the input variables. Such a circuit is shown in magnetic-core form in Figure P.4.3a, in which the variables are the currents I_1, I_2, and I_3. Binary values are defined as

$$0, \quad \text{if } I = 0,$$
$$1, \quad \text{if } I = 100 \text{ mA}.$$

The value of the function is taken to be

$$0, \quad \text{if the core is at } +\phi_s,$$
$$1, \quad \text{if the core is at } -\phi_s,$$

The core characteristic is given in Figure P.4.3b.
(a) List the function values in a truth table such as the following if $I_0 = 100$ mA (a constant). Note that the polarity of its winding is opposite to that of the other windings.

I_3	I_2	I_1	f
0	0	0	
0	0	1	
0	1	0	
0	1	1	
1	0	0	
1	0	1	
1	1	0	
1	1	1	

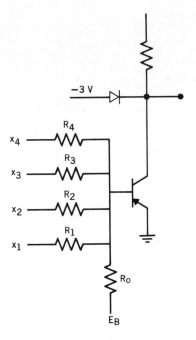

Figure P.4.1

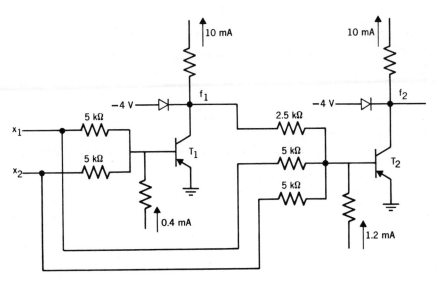

Figure P.4.2

167

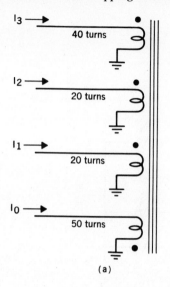

(a)

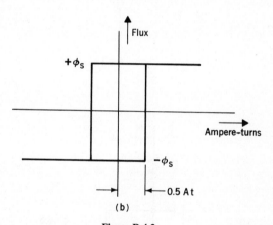

(b)

Figure P.4.3

(b) What are the tolerances on I_0 for correct operation of the circuit? (Assume all other quantities have zero tolerances).

4.4 Let the inputs to a three-input resistor-mapping transistor decision element be defined as follows: True $= -1$ V; False $= 0$ V. A bias voltage $V_B = \pm 1$ V is also available. Find the *relative* values of input resistors R_1, R_2, R_3, and bias resistor R_B to implement the following logic functions:

(a) $x_3\bar{x}_2 x_1$

(b) $\bar{x}_3\bar{x}_2\bar{x}_1 + x_2 x_1$

(c) $x_3\bar{x}_1 + \bar{x}_2\bar{x}_1$

(d) $\bar{x}_3 x_1 + x_3\bar{x}_1$

(e) $x_3\bar{x}_2 + x_3\bar{x}_1 + \bar{x}_2\bar{x}_1$

4.5 Calculate the rise-time of the circuit of Figure 4.17 in this chapter, under the following conditions:

(a) $I_{B2} = 2I_{B1} = 2\,\text{mA}$,
   $N_1 = N_2 = 10$,
   $I_R = 10\,\mu\text{A}$,
   $C = 100\,\text{pF}$.

(b) Replace the constant-current sources with bias voltages $E_B = \pm 12\,\text{V}$, and let $R_{B1} = 2R_{B2} = 12\,\text{k}\Omega$.

The input voltage swings between $+3$ and $0\,\text{V}$. Use both approximate and exact methods of estimation.

Chapter 5

Transistor Decision Elements

The major requirement imposed upon the decision component in a decision element is that it possess gain. This is certainly true of the transistor. The output of a decision component must, of course, be compatible with the inputs to other decision elements. In order to avoid the attenuation inherent in coupling networks, the ideal coupling arrangement illustrated in Figure 5.1 might therefore be constructed. Assume that the decision component is loaded by n identical components, each in the active region. To prevent signal deterioration through a network of such elements, it is necessary that any change in I_{bx} produce at least as large a change in each of its loads. Thus

$$nI_{by} + \beta I_{bx} - nI_B = 0,$$

or

$$\frac{dI_{by}}{dI_{bx}} = -\frac{\beta}{n}. \tag{5.1}$$

The limiting condition is $|dI_{by}/dI_{bx}| > 1$, and consequently $n < \beta$. This is the upper limit on the performance which can be obtained from single-stage transistor decision components.

For a variety of reasons, it is not practical to achieve maximum performance. First, allowance must be made for mapping-function networks in the coupling between elements. For example, the transistors in the second stage of Figure 5.1, must in general receive inputs from other sources. The value of the bias current I_B could be set individually for each coupling circuit to accomplish this purpose. However, it is apparent that this would result in undesirable coupling between decision components in the first stage. In order to avoid such crosstalk between driving sources it is desirable

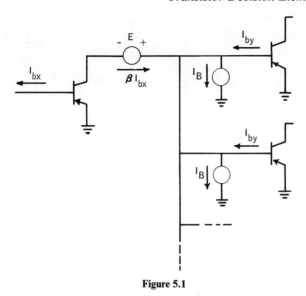

Figure 5.1

to employ either high-impedance mapping functions and ideal voltage-source decision components, or by duality, high-conductance mapping functions and ideal current-source decision components. Both techniques will be explored in this chapter. Another reason for failing to achieve maximum decision component performance lies in the common necessity for changing dc levels between outputs and inputs. It is usually expensive and inconvenient to provide floating voltage sources as indicated in Figure 5.1. However, they can be approximated by common voltage sources and a resistor network, usually in conjunction with the input mapping function. This amounts to a trade-off between cost and power loss. The coupling network of Figure 5.2 illustrates this approximation. The network can always

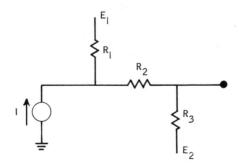

Figure 5.2

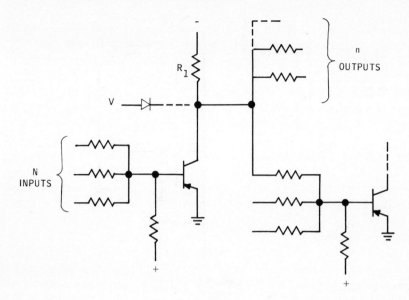

Figure 5.3

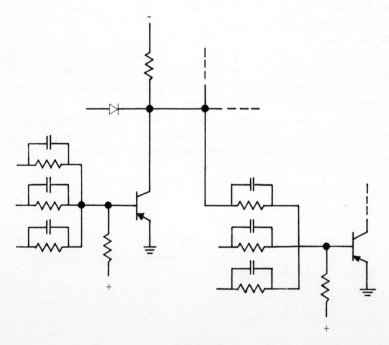

Figure 5.4

be replaced by an equivalent network consisting of a fixed voltage source, an active voltage source, and a resistor, which are functions of R_1, R_2, R_3, E_1, E_2, and I. A great deal of flexibility is therefore available in establishing dc levels and impedances.

The transistor decision elements are commonly classified on the basis of the components employed in the coupling network. Although a large number of variations exist, the basic types can be categorized as follows.

RTL. Resistor-transistor logic elements employ pure resistive coupling among elements, as shown in Figure 5.3. As pointed out in the previous chapter, all N input conductances may be regarded as equal, with the input bias selected to achieve the desired decision level. With binary signal levels of approximately 0 V and some signal voltage V (negative for the *p-n-p* component shown), operation is as follows. If m or more of the N inputs are negative, the transistor turns ON (usually to saturation), and its output signal is near zero (collector V_{ces}). If fewer than m inputs are negative, the transistor is turned OFF, and its output is negative. The output may or may not be clamped to the signal voltage V, as shown in Figure 5.3 by the dotted line to the collector clamp diode. When the transistor output is negative, the current through the resistor R_1 must be sufficient to supply the drive current to the decision element's n output loads.

RCTL. Resistor-capacitor-transistor logic elements can be generated by placing capacitors across the input conductances of the basic RTL circuit, in Figure 5.4. This has approximately the effect of converting the device to direct-coupled transistor logic during switching transients and to low-frequency (dc) RTL during other times. These points will be explored in subsequent analyses.

DTL. Diode-transistor logic is illustrated in two typical versions in Figure 5.5*a* and *b*. In these cases, if any input is positive (near 0 V), the diode gate is held positive, and the transistor is turned OFF by the bias current through R_3. If all inputs are negative, R_4 draws enough current to overcome the bias, and the transistor turns ON. Note that to the right of the diode gate (to the right of R_4), the circuit is equivalent to a one-input RTL element. As in the RTL case, a capacitor may be placed across the input conductance. The output may or may not be clamped to the signal voltage V, as before. Note that with the coupling arrangement shown in Figure 5.5, the equivalent of R_1 in Figure 5.3 has been split up and effectively distributed among the R_4 in the n diode gates driven by the element. In the version of Figure 5.5*b*, attenuation in the resistor network has been reduced somewhat by coupling with one or more series diodes that provide a nearly constant voltage difference between the base of the transistor and the diode gate. Zener diodes have also been used to perform this function.

DCTL. Direct-coupled transistor logic elements have been derived from diode logic gates on the basis of duality in the previous chapter. A representative network with both fan-in and fan-out is illustrated in Figure 5.6.

CML. Current-mode logic is illustrated in two versions in Figure 5.7a and *b*. In this case binary signal levels are taken to be positive and negative with respect to ground (or any other reference). Operation is as follows. An

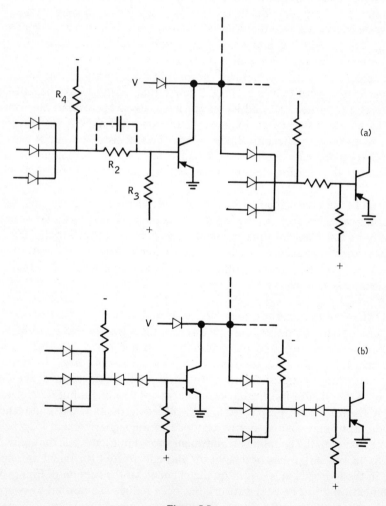

Figure 5.5

approximately constant current flows through R_E from the positive supply. If any one of the inputs is negative, the corresponding transistor is turned ON, and passes this current to the load, producing a relatively positive output. A zener-diode level shifter is shown as translating the output signal to the level required at the inputs to subsequent loads. If all inputs are positive, the constant current is shunted to ground via the diode D_1, thus clamping the emitters to ground, and all input transistors are turned OFF. The output is therefore relatively negative. The shunt diode can be replaced by an equivalent base-emitter junction, as in Figure 5.7b, with the advantage of obtaining simultaneously the decision function (f) and its complement ($\bar{f}$).

Basic Circuit Constraints

The number of constraints imposed on a decision element for proper operation is relatively modest, whereas the number of variables available is large. Commonly, therefore, a range of possible circuit values exists. In this

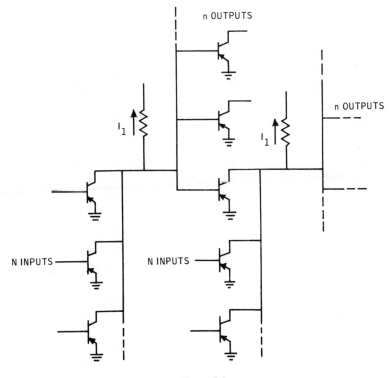

Figure 5.6

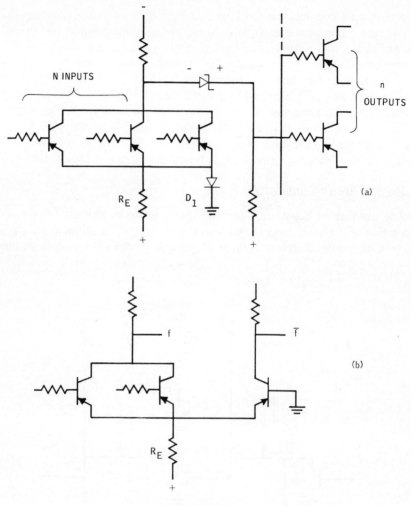

N INPUTS

R_E D_1 (a)

n OUTPUTS

f   f̄   (b)

R_E

Figure 5.7

section the basic circuit constraints are developed; in the next section the range of choice is narrowed by optimizing certain of the variables. The problem consists of designing a decision element having specified properties, based on a transistor component whose characteristics are known.

Decision element specifications include (a) the logic function to be mechanized and (b) the binary signal conditions to be fulfilled. The latter may or may not include power output levels. If already-existing devices are to be driven by the decision elements, then power requirements will be known. If, on the other hand, it is only necessary to derive a set of operable logic

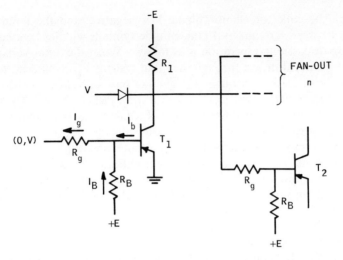

Figure 5.8

elements without other constraints, it is only required that the elements be compatible among themselves. It is the second problem which is discussed in this chapter.

Specifications on binary signal conditions will normally involve the required switching times. In this section attention is confined to simplified dc cases; full treatments, including switching-time specifications, are developed in conjunction with individual decision element types in later sections. Transistor characteristics, including β (or α_N), I_{co}, I_{eo}, τ_a, τ_s, C_c, C_e, and so on, are in all cases taken to be known or measurable.

The logic conditions that a decision element must fulfill are that it be ON for certain input combinations, OFF for others, and possess minimum fan-in and fan-out. In this section the ON condition is taken to be that condition in which the transistor is at least at the edge of or in the saturation region. The OFF condition is that condition in which the transistor is at least at the edge of or in the cutoff region. Fan-in and fan-out are considered as dependent variables (usually interrelated). The design procedure therefore consists of deriving the maximum values of fan-in and fan-out which can be reliably obtained with the given transistor. If these values are sufficient, the design is complete. If not, a different transistor must be chosen and the design repeated. In this section we shall consider fan-out alone in some highly idealized cases that serve to bring out the major dependencies among variables.

In particular, consider the simple one-input *RTL* decision element of Figure 5.8. For the *p-n-p* case shown, the collector supply voltage $|E|$ is negative, and the bias supply voltage (assumed equal in magnitude) is positive. The collector clamp voltage V is negative, and the input signal swing is between zero and $-|V|$ (magnitude symbols will not be used in the following discussion). Operation is as follows. When the input voltage is V (negative), current drawn through the input resistor R_g is sufficient to overcome the bias current through R_B. The net base current I_b is then in the turn-ON direction I_{b1}, driving T_1 to saturation. When the input voltage is zero, little or no current flows through R_g, and the bias current serves to turn the transistor OFF with a current directed *into* the base, I_{b2}. These are the input constraints. The corresponding output constraints are as follows. When the transistor is ON, its collector current must be sufficient to supply the shunt load current to R_1. When the transistor is OFF, the collector should be clamped through the clamp diode in order to present a low output impedance. The clamp-diode forward current is the difference between the current through R_1 and the total through the n loads being driven. This difference must exceed zero.

Assume for the moment that V_{beN}, the base-emitter voltage of an ON transistor, is sufficiently small relative to signal swings to be negligible. Then the four constraints above may be written

$$\frac{V}{R_g} - \frac{E}{R_B} \geq I_{b1}, \tag{5.2}$$

$$\frac{E}{R_B} \geq I_{b2}, \tag{5.3}$$

$$I_{b1} \geq \frac{1}{\beta} \frac{E}{R_1}, \tag{5.4}$$

$$\frac{E - V}{R_1} \geq n \frac{V}{R_g}. \tag{5.5}$$

The following quantities are defined

$$I_g \equiv \frac{V}{R_g},$$

$$I_B \equiv \frac{E}{R_B},$$

$$I_1 \equiv \frac{E}{R_1}. \tag{5.6}$$

Then (5.2) through (5.5) can be written as

$$I_g - I_B \geq \frac{I_1}{\beta}, \tag{5.7}$$

$$I_B \geq I_{b2}, \tag{5.8}$$

$$I_1\left(1 - \frac{V}{E}\right) \geq nI_g. \tag{5.9}$$

The current I_{b2} required to keep the transistor OFF under dc conditions is I_{co}. For the moment, let this be negligible relative to other current levels. Then (5.7) through (5.9) yield the limiting fan-out,

$$\frac{n}{\beta} \leq \left(1 - \frac{V}{E}\right). \tag{5.10}$$

Note that (5.10) is less than the ideal of (5.1) by the amount that E and R_1 depart from the behavior of a constant-current generator. Note further that the circuit values are not yet determined. It is only required that $I_g \geq I_1/\beta$, and neither current level is specified.

To be slightly more realistic, let us introduce the finite value of I_{co} at the maximum temperature of desired operation. Then $I_{b2} = I_{co}$ in (5.8). To be consistent, (5.9) should also be rewritten as

$$I_1\left(1 - \frac{V}{E}\right) \geq nI_g + I_{co}. \tag{5.11}$$

Under these conditions, the limiting fan-out becomes

$$\frac{n}{\beta} \leq \frac{1 - V/E - I_{co}/I_1}{1 + \beta I_{co}/I_1}. \tag{5.12}$$

The desirability of choosing operating current levels such that $I_1 \gg \beta I_{co}$ in order to maximize fan-out is apparent. With I_{co} specified, the inequalities (5.7), (5.8), and (5.11) require a minimum operating current level of

$$I_1 \geq \frac{(n + 1)I_{co}}{1 - (V/E + n/\beta)}. \tag{5.13}$$

It is seen that n cannot be arbitrarily chosen; that is, as $V/E + n/\beta \to 1$, $I_1 \to \infty$, and power dissipation must inevitably limit circuit realization. Far more serious constraints are imposed by power dissipated during switching transients.

Consider the effects of finite voltages at base input and collector output. When the transistor is ON, voltage V_{beN} (negative) appears at the base. If the transistor is just at the edge of saturation, by definition the collector junction voltage $V_c = 0$. Therefore under this limiting condition $V_{ces} = V_{beN}$ in magnitude. If the transistor is further into saturation, $V_{ces} < V_{beN}$ by the magnitude of the forward bias across the collector, V_c. Under these conditions the input drive currents become

$$\frac{V - V_{beN}}{R_g} - \frac{E + V_{beN}}{R_B} = I_g\left(1 - \frac{V_{beN}}{V}\right) - I_B\left(1 + \frac{V_{beN}}{E}\right) \geq \frac{I_1}{\beta}. \quad (5.14)$$

The voltage at the base of an OFF transistor will be very nearly zero (it will usually be slightly positive). Equation (5.14) is therefore conservative in the limit, because collector saturation current can be reduced by small currents in the loads R_g as a result of the voltage difference $V_{ces} + V_{beF} = V_{beN} + V_{beF}$. This refinement is not included. At the input of an OFF transistor, however, the bias must supply not only I_{co} but also this small difference current. Therefore

$$\frac{E - V_{beF}}{R_B} = I_B\left(1 - \frac{V_{beF}}{E}\right) \geq I_{co} + \frac{V_{beN} + V_{beF}}{R_g} = I_{co} + I_g\frac{V_{beN} + V_{beF}}{V}. \quad (5.15)$$

In order to simplify and to bring out the importance of V_{beN}, let $V_{beF} = 0$, and assume E is large enough so that $V_{beN}/E \ll 1$ and can be neglected. It is also conservative to use (5.11) for the fan-out relationship, even though an improvement can be realized by noting that the current in a driven load R_g is slightly less than I_g. Combining (5.11), (5.14), and (5.15) with these simplifications, the limiting fan-out becomes

$$\frac{n}{\beta} \leq \frac{(1 - V/E - I_{co}/I_1)(1 - 2V_{beN}/V)}{1 + \beta I_{co}/I_1}. \quad (5.16)$$

This result implies that it is desirable to make $I_1 \gg \beta I_{co}$, and $E \gg V \gg V_{beN}$. Opposing such a choice is the increased power dissipation in the circuit as the ratios become large. Some selection of circuit variables can therefore be made on the basis of a system optimization criterion.

Selection and Optimization

Given a particular device (β, I_{co}, and V_{beN} approximately established), the optimum values to select for I_1, E, and V will depend upon system requirements. If power is a precious commodity, it must be conserved at the expense of circuit performance, such as the fan-out given by (5.16). If minimizing the number of components is more important, each must handle as large a logic load as possible, and power consumption can be sacrificed. Because no single criterion will be satisfactory for all system requirements, only the general nature of the trade-off and some illustrative examples can be given in this section. The practicing circuit designer must define for his own application the functional form of the quantity to be optimized.

One simple criterion might be to minimize total system power. This is given by the product of power consumption per circuit and the number of circuits. The power consumed in a circuit of the type of Figure 5.8 can be divided into that consumed in the transistor, that consumed in the load resistor R_1, and that consumed in the input coupling network. These in turn are complex functions of fan-in, fan-out, and system clock rate.

To first obtain an idea of the relationships involved, let us consider each power loss independently, under dc conditions. For example, at the input the power dissipation under the worst possible combination of circumstances is, neglecting I_{co} and V_{beN} relative to E (but not to V), from (5.14) and (5.15)

$$W_i = I_g V \cong \frac{I_1}{\beta} \frac{V}{(1 - 2V_{beN}/V)}. \qquad (5.17)$$

Suppose that the total number of circuits in the system is inversely proportional to the fan-out n of each (note that this is only an illustrative supposition, and may not be valid for a particular system). With this criterion, the total system power is minimized when the ratio W_i/n is minimum. The ratio may be written using (5.16), with $I_{co} \ll I_1$, as

$$\frac{W_i/n}{(I_1 V_{beN}/\beta^2)(1 + \beta I_{co}/I_1)} \cong \frac{1}{(1 - 2V_{beN}/V)^2(V_{beN}/V - V_{beN}/E)}. \qquad (5.18)$$

The minimum value of this quantity occurs when the variable V_{beN}/V is given by

$$\frac{V}{V_{beN}} = \frac{6}{1 + 4V_{beN}/E} \cong 6, \qquad (5.19)$$

for $V_{beN} \ll E$. A plot of the inverse of (5.18) for V_{beN}/E negligible is given in Figure 5.9. The peak is reasonably broad. For example, any value of V within the range $4V_{beN} < V < 10 V_{beN}$ places power dissipation within 80 percent of its optimum value. Because of the desirable effect on fan-out, V is commonly selected near the upper end of this range.

Given that V is selected according to a criterion such as the above, a similar argument can be used for the choice of E. The maximum power dissipated in the load occurs when the transistor is ON and is approximately $W_L = EI_1$. Again assuming an inverse relationship between fan-out and number of circuits, we should minimize the ratio W_L/n. It can be written

$$\frac{W_L/n}{(I_1 V/\beta)(1 + \beta I_{co}/I_1)/(1 - 2V_{beN}/V)} = \frac{1}{(V/E)(1 - V/E)}, \qquad (5.20)$$

which possesses a minimum when $E = 2V$. Again, ratios in the range $1.5V < E < 3V$ lie within 80 percent of the peak, and it is common to select E near the high end of the range.

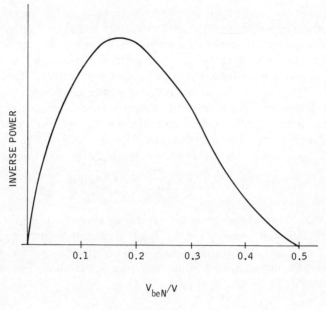

INVERSE POWER

0.1   0.2   0.3   0.4   0.5

V_{beN}/V

Figure 5.9

With V and E selected as above, I_1 may also be chosen to minimize power consumption. For example, the power consumed at the collector when the transistor is OFF is approximately $W_F \cong I_{co}V$, because the collector is clamped to V. In the worst ON case, the collector is at the edge of saturation, $V_c = 0$, and $V_{ces} \cong V_{beN}$. The power consumed is, therefore, $W_N \cong I_1 V_{beN}$. Either one of these losses may predominate. The circuit may also alternate approximately equally between the two states, in which case the total transistor power is proportional to

$$W_T = W_F + W_N = I_{co}V + I_1 V_{beN}. \qquad (5.21)$$

In this case the ratio of transistor power to fan-out can be written in the form, for $I_1 \gg I_{co}$,

$$\frac{W_T/n}{(VI_{co}/\beta)/(1 - V/E)(1 - 2V_{beN}/V)} \cong \left(1 + \frac{\beta V_{beN}}{V}\frac{I_1}{\beta I_{co}}\right)\left(1 + \frac{\beta I_{co}}{I_1}\right). \qquad (5.22)$$

The inverse of this function is plotted in Figure 5.10. Again, a rather broad peak is observed, and the selection of $I_1 \gg \beta I_{co}$ is usual.

Reliability. The foregoing discussion is highly oversimplified, because the optimization criterion illustrated is simply the minimization of total power,

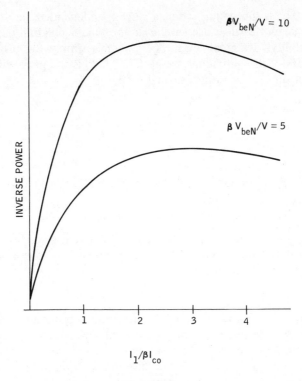

$$\beta V_{beN}/V = 10$$

$$\beta V_{beN}/V = 5$$

INVERSE POWER

1   2   3   4

$$I_1/\beta I_{co}$$

Figure 5.10

without any statement about whether that minimum is sufficiently good. Whether the minimum is satisfactory ultimately depends on other factors. An important one is the over-all reliability of the system. Specifying this reliability figure in any manner places additional constraints on the circuit design.

Typically, system reliability is specified in terms of mean time to failure or in terms of the probability of no failures within a given time. From this specification and additional data relating component failure rates to operating power level, it is possible to derive limits on circuit variables such as maximum collector current and signal voltage swing. These in turn limit the performance obtainable from the logic elements, in accordance with such relationships as (5.16). The many subtleties relating failure rate to power dissipation cannot be exhausted here. Nevertheless, the main line of argument can be outlined, and the data required for design can be identified.

The reliability specification is commonly related to individual component reliability requirements by means of two assumptions about the nature of the probability distributions that characterize component and system behavior. One is that failures are statistically independent and that a single component failure will cause system failure (chain reliability). In this case the probability of a joint event $P(A, B, \ldots)$ is simply the product of the probabilities of the individual events. Thus if P_s is the probability that the system will survive for some specified time,

$$P_s = P_1 \cdot P_2 \cdot P_3 \cdots = \prod_{i=1}^{N_T} P_i, \qquad (5.23)$$

in which

$$P_i = \text{probability the } i\text{th component will survive for}$$
$$\text{the specified time,}$$
$$N_T = \text{total number of individual components.}$$

The second assumption usually made with regard to component probability distributions is that survival is an exponentially decreasing function of time. Thus, the form of P_i is

$$P_i(t) = e^{-t/\tau_i}, \qquad (5.24)$$

in which τ_i is a "life-time-constant" characteristic of the particular component type in the particular operating environment. This form of the survival function is in turn the result of an assumption about the nature of the failure mechanism and the consequent rate of failures. Suppose that the environmental stress imposed on components can be described by some probability distribution. Suppose further that components possess some characteristic failure distribution in the presence of stress. When the stress happens to exceed the limit of a particular component, a catastrophic failure results. The average rate of such failures is given by the convolution of the two distribution functions and is therefore a constant, dependent only on the probability functions, provided the functions are not time-varying. In fact, failure rate does vary with time. A typical curve is given in Figure 5.11, showing an initially high system debugging or faulty component failure rate, and an end-of-life hump (the one-horse-shay effect) separated by a period of approximately constant failure rate. It is this intermediate, operating-life period that is being considered here.

If the average rate for the ith component is r_i, the average number of failures in time t is $r_i t$. The Poisson distribution describes the probability that exactly n events occur in an experiment when the average number of events in a given time is λ. It is

$$P_i^n = \frac{\lambda_i^n}{n!} e^{-\lambda_i}. \qquad (5.25)$$

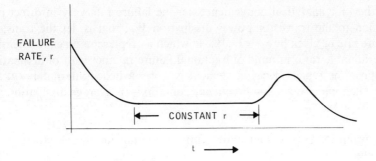

Figure 5.11

The probability of survival or that no failures occur ($n = 0$) is thus

$$P_i = e^{-\lambda_i} = e^{-r_i t}, \tag{5.26}$$

which, by comparison with (5.24), defines $\tau_i \equiv 1/r_i$.

Let the components be divided into types i and corresponding numbers N_i, such that $\Sigma N_i = N_T$. Given a particular form of decision element circuit design, the relative number of components of each type is determined (e.g., three resistors, one diode, and one transistor for the circuit of Figure 5.8). Let the relative number of each type be a_i, and let the number of decision elements in the system be N_D, such that $N_i = a_i N_D$, and $N_T = N_D \Sigma a_i$. Then, using (5.23) and (5.26), we can write the probability of survival of the complete system as

$$P = \prod_i (e^{-r_i t})^{N_i} = e^{-N_D t (\Sigma a_i r_i)}. \tag{5.27}$$

The time to failure, T, is generally defined as that time at which the probability of survival has dropped to some specified value (e.g., 99 percent, 90 percent, or $1/e$). Let this probability be defined by $P(T) \equiv e^{-k}$. Then in terms of system specifications, the constraint imposed upon the individual component failure rates is

$$\sum a_i r_i \leq \frac{k}{N_D T}. \tag{5.28}$$

With r_i restricted by (5.28), the remaining relationships to be determined are those between r_i and circuit variables. A number of factors, such as breakdown voltages and maximum current limits, enter into these relationships. Commonly, however, the most important single factor is power dissipation in the component. In order to explore the effect of this factor consider a highly simplified situation in which failures are dominated by those due to transistors in a single-transistor ($a_i = 1$) decision element system.

Further, for analytical convenience, let the failure rate vary in direct proportion to the transistor power dissipation W_T; that is, let the transistor failure rate be given by $r_T = w_T W_T$, in which w_T represents an experimentally determinable ratio in units of fractional failure rate per watt of dissipation. Suppose the typical decision element is to be switched alternately ON and OFF. Then under dc or low-frequency conditions its average dissipation is

$$W_T = \tfrac{1}{2}(V_{ces}I_1 + VI_{co}). \qquad (5.29)$$

Constraint (5.28) places an upper limit on the maximum collector current, namely,

$$\frac{I_{co}}{I_1} \geq \frac{1}{(W_T'/W_0)(2 - VI_{co}/W_T')}, \qquad (5.30)$$

in which $W_0 \equiv I_{co}V_{ces}$ may be regarded as a parameter of the transistor, and $W_T' \equiv k/w_T N_D T$ may be regarded as a "characteristic power" related to the transistor failure rate, system size, and lifetime. Although circuit realization requires that the denominator of (5.30) not be permitted to approach zero,

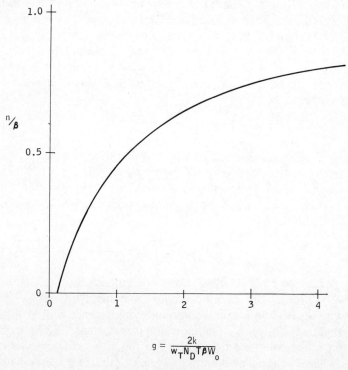

$$g = \frac{2k}{w_T N_D T \beta W_0}$$

Figure 5.12

a more serious limitation on fan-out usually occurs before realizability is violated. For example, the sample *RTL* circuit exhibits the fan-out characteristic of (5.16). The fan-out of this circuit is therefore ultimately restricted by system reliability specifications via the limitation imposed on the collector current by (5.30). The relationship between fan-out and system lifetime, for $I_1 \gg I_{co}$ and $E \gg V \gg V_{beN}$, can be put in the form

$$\frac{n}{\beta} \le \frac{1}{1 + 1/(g - V/\beta V_{ces})}, \tag{5.31}$$

in which $g \equiv 2W'_T/\beta W_0$. This function is plotted in Figure 5.12 for a typical value of $V/\beta V_{ces} = 0.1$. Small values of g correspond to large values of system lifetime, number of components, failure rate, or W_0. It is clear that contradictory specifications are possible if system reliability forces achievable fan-out to intolerably low values.

The transistor is not the only component subject to failure, and a complete treatment must also take into account the effect on system failure rate of passive components. For example, for the simple *RTL* circuit of Figure 5.8, maximum power dissipation usually occurs when the transistor is ON. In this state the dc power consumed in the various components is a minimum of

$$W_B = EI_{co},$$

$$W_g = V\left(\frac{I_1}{\beta} + I_{co}\right),$$

$$W_1 = EI_1,$$

$$W_T = V_{ces}I_1,$$

$$W_D = VI_R, \tag{5.32}$$

in which I_R is the maximum clamp-diode reverse current, and W_D is the diode dissipation. A limitation on I_1 similar to that of (5.30) can be obtained from the above, together with (5.28). For simplicity, let $w_B = w_D = w_g$, and define $W_{co} \equiv EI_{co}$, and $W'_i \equiv k/w_i N_D T$. Then the limiting value of collector current can be put in the form

$$\frac{I_{co}}{I_1} \ge \frac{1 + (W'_T/W'_g)(V/\beta V_{ces})[1 + (\beta E/V)(W'_g/W'_1)]}{(W'_T/W_0)[1 - (1 + 2V/E)(W_{co}/W'_g)]}. \tag{5.33}$$

Note the increase in (5.33) over (5.30), imposing smaller values of I_1 and consequently smaller fan-out.

Steady-state power dissipation is seldom a sufficient measure of circuit losses, particularly at the higher switching frequencies. The worst case occurs when a circuit such as that of Figure 5.8 has been designed to drive a large

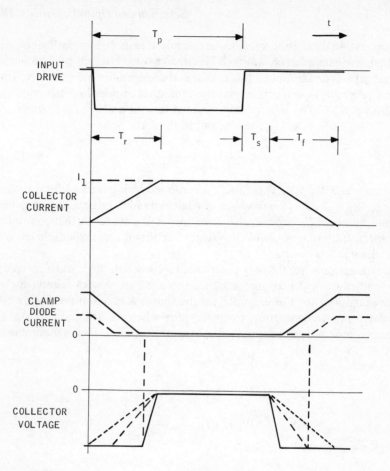

Figure 5.13

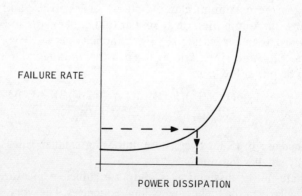

Figure 5.14

188

number of loads, but in fact the particular decision element is lightly loaded. The individual loading is a function of system logic. Although a decision element would presumably not be put into a system if its output were totally unused, we can take such an unloaded circuit as the limiting case (the practical limit being a load of 1). Under these conditions, the collector current and voltage waveforms are as shown by the solid lines in Figure 5.13 (waveforms for a circuit with intermediate load are shown dashed). A conservative estimate of energy dissipation during the switching transient is obtained if it is assumed that the collector voltage remains clamped during most of the transient but rises (falls) suddenly near the end (beginning). Neglecting dc power dissipation and assuming high-speed switching (approximately linear rise of collector current), we find that the energy expended during the transient is

$$\int_0^{T_r} e_c i_c \, dt \cong \frac{VI_1 T_r}{2}. \tag{5.34}$$

A similar loss occurs during the fall time T_f. If a timing period T_p is taken as the time for either turn ON or turn OFF followed by a quiescent state, then, neglecting the dc contribution, the average power dissipated is $W_T = VI_1 T_r / 2T_p$. This power must satisfy the reliability condition (5.28).

It is important to observe that loading can cause power dissipation to vary over wide ranges. For example, at one extreme, in a decision element of the above type with maximum loading, the collector-clamp current approaches zero, and collector voltage rises early in the switching transient as indicated by the dotted line in the collector voltage waveform of Figure 5.13. In this case, average power dissipation is reduced to $VI_1 T_r / 6T_p$. On the other extreme is a decision element with capacitive loading. In this case the collector current may temporarily exceed its quiescent value, whereas collector voltage is prevented from changing rapidly by a slow capacitance-charging process.

Although the evaluation of circuit power dissipation can itself become a difficult problem, relating such losses to failure rates is by no means as simple as suggested by the linear model previously employed. In practice, failure rate versus power dissipation tends to follow a curve more nearly like that of Figure 5.14, with rapidly increasing failure probability as destruction is approached. Thus, in general, r is some arbitrary function of W, $r_i = f_i(W_i)$, which may be different for each component, and the system reliability specification should be stated in the form

$$\sum a_i f_i(W_i) \le \frac{k}{N_D T}. \tag{5.35}$$

Discussion of techniques for solving inequalities of this type, involving failure-rate dissipation functions f_i that may be entirely empirical, is beyond the scope of this text. The characteristics of an ideal solution are, however, straightforward. What is usually desired is a set of values for I_1, E, and V (given a particular component with known parameters such as β, I_{co}, and V_{ces}) that will maximize some quantity such as fan-out [as given, for example, by (5.16)] while at the same time staying within the constraints of (5.35). In practice this ideal procedure is not always followed, because relevant parameters can often be freely manipulated (e.g., resistor power ratings and provisions for refrigeration), and an early trial-and-error process may rapidly converge to a usable region of performance. Therefore, in the following discussions, it is assumed that certain circuit variables are chosen or can be estimated on a simple but rational basis. For example, signal voltage V can be selected much larger than junction potentials to overcome attenuation and noise but need not be chosen much above its optimum because the effect upon fan-out diminishes while the effect upon system power increases rapidly (e.g., Figure 5.9). Similar considerations apply to the choice of bias supply E. Limiting collector current can be estimated from considerations such as those leading to (5.30) or (5.33) for low frequencies, or from transient power losses such as that of (5.34). In the following discussions it is assumed that these choices have been made, and only the remaining relationships among circuit performance, device parameters, and switching specifications are explored for the different forms of decision element circuit.

Resistor Transistor Logic

The most straightforward decision element circuit is the *RTL* device of Figure 5.15. It is shown including a collector-clamp diode to the signal voltage V. Unclamped *RTL* circuits are also widely used but present more difficult analytical problems. Because the behavior of unclamped circuits frequently merges into that of other forms of decision element, particularly *DCTL* and *DTL*, consideration of that form is deferred to later sections.

DC constraints. The dc behavior of the one-input RTL circuit has been analyzed in (5.2) through (5.16). To generalize, the requirements imposed upon an N-input circuit can be stated as follows. When m or more (out of N) inputs are true (negative for the *p-n-p* case), the transistor turns ON (in the worst case, at the edge of saturation). When fewer than m inputs are true, the transistor turns OFF (passes I_{co}). Finally, when the transistor is OFF, the collector voltage must be capable of remaining true (negative) even when up to n other circuits are being driven. This last requirement is equivalent to demanding some forward current in the clamp diode—down to a worst-case limit $I_d \to 0$ for maximum loading.

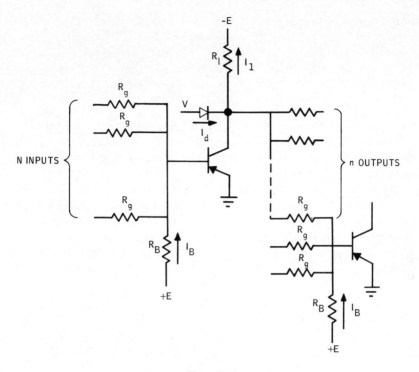

Figure 5.15

In order to bring out first-order properties the circuit can be idealized in the following ways. Let the base-emitter voltage of an ON transistor be negligible relative to the signal voltage swing. Because the base-emitter voltage of an OFF transistor need only be zero or slightly positive, let this also be negligible. Assume that V_{ces} and I_{co}, and diode forward drop and leakage current are also negligible. Thus the signal inputs to a decision element possess the two voltage values 0 and V. The requirements on the operation of this idealized circuit can be written, using definitions (5.6), as

$$\text{ON}: mI_g - I_B \geq I_{b1} = \frac{I_1}{\beta},$$

$$\text{OFF}: I_B - (m - 1)I_g \geq I_{b2} = 0,$$

$$\text{LOAD}: I_1\left(1 - \frac{V}{E}\right) \geq nI_g. \tag{5.36}$$

Current gain β is always to be taken at its smallest worst-case value. Limiting solutions for this case are

$$I_g = \frac{I_1}{\beta},$$

$$I_B = (m - 1)\frac{I_1}{\beta}, \tag{5.37}$$

from which, by definitions (5.6), R_g and R_B are determined. Circuit fan-out is given by (5.10). Thus the ideal N-input circuit is identical to the one-input circuit with a shift in the bias level corresponding to the m (rather than one) true inputs required for turn ON.

One of the major disadvantages of the multi-input RTL decision element is its sensitivity to tolerance effects. Consider, for example, the effect of finite tolerances on power supplies and resistors while maintaining other idealizations. Let $r \equiv$ fractional resistor tolerance, and let $v \equiv$ fractional voltage tolerance. The above circuit requirements can then be written in the form

$$mI_g\frac{1 - v}{1 + r} - I_B\frac{1 + v}{1 - r} \geq \frac{I_1}{\beta}\frac{1 + v}{1 - r},$$

$$I_B\frac{1 - v}{1 + r} - (m - 1)I_g\frac{1 + v}{1 - r} \geq 0,$$

$$I_1\frac{1 - v}{1 + r}\left(1 - \frac{V}{E}\frac{1 + v}{1 - r}\right) \geq nI_g\frac{1 + v}{1 - r}. \tag{5.38}$$

Note that the maximum collector current is no longer I_1 but $I_1(1 + v)/(1 - r)$. Manipulating (5.38), we obtain the following constraint on decision element fan-out:

$$\frac{n}{\beta} \leq \left(1 - \frac{V}{E}\frac{1 + v}{1 - v}\right)\left\{1 - m\left[1 - \left(\frac{1 - v}{1 + v}\right)^2\left(\frac{1 - r}{1 + r}\right)^2\right]\right\}. \tag{5.39}$$

This relationship means that there exists a significant trade-off between fan-in and fan-out. Given finite tolerances, as m increases the maximum n must correspondingly decrease. It is easy to find simple cases in which circuit realizability conditions can be violated. For example with tolerances on both resistances and power supplies of ± 5 percent, the term in brackets in (5.39) is negative for $m > 3$, indicating an impossible condition. The function (5.39) is plotted in Figure 5.16 for a typical value of $E = 3V$ and $r = v$.

The effect of a finite collector leakage current I_{co} upon performance of the one-input RTL circuit has been derived (5.12). For a multiple-input circuit

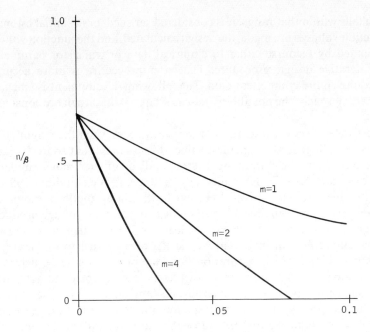

FRACTIONAL TOLERANCES

Figure 5.16

let $I_{b2} = I_{c0}$ in (5.36). The resulting fan-out limit is

$$\frac{n}{\beta} \leq \frac{\left(1 - \dfrac{V}{E}\dfrac{1+v}{1-v} - \dfrac{I_{c0}}{I_c}\dfrac{1+r}{1-v}\right)\{B\}}{1 + \dfrac{\beta I_{c0}}{I_c}\dfrac{1+r}{1-v}}, \tag{5.40}$$

in which $\{B\}$ is the same as the term in braces in (5.39), and $I_c \neq I_1$ is the selected maximum collector current. Compare this with (5.12) and note the further departure from ideal behavior.

With the exception of tolerance effects, the foregoing analysis of circuit performance has not brought to light any interaction between fan-in and fan-out. This is a result of the lack of coupling imposed by the assumed negligible base-emitter, collector-emitter, and forward diode drops. The introduction of these variables introduces serious nonlinearities into the design equations, because junction voltages are nonlinear functions of junction currents. This problem is usually handled in one of two ways: (a) maximum values of junction voltages are taken from empirical data covering the anticipated range of operation and, provided the resulting design falls within that range, it is considered satisfied, or (b) initial estimates of junction voltages are made, the design calculated, and the junction voltages reestimated by recourse either to empirical data or transistor parameters, in an iterative design procedure. The latter procedure is more accurate, but requires more component data. The following development is amenable to either approach; the possible worst-case states of the circuit are considered in turn.

ON: The least ON drive to an input resistor occurs when the magnitude of the source voltage is at its smallest value. It is assumed that source voltages are derived from the outputs (collector load) of other circuits similar to those of Figure 5.15. Such circuits may be drawing their limiting collector load, in which case the forward current in the clamp diode can approach zero. If so, there is negligible forward diode drop, and the voltage producing the input signals approaches V. For the least ON drive, m such inputs are at V. The remainder $N - m$ inputs are at the V_{ces} of ON transistors. These transistors may be heavily in saturation (for example, they may be driven ON by N inputs). The V_{ces} of such transistors may be correspondingly small, and it is conservative to regard it as approaching zero. At the base summing junction, the voltage is $V_{be(m)}$, corresponding to a turn-ON drive of m units. The collector voltage of the ON transistor must not exceed the maximum value of V_{ces} if it is to drive subsequent loads correctly. However, only one such critical load may be present, the remainder being driven heavily ON by their other $N - 1$ inputs. This worst-case situation is depicted in Figure

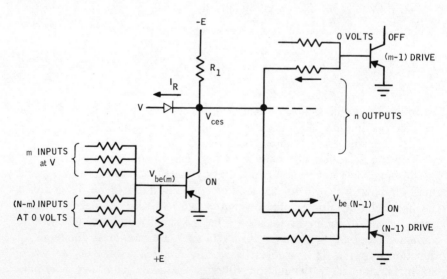

Figure 5.17

5.17. The single OFF load is taken to have a limiting base-emitter voltage of zero, and the remaining $n - 1$ loads have a base-emitter voltage of $V_{be(N-1)}$ due to $N - 1$ true inputs. The transistor input-output requirements can be written as

$$m\frac{V - V_{be(m)}}{R_g} - \frac{E + V_{be(m)}}{R_B} - (N - m)\frac{V_{be(m)}}{R_g} \geq \frac{I_c}{\beta}$$

$$= \frac{1}{\beta}\left[\frac{E - V_{ces}}{R_1} + I_R - \frac{V_{ces}}{R_g} + (n - 1)\frac{V_{be(N-1)} - V_{ces}}{R_g}\right]. \qquad (5.41)$$

OFF: The transistor must be turned OFF even when $m - 1$ inputs are true. These inputs could come from decision elements that, although designed to supply n loads, are in fact lightly loaded. The forward diode drop could therefore add to the signal supply voltage for a total drive of $V + V_D$. Similarly, the remaining $N - m + 1$ inputs which are nominally at 0 V could in fact be barely ON, with a maximum collector voltage of V_{ces}. We shall assume that it is only necessary to drive the base voltage to zero in order that the transistor be OFF. Thus

$$\frac{E}{R_B} - (m - 1)\frac{V + V_D}{R_g} - (N - m + 1)\frac{V_{ces}}{R_g} \geq I_{co}. \qquad (5.42)$$

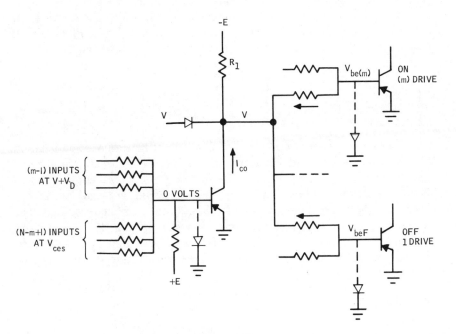

Figure 5.18

LOAD: The worst-case condition of an OFF decision element is depicted in Figure 5.18. In the limit, clamp-diode current can be allowed to approach zero, and the output voltage can be allowed to approach V. This signal must provide the mth true input to at least one load circuit (if other decision element states are independent of this signal, the logic configuration is not the worst possible). However, the remaining $n - 1$ driven elements may have all other inputs at 0 V and a positive OFF base voltage V_{beF}. These constraints can be written in the form

$$\frac{E - V}{R_1} \geq (n - 1)\frac{V + V_{beF}}{R_g} + \frac{V - V_{be(m)}}{R_g} + I_{co}. \tag{5.43}$$

The value of V_{beF} is a function of m and N as well as a function of other circuit variables. Some idea of its limiting value can be obtained by considering R_B to be a current source I_B, by letting $m = N \gg 1$, and by lumping all parallel input resistors together into a single equivalent R. If the bias is such that all inputs are required to be true to turn the transistor ON and the base voltage under this condition is taken to be near zero, then V/R must draw $I_B + I_c/\beta$. When all inputs are at 0 V, neglecting I_{co}, then $I_B R = V - (I_c/\beta)R$ is the voltage at the base. For large N, $R = R_g/N$ can become small, and $V_{beF} = I_B R$ approaches V. Because coupling of this type can substantially reduce circuit performance, an input-limiting diode is sometimes placed across the input to the base, as indicated by the dotted connections in Figure 5.18. In this case $V_{beF} = V_D$ as an upper limit. If the diode is omitted, an estimate of V_{beF} can be obtained by neglecting I_{co} and letting

$$\frac{E - V_{beF}}{R_B} - N\frac{V_{beF}}{R_g} = 0, \tag{5.44}$$

in which N rather than $N - 1$ is used in order that the result be valid for small values of N including $N = 1$.

The above expressions completely define circuit performance and can be solved for R_1, R_g, R_B, and n in terms of I_c (or I_1), E, V, transistor parameters, and desired fan-in N and m. In order to bring out the salient features of circuit performance let us simplify by making several practical approximations. Let all forward-biased junction voltages be approximately equal, that is, let $V_{be(m)} \cong V_{be(N)} \cong V_{ces} \cong V_D \equiv V_N$. Neglect I_R and V_{ces}/R_g in (5.41) relative to $I_1 \equiv E/R_1$, and let E be large enough so that $V_N/E \ll 1$. Then limiting solutions for the input coupling network, using definitions (5.6), are

$$I_g = \frac{I_1}{\beta}\frac{1 + \beta I_{co}/I_1}{1 - 2N(V_N/V)},$$

$$I_B = \frac{I_1}{\beta}\left[m - 1 + N\frac{V_N}{V} + \left(m + N\frac{V_N}{V}\right)\frac{\beta I_{co}}{I_1}\right]. \tag{5.45}$$

If a base-input limiting diode is used in the circuit such that $V_{beF} \equiv V_F = V_N$ does not exceed a forward-biased junction voltage and $\beta \gg 1$, the fan-out is

$$\frac{n}{\beta} \le \frac{(1 - V/E)[1 - 2N(V_N/V)]}{(1 + \beta I_{co}/I_1)(1 + V_F/V)}. \qquad (5.46)$$

If a limiting diode is not used, the ratio V_F/V in (5.46) can be replaced by the value obtainable from (5.42) and (5.44), namely,

$$\frac{V_F}{V} \cong \frac{1}{N/[m - 1 + N(V_N/V)] + (V/E)}. \qquad (5.47)$$

It can be seen that $1 < 1 + V_F/V < 2$ in (5.46) over the range $1 < N = m < \infty$. Note the importance of selecting $V \gg V_N$ for good circuit performance, because realizability requires $N < V/2V_N$. It is worth pointing out that the NOR decision element, with $m = 1$, takes maximum advantage of (5.47) by forcing $1 + V_F/V$ to approach its lower limit, whereas the NAND, with $m = N$, causes the same quantity to approach its upper limit for $N \gg 1$.

When both tolerance and finite impedance effects are considered, further departure from idealized circuit performance results. For notational convenience, let tolerance ratios be defined as $(1 - v)/(1 + v) \equiv U_v$; and $(1 - r)/(1 + r) \equiv U_r$. Repeating the previous development, an approximate limit on circuit performance is found to be

$$\frac{n}{\beta} \lesssim \frac{[1 - (V/E)(1/U_v)]\{1 - m[1 - (U_vU_r)^2] - N(V_N/V)[(1 + U_vU_r)/(1 + v)]\}}{[1 + (\beta I_{co}/I_1)(1 + r)/(1 - v)]\{1 + (V_F/V)[1/(1 + v)]\}}, \qquad (5.48)$$

in which second-order product terms (e.g., $rV_N/V \ll 1$) have been dropped, and $\beta \gg 1$ is assumed. The term in braces in the numerator is the major criterion of circuit realizability. The function (5.48) is plotted in Figure 5.19 for the NOR decision element ($m - 1$) and the typical conditions $E = 3V$, $V = 6V_N$, $V_F = V_N$ (i.e., base limiting diode), $I_1 \gg \beta I_{co}$, and equal resistor and voltage tolerances, $v = r$. Observe the further degradation in performance over the idealization depicted in Figure 5.16 ($m = 1$), which included tolerance effects alone.

AC constraints. In the previous section constraints imposed on the coupling network by the two states of the transistor, ON and OFF, have been discussed from the dc standpoint. For example, the turn-ON drive has been taken as $I_{b1} = I_1/\beta$ and the turn-OFF drive as $I_{b2} = I_{co}$. When the required switching-time specifications are also attached to the transistor ON and OFF states, however, additional source drive is required. Since input drive and switching time are functionally related, as outlined in Chapter 2, the corresponding circuit performance can be obtained directly in terms of such specifications. In this section it is assumed throughout that desired turn-ON and turn-OFF times are given as part of the system specifications and therefore constitute known design parameters.

Some care is necessary in defining precisely which switching times are of interest. In turning ON, is the important factor the time for the collector current to rise from zero to saturation, or the time for the collector voltage to change from clamp to V_{ces}? In turning OFF, is storage time, fall time, or collector voltage swing the crucial parameter? Reference to Figure 5.13 shows that these are not necessarily equivalent criteria, and they can vary widely depending upon load conditions. In general, however, we are interested in limiting conditions, in which case only certain criteria are appropriate for particular forms of decision element. For RTL with maximum loading, Figure 5.13 indicates that the changes of collector voltage and current between cutoff and saturation are concurrent. Therefore, the turn-ON time will be taken to be equal to the rise time, and the turn-OFF time will be taken equal to storage plus fall time. For notational symmetry these quantities will be denoted as turn-ON $= T_N = T_r$ and turn-OFF $= T_F = T_s + T_f$.

The switching-time expressions developed in Chapter 2 were based upon initial quiescent conditions involving steady-state base drive. Although external variables (collector current and voltage) will always be assumed to have reached a final state before switching is initiated, under conditions

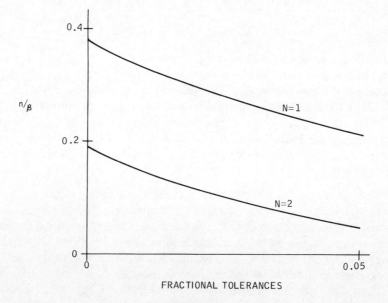

FRACTIONAL TOLERANCES

Figure 5.19

of repeated switching this may not hold true of internal transistor variables such as base charge. It is convenient, therefore, to state initial conditions in terms of base charge and to relate the latter to quiescent drive if the time between successive transients is sufficiently long.

Assuming purely resistive loads and $C_c = 0$, the simplest forms for relating base drive to specified switching times are obtained by inverting the relationships derived in Chapter 2. These are

$$\text{ON}: I_{b1} = \frac{Q(T_N)}{\tau_a}\left(\frac{1}{1 - e^{-T_N/\tau_a}}\right),$$

$$\text{OFF}: I_{b2} = \frac{Q(0)}{\tau_a}\left(\frac{e^{-T_F/\tau_a}}{1 - e^{-T_F/\tau_a}}\right). \tag{5.49}$$

$Q(T_N)$ is the base charge at the completion of the transient and is $\tau_a(I_c/\beta)$

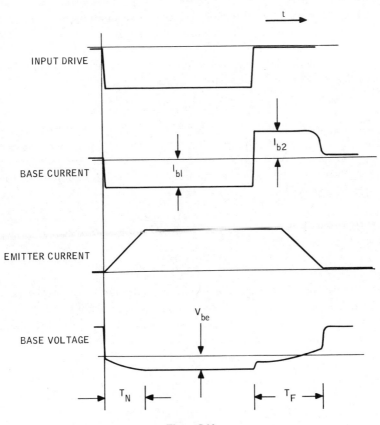

Figure 5.20

for the resistive load case. $Q(0)$ is the initial base charge at the beginning of the turn-OFF transient. Because some overdrive is required to achieve finite switching times, it will normally not be the same as $Q(T_N)$.

In order to emphasize the most important results of specifying frequency response, let us again idealize the RTL circuit as in the dc analysis. Let V_{ces} and V_D be negligible compared to signal voltage, and let I_{co} and I_R be negligible compared to current drive levels. Consider all tolerances to be zero. Finally, let base input impedance be negligible during switching times. Note that the base impedance becomes very high at the end of a turn-OFF transient, but that during the transient the base-emitter junction remains forward-biased until emitter current drops to nearly zero. The sequence of events at the base is illustrated in Figure 5.20.

During turn-OFF, the voltage $r_b I_{b2}$ subtracts from the forward-junction voltage, which decays toward zero as cutoff is approached. When cutoff is reached, base impedance becomes that of a reverse-biased junction, and I_{b2} drops to near zero (I_{co}). The idealization that base impedance is negligible amounts to the condition $V_{be} \ll V$.

Worst-case switching occurs as follows. For turn-ON, m true inputs must be capable of turning the transistor ON within T_N. Thus

$$mI_g - I_B \geq I_{b1}, \tag{5.50}$$

in which I_{b1} is given by (5.49) with $Q(T_N)/\tau_a = I_1/\beta$. For turn-OFF, a more serious worst-case situation arises. The transistor may previously have been turned ON by as many as N true inputs. It is specified that the transistor must turn OFF within T_F when as few as $N - m + 1$ of them (remaining $m - 1$ still true) return to 0 V. Thus

$$I_B - (m - 1)I_g \geq I_{b2}, \tag{5.51}$$

in which I_{b2} is given by (5.49) with

$$Q(0) = \tau_a(NI_g - I_B). \tag{5.52}$$

This value of initial base charge implies that the interval between turn-ON and turn-OFF is sufficiently long for base charge to reach its quiescent value following turn-ON. Combining the above relationships, we find that the limiting (minimum-power) solutions for circuit variables are

$$I_g = \frac{I_1}{\beta} \frac{1}{(1 - e^{-T_N/\tau_a})[1 - (N - m + 1)e^{-T_F/\tau_a}]},$$

$$I_B = \frac{I_1}{\beta} \frac{m - 1 + (N - m + 1)e^{-T_F/\tau_a}}{(1 - e^{-T_N/\tau_a})[1 - (N - m + 1)e^{-T_F/\tau_a}]}. \tag{5.53}$$

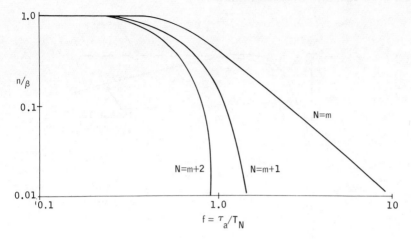

Figure 5.21

Compare this result with (5.37), and observe the large effect of fan-in on required drive current levels. It is commonly asserted that RTL circuits are low-frequency devices, reflecting the realizability constraint that $T_F \gg \tau_a$ if fan-in is at all substantial. For this idealized circuit, the LOAD condition of (5.36) applies. The resulting measure of circuit performance is

$$\frac{n}{\beta} \le \left(1 - \frac{V}{E}\right)(1 - e^{-T_N/\tau_a})[1 - (N - m + 1)e^{-T_F/\tau_a}]. \tag{5.54}$$

Note that the NAND decision element ($N = m$) possesses a significant advantage over other RTL circuits for higher-speed applications (small T_F). The effect of fan-in on the circuit performance given by (5.54) is shown in Figure 5.21 for several values of N, based on the symmetrical switching specification, $T_N = T_F$, and $E \gg V$. A relative frequency is defined by $f \equiv \tau_a/T_N$. At high frequencies, performance for $N = m$ drops off at the rate of 12 dB/octave.

The performance figure (5.54) and minimum-power circuit solutions (5.53) imply that within the constraints of fan-in and switching times, any desired system reliability can be achieved by operating at sufficiently low power levels. Thus, by reducing I_1 in (5.53), we can make resistor and transistor power dissipation vanishingly small, and (5.35) could always be satisfied. This illusion is, of course, a consequence of the simplified treatment of circuit behavior. In fact, when all factors are taken into account, a definite optimum choice of I_1 or V can always be made to simultaneously satisfy reliability requirements while maximizing performance. This has already been pointed out in connection with selection and optimization of circuit variables under dc conditions.

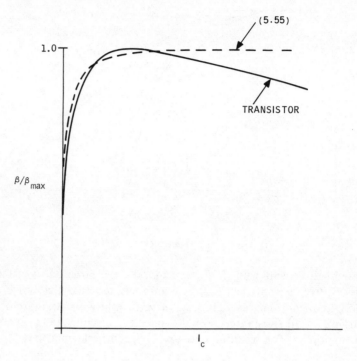

Figure 5.22

One of the important effects is that current gain β is a function of I_c. A typical relationship between current gain and collector current is given in Figure 5.22. Because operation at the lowest possible current is desirable from the standpoint of power dissipation, the low-I_c end of the curve is of greatest interest. Although the actual relationship is complex, a satisfactory approximation for β versus I_c in this region is

$$\beta \cong \frac{\beta_{\max}}{1 + I_h/I_c},$$ (5.55)

as indicated by the dotted curve of Figure 5.22, in which I_h is the collector current at which β falls to half its maximum value. Introducing this correction for current gain into the design equations has the effect of replacing the quantity βI_{co} with the quantity $I_h + \beta I_{co}$, thus shifting the optimum operating point to somewhat higher values of collector current.

Another idealization which deserves attention is the assumption that the transistor time constant dominates circuit response. In general, other factors must be taken into account. For example, turn-ON delay time and the charging of base-emitter and base-collector junction capacitances will delay the beginning of the rise time. The approximate expression (2.85) describes T_N when these factors are included. Rearranging this expression, we can see that the required value of ON drive, I_{b1} in (5.49), should be replaced by,

$$I_{b1} = \frac{Q(T_N)}{\tau_a} \left\{ \frac{1 + [\Delta Q / Q(T_N)] e^{-(T_N - T_D)/\tau_a}}{1 - e^{-(T_N - T_D)/\tau_a}} \right\}, \tag{5.56}$$

in which ΔQ is the charge transferred to the base-emitter junction capacitance in going from OFF to ON. For symmetrical base voltages $(V_F = V_N)$ it is

$$\Delta Q = 2V_N(C_e + C_c).$$

This replacement has a corresponding effect on the performance measure (5.54).

Finite collector and load capacitance may have substantial effects upon switching times. As pointed out in Chapter 2, a first approximation to circuit response under these conditions involves a single-pole transfer function whose equivalent time constant is the sum of the individual effects; thus $\tau_T = \tau_a + \beta\tau_c + \tau_L$. The duration of the turn-ON transient is related to the collector voltage swing between clamp and forward-biased collector junction, rather than to collector current. From (2.81) changes in collector voltage in terms of step changes in base drive with $k \ll 1$ are given by

$$\Delta e_c(t) \cong \beta R \, \Delta I_b (1 - e^{-t/\tau_T}), \tag{5.57}$$

in which R is the equivalent load resistance. The collector circuit with maximum fan-out and load capacitance C_L is shown in Figure 5.23a. The value of the equivalent load resistor R may be estimated from the OFF condition (Figure 5.23b), in which it is assumed that the base voltages of the n driven elements are near zero, and the equivalent load R_g/n is therefore connected to ground. The quiescent OFF requirement is that

$$\frac{E - V}{R_1} \geq \frac{V}{R_g/n}, \tag{5.58}$$

from which, with limiting equality prevailing, $R = R_1(V/E)$. The ac equivalent load circuit is shown in Figure 5.23c. Because the transient is complete when e_c has changed by V, the required drive current is

$$I_{b1} = \frac{V}{\beta R}(1 - e^{-T_N/\tau_T}) = \frac{I_1}{\beta}(1 - e^{-T_N/\tau_T}). \tag{5.59}$$

With $\tau_L = RC_L$ and $\tau_c = 0$,

$$\tau_T = \tau_a \left(1 + \frac{V}{E} \frac{R_1 C_L}{\tau_a} \right). \tag{5.60}$$

The above discussion is unchanged if $\tau_c = RC_c$ is also included.

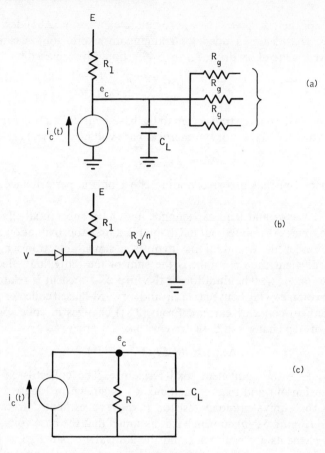

Figure 5.23

Turn-OFF under nonresistive load conditions also deserves careful consideration. In the active region, with the collector voltage changing, the single-pole form of (2.80) with $k \ll 1$ is a fair first-order estimate of behavior. Inserting initial and final conditions on collector voltage, we obtain the OFF drive required to produce a specified fall time T_f as

$$ I_{b2} = \frac{I_1}{\beta} \left(\frac{e^{-T_f/\tau_T}}{1 - e^{-T_f/\tau_T}} \right). \tag{5.61} $$

However, the total turn-OFF time is $T_F = T_s + T_f$, and storage time is dependent only upon transistor storage time constant τ_s. (It will be assumed for convenience in the following discussion that $\tau_s = \tau_a$, reflecting the basic recombination-rate mechanism. Similar results can readily be derived

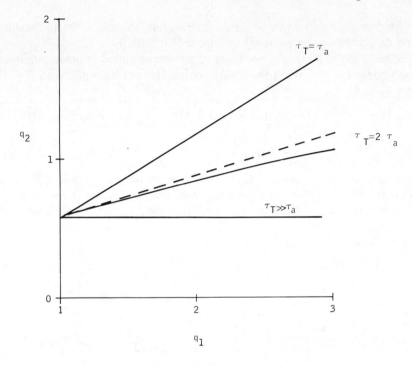

Figure 5.24

without this restriction.) Using the switching-time equations of Chapter 2, we obtain the following expression relating total turn-OFF time and current drive:

$$e^{T_F/\tau_T} = \left(1 + \frac{I_1}{\beta I_{b2}}\right)^{1-\tau_a/\tau_T}\left(1 + \frac{Q(0)}{\tau_a I_{b2}}\right)^{\tau_a/\tau_T}. \tag{5.62}$$

Although (5.62) determines I_{b2} in terms of initial conditions and system parameters, the presence of the exponents makes solution awkward. Therefore, recourse is often made to alternative estimates. If we define

$$q_1 \equiv Q(0)/\tau_N I_1 \geq 1$$

as the ON overdrive ratio and $q_2 \equiv \beta I_{b2}/I_1$ as the OFF drive ratio, one reasonable accurate form of approximation is

$$q_2 \cong \left[1 - (1 - q_1)\frac{\tau_a}{\tau_T}\right]\left(\frac{e^{-T_F/\tau_T}}{1 - e^{-T_F/\tau_T}}\right). \tag{5.63}$$

Functions (5.62) and (5.63) coincide for either $\tau_T = \tau_a$ or $\tau_T \gg \tau_a$. They are plotted for these extremes and for the intermediate value $\tau_T = 2\tau_a$ in the solid and dashed curves, respectively, in Figure 5.24, with $T_F = \tau_T$.

When load and input capacitance effects are included in device analysis, the performance figure of the ideal decision element becomes, with $E \gg V$ and $T_D = 0$,

$$\frac{n}{\beta} \leq \frac{(1 - e^{-T_N/\tau_T})[1 - (\tau_a/\tau_T)(N - m + 1)\,e^{-T_F/\tau_T} - (1 - \tau_a/\tau_T)\,e^{-T_F/\tau_T}]}{(1 + \Delta q\,e^{-T_N/\tau_T})\{1 - (1 - \tau_a/\tau_T)}$$
$$\times\ [(1 + \Delta q)/(1 + \Delta q\,e^{-T_N/\tau_T})]\,e^{-(T_N + T_F)/\tau_T}\}$$, \hfill (5.64)

in which $\Delta q \equiv \Delta Q/Q(T_N)$. This function is plotted in Figure 5.25 with $\Delta q = 0$ and $T_N = T_F = \tau_T$. Note the improved performance for $\tau_T \gg \tau_a$. This means that given a specified switching time and therefore a fixed total

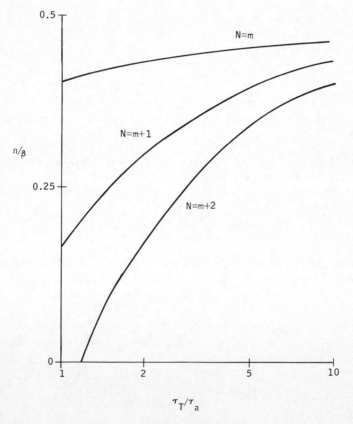

Figure 5.25

circuit time constant that can be distributed in any way between the transistor and the remainder of the circuit, it is best to select a transistor with $\tau_a \ll \tau_T$. In this case, virtually all of the switching transient occurs with the transistor in the active region. Thus as $\tau_a \to 0$, (5.64) becomes independent of fan-in, illustrating the fact that (to a first order) ON overdrive exhibits its detrimental effects primarily through increased storage time, and can be minimized with a sufficiently fast active device.

With $\tau_T = \tau_a$ and $\Delta q \to 0$, (5.64) reduces to (5.54) with its characteristic frequency drop-off of 12 dB/octave even under the best of conditions ($N = m$). With $\tau_T \gg \tau_a$, under the same conditions (5.64) reduces to

$$\frac{n}{\beta} \le \frac{(1 - e^{-T_N/\tau_T})(1 - e^{-T_F/\tau_T})}{1 - e^{-(T_N + T_F)/\tau_T}}. \tag{5.65}$$

This function is plotted in Figure 5.26 for $T_N = T_F$. It can be taken as the characteristic frequency response of an ideal nonsaturating decision element, because with $\tau_a \ll \tau_T$, storage time is a negligible portion of total switching time. Note that the response drops off at 6 dB/octave, being proportional to $T_N/2\tau_T$ for $T_N = T_F \ll \tau_T$.

Finite tolerances, base voltages, and other degrading effects produce approximately the same effects on performance under specified switching conditions as those previously noted under dc conditions. For example, the effect of tolerances alone on the simplified performance measure (5.54)

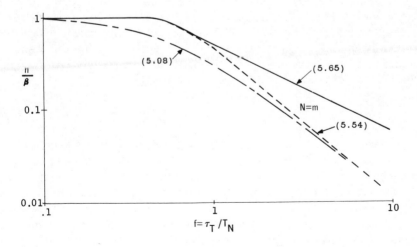

Figure 5.26

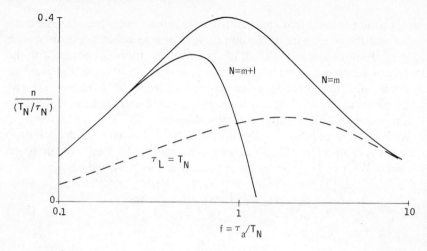

Figure 5.27

is to multiply (5.54) by $U_v^2 U_r^2 < 1$ and replace the last factor in (5.54) by

$$[1 - m(1 - U_v^2 U_r^2) - (N - m + 1)e^{-T_F/\tau_a}]. \qquad (5.66)$$

Compare this with (5.39). Other quantities similar to those in (5.48) appear in the performance measure when finite transistor impedances are included.

It has been implied by the form of performance–switching-time expressions that β and τ_a are independent parameters of the transistor. In fact, although a wide latitude exists in manufacture, the two quantities tend to be correlated. The dependency follows from the first-order charge model, in which collector current is related to base charge by $I_c = Q/\tau_N$, and base charge is related to recombination rate by $Q = \tau_a I_b$. Thus $\beta = \tau_a/\tau_N$. If switching depends primarily upon transistor parameters (i.e., resistive loading), a value of β and τ_a exists that will optimize the performance measure. For example, for $T_N = T_F$, (5.54) can be put in the form with $(E \gg V)$,

$$\frac{n}{T_N/\tau_N} = f(1 - e^{-1/f})[1 - (N - m + 1)e^{-1/f}], \qquad (5.67)$$

in which $f \equiv \tau_a/T_N$. This function is plotted in Figure 5.27 (solid curves) for two values of fan-in. Because τ_a (and β) will vary from one component to another, performance is degraded to that corresponding to the limits of the range, even though the nominal value is selected to coincide with the peak. As nonresistive loading becomes the dominant factor in frequency response, performance is less dependent on τ_a and is simply proportional to minimum β, as in (5.65). At intermediate loading values, a distinct performance peak still remains, but it does not coincide with that of (5.67). For example, with

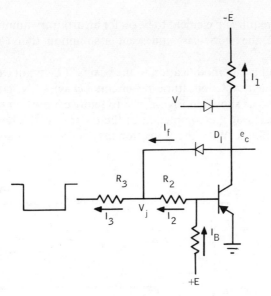

Figure 5.28

$\Delta q \to 0$, $T_N = T_F$, and $\tau_T = \tau_a + \tau_L$, (5.64) can be written as

$$\frac{n}{T_N/\tau_N} = \frac{f(1 - e^{-1/f[1 + (\tau_L/T_N)/f]})^2}{1 - [(\tau_L/T_N)/f]\{1/[1 + (\tau_L/T_N)/f]\} e^{-2/f[1 + (\tau_L/T_N)/f]}} \quad (5.68)$$

for the NAND case ($N = m$). The dotted curve of Figure 5.27 represents (5.68) for the intermediate value, $\tau_L = T_N$. Note that the peak has shifted from approximately $\tau_a = T_N$ for $\tau_L = 0$ to $\tau_a = T_N + \tau_L$. In other words, no advantage is to be gained by reducing the transistor time constant to values much less than that of the load time constant if dc current gain must be sacrificed in the process.

Performance measures have directly or indirectly incorporated the quantity $Q(0)$, the stored base charge at the initiation of the OFF transient. If quiescent conditions have been reached, $Q(0) = \tau_a I_{b1}$. It is worth observing that this may not be the case. Particularly, as the duration of the period between transients approaches the switching time ($T_P \to T_N$), base charge may not have risen to its final value. In the limit ($T_P = T_N$) base charge has just reached the value corresponding to the edge of saturation at the instant OFF switching begins. Under these conditions the decision element behaves as a nonsaturating circuit, whose performance is given by (5.65), even for $\tau_T = \tau_a$. Although it is possible to take advantage of this improved performance in special cases, in the general computing situation there is no guarantee that a given element will remain ON for a single period only. In general, the logic may require an element to be ON for an arbitrary number of periods, in which case the worst-case quiescent assumption that $Q(0) = \tau_a I_{b1}$ is valid.

A simple and useful modification to the basic RTL circuit can be made to obtain the benefits of nonsaturating circuit behavior. Negative feedback can be provided as shown in Figure 5.28 to remove a portion of the turn-ON drive at or nearly at the completion of the ON transient. Operation of the circuit is as follows. With the transistor OFF, collector voltage is at V, the

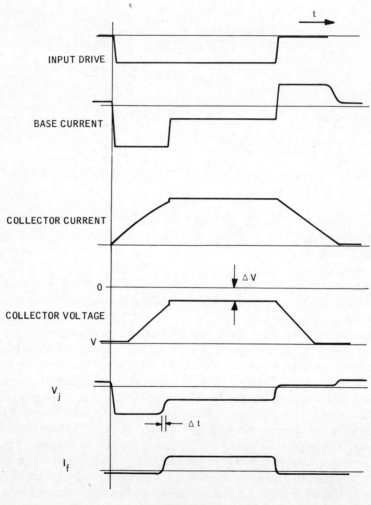

Figure 5.29

input is near 0 V, and the junction point V_j between R_2 and R_3 is at zero or is slightly positive. Thus D_1 is reverse-biased. When the input goes negative by V, V_j assumes some intermediate value between V and V_{beN}. Until the collector voltage comes off clamp, however, D_1 remains reverse-biased. The design can be arranged so that D_1 becomes forward-biased when e_c rises to a voltage near zero, but one sufficiently different from zero to prevent forward-biasing the collector junction. At this point I_f flows, increasing V_j such that I_2 is reduced to just that value required to keep the transistor at the edge of saturation. Waveforms are shown in Figure 5.29. Because the collector junction never becomes forward-biased, saturation is avoided, and $T_F \cong T_f$. When the input goes positive again, D_1 is immediately reverse-biased, and collector current turn-OFF occurs in the active region under drive I_B. The voltage ΔV at which the collector voltage "catches" can be selected as a design parameter. It should be chosen larger than the maximum base-emitter voltage anticipated, because the reverse bias across the collector junction is $-V_c = \Delta V - V_{beN} > 0$. It should be chosen as small as possible in order to minimize transistor power because an ON transistor is dissipating $W_T = \Delta V I_1 \gg V_{ces} I_1$. This additional power, as reflected in system lifetime and in the cost of the additional components versus improved performance,

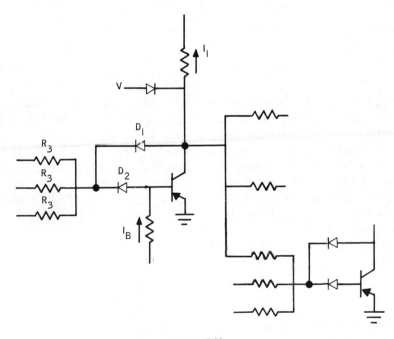

Figure 5.30

constitutes the trade-off between saturating and nonsaturating RTL decision elements.

An idea of the improved performance can be obtained by analyzing an idealized circuit of the type of Figure 5.28 and by comparing it with the saturating case. For this purpose the system of Figure 5.30, with R_2 replaced by diode D_2, serves to simplify the discussion. The requirement imposed on the diodes is that the forward drop across D_2 be greater than that across D_1, because the reverse bias across the collector junction is

$$-V_c = V_{D2} - V_{D1} > 0.$$

The diodes and transistor determine the collector catch voltage,

$$\Delta V = V_{beN} + V_{D2} - V_{D1}.$$

Let $E \gg V \gg \Delta V$ so that I_1 and I_B may be regarded as current sources. The final ON collector current is $I_1 + I_f$. There are two reasons for neglecting I_f relative to I_1. A first estimate for I_f can be obtained from Figure 5.28 by noting that the current through R_3 during the ON transient must be $I_{b1} + I_{b2}$, but that after turn-ON is complete, I_f must reduce it to

$$I_{b2} + (I_1 + I_f)/\beta.$$

Therefore, using (5.49) for I_{b1} with $Q(T_N) \cong \tau_N I_1$, we obtain

$$\frac{I_f}{I_1} \cong \frac{1}{1 + \beta} \frac{e^{-T_r}}{1 - e^{-T_r}} \ll 1, \tag{5.69}$$

particularly at lower switching rates. The second reason for neglecting I_f is that the time for the collector current to rise from I_1 to $I_1 + I_f$ is under control of a time constant much smaller than the active-region time constant τ_a. Because forward-biasing the diode D_1 has the effect of placing a negative feedback network with feedback ratio approaching unity across the amplifier, the frequency response is increased by the gain of the transistor, and the time constant for this transition approaches $\tau_N = \tau_a/\beta$. The two input signal levels available for drive are V and ΔV. We now define $I_3 \equiv V/R_3$. For turn-ON, m inputs are at V and $N - m$ are at ΔV. The voltage at the input to D_2 is $V_j = V_{beN} + V_{D2}$. The ON specification can therefore be written as

$$mI_3\left(1 - \frac{V_{beN} + V_{D2}}{V}\right) - (N - m)I_3\frac{V_{D1}}{V} - I_B \geq I_{b1}, \tag{5.70}$$

with I_{b1} given by (5.49) and $Q(T_N) \cong \tau_N I_1$. Worst-case turn-OFF occurs when $m - 1$ inputs are at V and the remainder at ΔV. Diode D_2 remains forward-biased under these conditions. The required drive is

$$I_B - (m - 1)I_3\left(1 - \frac{V_{beN} + V_{D2}}{V}\right) + (N - m + 1)I_3\frac{V_{D1}}{V} \geq I_{b2}. \quad (5.71)$$

Because feedback current I_f prevents saturation, I_{b2} is also given by (5.49) with $Q(0) \cong \tau_N I_1$. Worst-LOAD conditions occur when the given decision element constitutes the critical input to one load element and the remaining $n - 1$ load elements are OFF. For simplicity, assume that all n loads are OFF elements, with maximum base-emitter voltage V_F given either by a diode drop (base limiter) or by the estimate (5.47). Because D_2 remains forward-biased, fan-out is determined by

$$I_1 \geq nI_3\left(1 + \frac{V_F - V_{D2}}{V}\right). \quad (5.72)$$

From (5.70), (5.71), and (5.72), the limiting performance of this element is

$$\frac{n}{\beta} \leq \frac{(1 - \Delta V/V)(1 - e^{-T_N/\tau_T})(1 - e^{-T_T/\tau_T})}{[1 + (V_F - V_{D2})/V](1 - e^{-(T_N + T_F)/\tau_T})}. \quad (5.73)$$

Note that the first-order performance is independent of fan-in, and the frequency response is identical to that of the ideal nonsaturating circuit described by (5.65) and plotted in Figure 5.26.

Diode Transistor Logic

The performance constraints imposed by fan-in on RTL decision elements constitute the most serious disadvantage of that class of circuit. Fan-in restrictions can be greatly relieved by replacing the linear coupling between elements with nonlinear coupling. Although a variety of components have been tried with some success, the diode has proven to be by far the most convenient and satisfactory device for this purpose. The resulting forms of circuitry have been dubbed DTL, even though resistors still retain a prominent role.

Voltage switching. A great number of decision elements based upon diode coupling have been invented. Perhaps the simplest form—and the one most closely related to RTL operation—results when the active element is conceptually separated from the coupling network. Thus, the logic is thought of in terms of AND or OR diode gates, and NOT or inverter devices. This viewpoint is illustrated in Figure 5.31, in which a standard inverter (*a*) and a single-stage diode gate (*b*) are combined to generate the representative logic network of (*c*). Operation is as follows. If one or more of the N inputs to the diode gate is true ($- V$ for the *p-n-p* case illustrated), the input to the inverter is also true, and the inverter turns ON. If all inputs are false, the inverter turns OFF, producing a true $(-V)$ output. This system has also been referred to as voltage-operated or voltage-switching DTL.

Provided that the forward voltage drop and reverse leakage current of the diodes are sufficiently small relative to signal swing V and collector current level I_1, the performance of this system is identical in every way to that of a one-input RTL decision element; that is, if the coupling diodes are "perfect,"

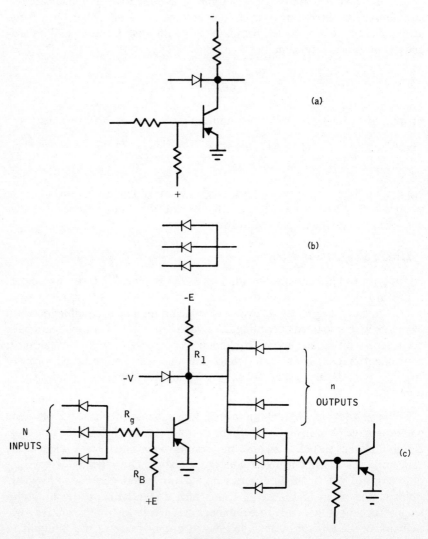

Figure 5.31

as far as the inverter is concerned, $N = m = 1$. Thus all the results of the preceding section can be applied by making that substitution in the general *RTL* expressions. In summary, the system is characterized by a frequency response flat to $T_N = T_F = 2\tau_T$, dropping off at 12 dB/octave thereafter (see Figure 5.26, $N = m$), and subject to the same degrading effects due to finite tolerances, device impedances, etc., as the one-input saturating *RTL* element. Feedback may be provided to prevent saturation. More than one level of diode logic may be performed between inverter stages without affecting idealized performance. Such a system is illustrated in Figure 5.32, in which the n individual gate resistors R_2 may be regarded as equivalent to the lumped load R_1 in the single-level logic system of Figure 5.31c.

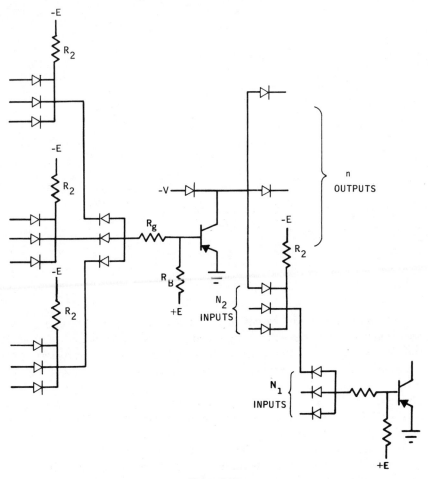

Figure 5.32

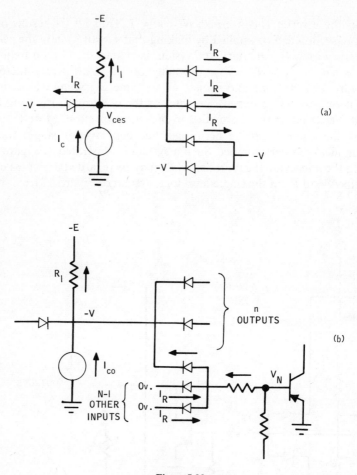

Figure 5.33

The above statements essentially complete the characterization of the idealized (perfect-diode) voltage-operated *DTL* system. It remains to explore the degrading effects attributable to real coupling diode parameters. The parameters of interest are invariably taken to be maximum forward voltage drop V_D and maximum reverse leakage current I_R. These quantities are not independent of operating-current levels and voltages. However, they tend to be relatively insensitive and can be estimated conservatively, or the design calculations can be iterated with rapid convergence. For the single-stage system of Figure 5.31c, the minimum ON drive voltage across R_g is $V - V_D - V_N$, with V_N the maximum ON base-emitter voltage. The worst-case ON collector load is illustrated in Figure 5.33a, in which the collector may be required to supply I_1 and, including the clamp diode, $(n + 1)I_R$ if at least one input to each driven element is negative. The ON specification is then

$$I_g\left(1 - \frac{V_D + V_N}{V}\right) - I_B\left(1 + \frac{V_N}{E}\right) \geq I_{b1},$$

$$I_{b1} = \frac{[I_1 + (n + 1)I_R]}{\beta} f_N(T_N), \tag{5.74}$$

in which $f_N(T_N)$ is the time-dependent portion of either (5.49) or (5.56), with τ_a replaced by τ_T, depending on which additional factors such as base capacitance charging or collector capacitance must be accounted for. Before turn-OFF, more than the minimum ON drive may have been applied. In particular, the collector clamp diode of the source signal may be forward-biased, thus canceling the drop across the coupling diode. During the OFF transient, while the base-emitter junction is still forward-biased, the inputs to the diodes are all at the saturated collector voltage level V_{ces}. For any significant current to flow in the diode, base resistance must be large enough to produce a positive voltage $r_b I_{b2}$ approaching V_D.

Diode capacitance has the effect of impressing a transient positive OFF voltage at the input to R_g as a result of the positive-going signal. A highly conservative worst-case assumption is that during the OFF transient the base voltage goes to some maximum positive OFF value V_F, diode recovery is slower than turn-OFF time, and the signal inputs correspond to transistors at the edge of saturation, with $V_{ces} \cong V_N$. The OFF specifications can then be written as

$$I_B\left(1 - \frac{V_F}{E}\right) - I_g\left(\frac{V_F - V_D + V_N}{V}\right) \geq I_{b2},$$

$$I_{b2} = \left[I_g\left(1 - \frac{V_N}{V}\right) - I_B\left(1 + \frac{V_N}{E}\right)\right] f_F(T_F), \tag{5.75}$$

in which $f_F(T_F)$ is the time-dependent portion of (5.49), (5.61), or the complete expression (5.63), depending upon which factors predominate. The worst-case LOAD condition is shown in Figure 5.33b. In the limit the forward current through the clamp diode can approach zero, and the current through R_1 must supply input drive to n loads plus leakage currents in the diode gates. Thus

$$I_1\left(1 - \frac{V}{E}\right) \geq n\left[I_g\left(1 - \frac{V_N}{V}\right) + (N - 1)I_R + I_{co}\right]. \tag{5.76}$$

The first-order effect of fan-in on performance can be most readily seen by

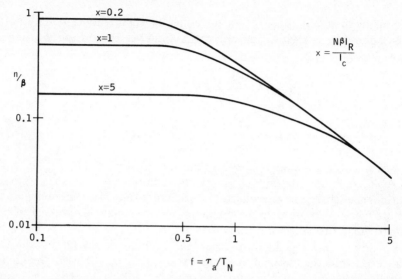

Figure 5.34

taking a signal swing V much larger than junction voltage drops (V_N, V_F, or V_D), constant current bias sources ($E \gg V$), $I_1 \gg I_R$ (but not necessarily greater than βI_R), and $I_{co} \cong I_R$. With these simplifications, and the frequency-dependent form of (5.49), the limiting fan-out becomes

$$\frac{n}{\beta} \lesssim \frac{(1 - e^{-T_N/\tau_a})(1 - e^{-T_F/\tau_a})}{[1 + (N\beta I_R/I_c)(1 - e^{-T_N/\tau_a})(1 - e^{-T_F/\tau_a})]}, \qquad (5.77)$$

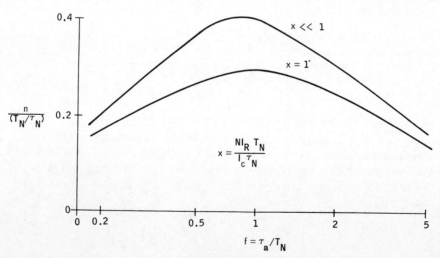

Figure 5.35

in which $I_c \neq I_1$ is the selected maximum dc collector current $I_1 + (n + 1)I_R$, and n and $N \gg 1$. The importance of choosing I_c as large as possible within the constraints of power dissipation (i.e., lifetime) is apparent. The function (5.77) is plotted in Figure 5.34 for several values of $N\beta I_R/I_c$ with specified $T_N = T_F$. If τ_a (rather than τ_L) indeed dominates frequency response and if transistor parameters β and τ_a are not independent but are correlated by $\beta = \tau_a/\tau_N$, (5.77) determines an optimum choice of $\tau_a \cong T_N = T_F$ for maximum performance. The optimum is, however, not particularly sensitive to fan-in, as shown in Figure 5.35, and either for large fan-in or $\tau_L > \tau_a$ it is virtually independent of τ_a.

Current switching. Although more than one stage of logic may be performed between inverters of the voltage-switching variety, successively more serious performance degredation occurs as a result of the nonideal current

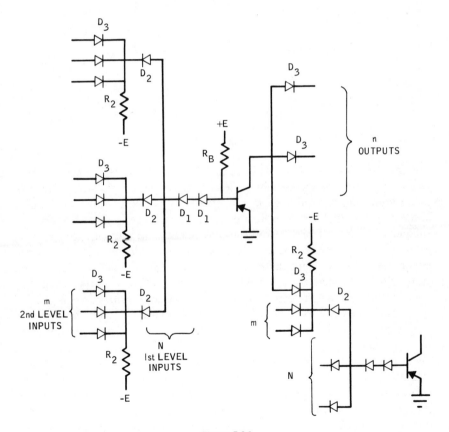

Figure 5.36

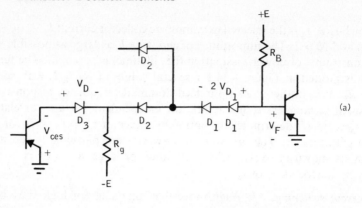

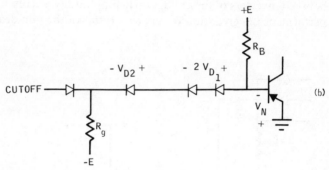

Figure 5.37

sources used for gate bias. The term $1 - V/E$ occurring in performance measures of all voltage-switching devices is further amplified by multiple diode-logic stages, tolerances, etc. In efforts to overcome this factor, many successful designs have been based on signal voltage swings that are just sufficient to forward-bias or reverse-bias the appropriate junctions and on current sources that approach ideal. Low-voltage swings possess the added advantage of minimizing the charge that must be supplied to diodes and stray circuit capacitances during switching. Designs incorporating this concept are variously referred to as *current-switching* or *low-level logic*.

A straightforward example of this type of decision element is shown in Figure 5.36. It will serve to illustrate principles of operation, as well as to illustrate the pitfalls that can be encountered with this type of circuit. Operation is as follows. If all m inputs to any one negative AND gate (diodes D_3, bias R_2) are true (negative, or cutoff), the corresponding R_2 draws current through its OR diode D_2. This is sufficient to overcome the OFF bias through R_B and turn the transistor ON. If at least one input to each of the m-input AND gates is at or near 0 V, all N diodes D_2 are temporarily reverse-biased, and the transistor turns OFF under the reverse drive from R_B. After the transistor turns OFF and its base input impedance becomes large, diodes D_2 again conduct, sharing the current from R_B. Diodes D_1 are present to prevent turn-OFF current from flowing into the input gates. The situation during the turn-OFF transient is illustrated in Figure 5.37a. The source saturated collector voltage V_{ces} and forward drop V_{D3} across D_3 provide a negative voltage at the OR gate input of $V_{ces} + V_{D3}$. The OFF base voltage V_F during the turn-OFF transient is determined by the difference between the changing forward-biased base-emitter junction voltage (negative) and the reverse drive times the base resistance, $I_{b2}r_b$. The net effect may be positive. Because diode D_2 may represent many diodes in parallel, it should conservatively be regarded as a net low impedance. Diodes D_1 must be selected so that $V_F + V_{D3} + V_{ces} < 2V_{D1}$. Otherwise, a significant fraction of the current through R_B is shunted through D_1 and is not available for turn-OFF. The maximum voltage swing in this circuit is at the AND gate. The ON condition is pictured in Figure 5.37b, in which the AND gate voltage is negative by $V_N + 2V_{D1} + V_{D2}$. The net voltage change between ON and OFF, with $V_{D2} = V_{D3}$, is given by $V_N + 2V_{D1} - V_{ces}$. Even for $V_{ces} \to 0$, this is no more than several diode forward voltage drops. Note that a collector clamp is not required for proper circuit operation, the AND gate or the base-emitter circuit of the succeeding stage serving as an effective clamp. A base limiter at the input to the transistor is also unnecessary, because the voltage there for $V_{D2} = V_{D3}$ cannot exceed the positive value $2V_{D1}$ even as $V_{ces} \to 0$.

Minimum ON, minimum OFF, and LOAD conditions determine the design of this circuit, as in previous analyses. To highlight principal circuit performance factors and avoid carrying excessively detailed notation, we shall assume that parameters are such that all diode forward drops and reverse currents can be regarded as equal. This has the effect of requiring $V_F < V_D$. The quantities $I_B \equiv E/R_B$ and $I_g \equiv E/R_2$ are defined as before. A preliminary gross picture of circuit performance can be obtained by making the following idealizations. Let R_B and R_g resemble current sources ($E \gg V_D$), neglect diode drops and reverse leakage, and assume all components purely resistive, so that the transistor determines the frequency response. Minimum turn-ON occurs when one AND gate goes true, the others remaining at 0 V. The minimum net ON drive is $I_g - I_B$. However, as many as N AND inputs may be true, and the worst OFF case occurs when the quiescent drive prior to the switching transient is $NI_g - I_B$. When all AND inputs go to near 0 V, diodes D_1 cease to conduct, and the OFF drive is I_B. The collector must be capable of delivering nI_g. Combining these ideal operating conditions, the performance measure becomes

$$\frac{n}{\beta} \le (1 - e^{-T_N/\tau_a})(1 - N e^{-T_F/\tau_a}). \tag{5.78}$$

We see that this circuit arrangement is similar to the saturating *RTL* with $m = 1$, in terms of its sensitivity to fan-in (N). Despite this drawback, a number of successful designs have been based upon such circuits incorporating transistors whose switching times are not the only important factors in circuit response. In the limit of large fan-in and fan-out, diode recovery time and diode capacitance charging times can become predominant. These effects deserve detailed consideration.

$t < 0$

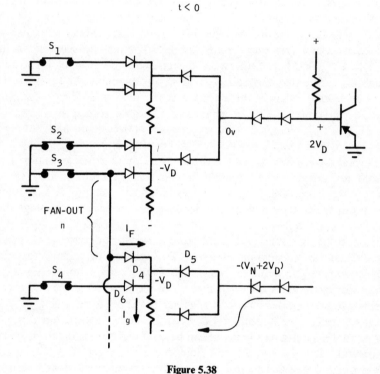

Figure 5.38

The turn-ON conditions at the AND gates are illustrated in Figures 5.38 and 5.39. The switches S_i are taken to be ideal, and the worst case occurs when S_2 and S_3 open at $t = 0$, while the others remain closed. The quiescent forward current through one of the n load diodes (e.g., D_4) can be as much as I_g if it does not share the AND load bias with other inputs (S_4 in this case) and if the OR diode (e.g., D_5) is reverse-biased because some other OR input is true (negative). After $t = 0$, two events occur at the AND gates. One is the recovery of D_4, which must occur before the AND summing point (S_2 AND S_3) can go negative. Transient reverse current I_R flows during this time into the S_3 bus. Its source is the net turn-ON drive to the decision element being switched, and it must also flow either as added ON drive to the OR gate attached to D_4, or from source S_4 through D_6 (see Figure 5.39). Note that the latter represents an additional transient collector load. Because there are m inputs to the AND gate and as many as $n-1$ recovering diodes on each input bus, the total reverse current flow into the turn-ON network can be as large as $m(n-1)I_R$.

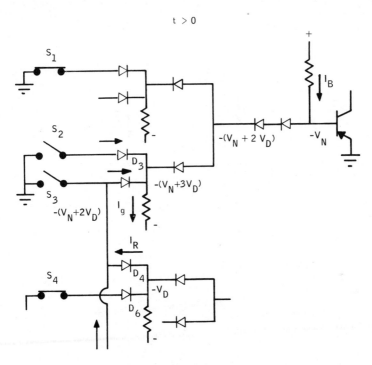

Figure 5.39

The net ON drive at the input of an element is $I_g - I_B$. Because this current supplies the recovery current I_R, for the AND gate the latter is given by

$$I_{RA} \cong \frac{I_g - I_B}{m(n-1)}. \tag{5.79}$$

Following diode recovery, the second event to occur at the AND gates is the charging of the diode capacitances (e.g., D_4) to a new voltage approximately $V_N + V_D$ (negative), shown as a final voltage in Figure 5.39. Again, the buses on each of the AND inputs must be so charged before the AND summing point can become sufficiently negative to turn on the driven circuit. The total charge required on the AND diodes is

$$Q_A \equiv m(n - 1)C_D(V_N + V_D), \tag{5.80}$$

in which C_D is the maximum diode depletion-layer capacitance averaged over the voltage swing.

Following AND turn-ON, the remaining $N - 1$ OR diodes must also recover and charge. The before and after situations are depicted in Figure 5.40. For

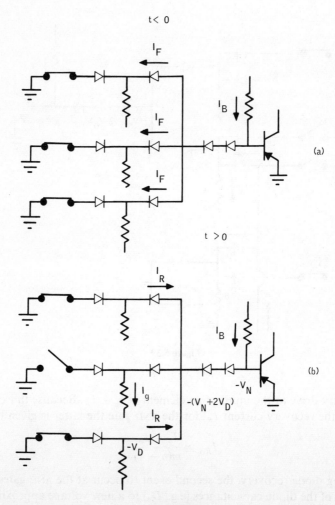

Figure 5.40

$t < 0$, the forward current through OR diodes D_2 is $I_F \cong I_B/N$. For $t > 0$, the initial transient reverse current is the net ON drive divided among the $N - 1$ OR diodes whose AND inputs remain near zero. Thus

$$I_{RO} = \frac{I_g - I_B}{N - 1}. \tag{5.81}$$

The required charge, for the quiescent reverse voltage across the OR diodes of $V_N + V_D$ is

$$Q_O = (N - 1)C_D(V_N + V_D). \tag{5.82}$$

Finally, the base capacitance of the transistor must be charged, requiring

$$Q_B = C_B(V_N + 2V_D). \tag{5.83}$$

With two junctions at the transistor base, and for N, m, or $n \gg 1$, no serious error is made by letting $C_B = 2C_D$, and the total charge

$$Q_T = Q_A + Q_O + Q_B \cong [m(n - 1) + N + 1]Q_{ND}, \tag{5.84}$$

with

$$Q_{ND} \equiv C_D(V_N + V_D).$$

The time for charging the diode and transistor capacitances is approximately

$$T_c \cong \frac{Q_T}{I_g - I_B}. \tag{5.85}$$

As discussed in Chapter 2, a first-order estimate of diode recovery time to a reverse current that is a fraction k of the initial reverse current I_R is proportional to the diode recombination rate τ_R and the logarithm of the ratio I_F/kI_R. The AND gate recovery time can therefore be written, using (5.79), as

$$T_{RA} = \tau_R \ln \frac{m(n - 1)}{k(1 - I_B/I_g)}, \tag{5.86}$$

and OR gate recovery can be written, using (5.81), as

$$T_{RO} = \tau_R \ln \frac{(N - 1)I_B/I_g}{kN(1 - I_B/I_g)}. \tag{5.87}$$

The total turn-ON time is

$$T_N \geq T_{RA} + T_{RO} + T_c + T_r. \tag{5.88}$$

Turn-OFF conditions are shown in Figure 5.41. The worst case occurs when all inputs are open for $t < 0$, and close at $t = 0$. As can be seen in Figure 5.39, diodes D_3 are slightly forward-biased prior to the transient and

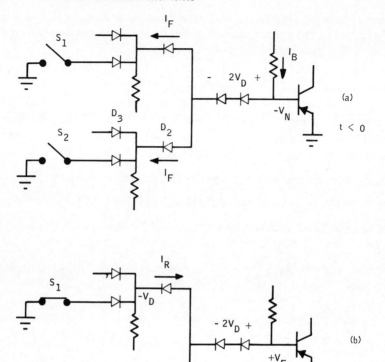

Figure 5.41

heavily forward-biased thereafter. Because forward recovery is rapid and capacitance charging is supplied by the source, the AND diodes can be neglected. Before the transistor turns OFF and its base voltage rises to V_F, diodes D_2 may be slightly reverse-biased. However, the reverse current is in a direction to aid switching, and it is on the conservative side to neglect it. Charging of the base junctions occurs at the completion of the transient. Because the transistor is already OFF ($I_c \rightarrow I_{co}$) during base-charging time, this period can also be ignored. Thus I_B is the normal transistor turn-OFF drive given by (5.49) and (5.61).

If $T_F/\tau_a \gg 1$ is not the case so that I_B is significant, T_{R0} must be accounted for. Combining (5.85), (5.86), (5.87), and transistor rise time under this condition in (5.88), a constraint involving ON drive, circuit parameters, and switching times is

$$e \geq \frac{[Nmn \, e^{-T_F/\tau_a}/k^2(1 - N \, e^{-T_F/\tau_a})^2]^{\tau_R/T_N} \, e^{T_c/T_N}}{[1 - I_c/\beta I_g(1 - N \, e^{-T_F/\tau_a})]^{\tau_a/T_N}}, \tag{5.89}$$

in which N, m, and $n \gg 1$ has been assumed for simplicity. Because fan-in and fan-out occur as the product Nmn in terms of diode recovery relative to switching time τ_R/T_N, the extreme importance of a small diode recovery time constant is thereby emphasized. Note that the individual factors entering into (5.89) represent the relative portion of the total switching time consumed by each, in accordance with the corresponding exponent, namely, diode recovery τ_R, capacitance charging T_c, and transistor turn-ON τ_a.

It is possible to obtain explicit solutions to (5.89) or to the set of design conditions leading up to it, usually with the aid of iterative algorithms and a computer. Goldstick and Mackie [15] have described such a procedure using the simplex algorithm, starting with an arbitrary initial assignment of the fraction of the total switching time to be occupied by each phenomenon. The program was designed to minimize resistor power dissipation. To achieve an optimum design, the procedure must be repeated for other apportionments of the switching times. For purposes of illuminating basic circuit behavior, it is sufficient here to approximate (5.89) over a range covering the performance of interest. Two approximations will be used. Providing T_c does not occupy more than half the total switching time, $e^{T_c/T_N} \cong 1 + T_c/T_N$ is valid within 10 percent. It can be seen from the factors in (5.89) that $T_N = T_F \gg \tau_a$ is virtually mandatory for useful circuit performance. In addition, the turn-ON drive ratio $(I_g - I_B)/(I_c/\beta)$ must exceed unity by a substantial amount. Under these conditions, another approximation that can be used is $(1 - x)^a \cong 1 - ax$.

LOAD conditions also require careful consideration. The driving source must supply not only nI_g, but also the recovery and charging currents through the $m - 1$ AND diodes and one OR diode connected to each of its n loads, as shown in Figure 5.42 with respect to S_1 (which closes at $t = 0$). In the worst case the forward current in the AND diodes approaches I_g/m. For recovery and charging times equal to the turn-ON time, the required switch (collector) current can therefore be written as

$$I_c \cong nI_g\left(1 + \frac{1}{k}e^{-T_N/\tau_R} + m\frac{Q_{ND}}{I_g T_N}\right). \tag{5.90}$$

Note that these recovery and charging currents may flow into other gates, imposing a further requirement on I_g; that is, if S_2 in Figure 5.42 is open and at least one input to each of its $n - 1$ other loads goes positive at $t = 0$

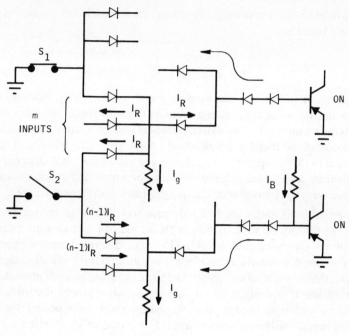

Figure 5.42

(S_1 is one such input), the AND gate current I_g at the remaining load must carry $m(n-1)$ times the reverse and charging currents, in addition to I_B and the dc ON current of the transistor driven by that gate. Otherwise, the gate may go positive, and a false transient OFF signal occurs at the input to the next stage.

Making the above approximations in (5.89), reinserting dc conditions, and using (5.90), we may invert the performance measure to obtain the necessary transistor current gain explicitly in terms of circuit parameters and specifications in the form

$$\beta \geq \frac{n[1 + (1/k)(e^{-\tau_a/\tau_R})^{1/f}]\{1 + f[1 + (mn + N)qr]/[(1 - N\,e^{-1/f})(1 - r)]\}}{1 - (mnq/\beta)f},$$

(5.91)

in which

$$f \equiv \frac{\tau_a}{T_N} = \frac{\tau_a}{T_F},$$

$$q = \frac{Q_{ND}}{\tau_N I_c},$$

$$r \equiv \frac{[(Nmn/k^2)^{\tau_R/\tau_a}]^f \, e^{-(1+\tau_R/\tau_a)}}{1 - 2Nf(\tau_R/\tau_a)\, e^{-1/f}}.$$

Special cases of interest can be illustrated by means of (5.91). For example, it is often desirable to make N, m, and n simultaneously as large as possible. For this purpose, let $N = m = n$. Let the recovery ratio be $k = 0.1$. The performance, $\beta\,(\mathrm{dc})/\beta(T_N)$, is plotted for several values of relative recovery time constant in Figure 5.43, for the reasonable choices $q = k = 0.1$ and $n = 5$. Similarly, the effect of q on performance is illustrated in Figure 5.44 for a fixed diode recovery time constant. Often fan-out is required to be greater than fan-in. A special case of this type is illustrated in Figure 5.45, based on $N = m = n/2$ for several values of fan-out. Observe in all cases that performance begins to drop off rapidly as specified switching time approaches within about an order of magnitude of circuit time constants.

Component tolerances, nonideal current sources, and so on, further degrade circuit performance. The effects on performance are similar to those discussed in previous treatments.

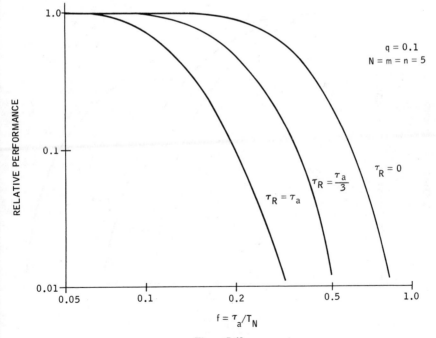

Figure 5.43

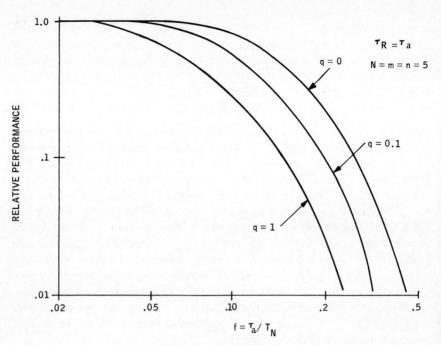

Figure 5.44

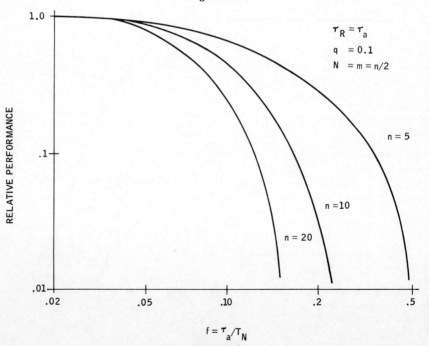

Figure 5.45

230

Much of the loss in high-frequency performance of the above circuit can be ascribed to the worst-case ON overdrive situation in which an element can be turned ON by more than one input gate. Some of this loss can be regained if the gating arrangement is reversed so that overdrive can only occur in the OFF case. A transistor turned OFF by more drive than required can still only be OFF, whereas a transistor turned ON by more than required goes into saturation. Generally the additional time required to charge an overly reverse-biased OFF base capacitance is much less than the time required to bring the overly saturated ON transistor back to the active region. If this situation holds, then the circuit of Figure 5.46 can provide improved performance. The operation is as follows. If at least one input to each AND gate (diodes D_3) is negative, I_3 flows into this sink, and OR diodes D_1 are reverse-biased. Current I_2 then turns the transistor ON. If all inputs to one AND gate go positive I_3 flows through D_1, and the net current $I_3 - I_2$

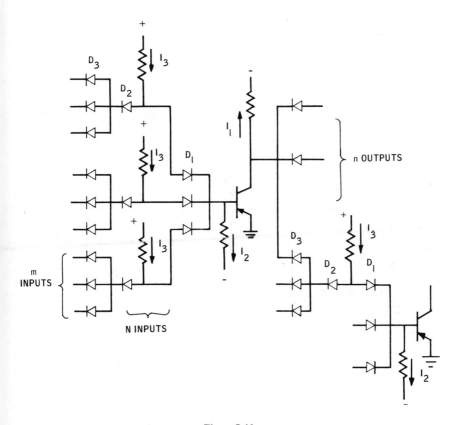

Figure 5.46

turns the transistor OFF. Note that if all inputs go positive, the OFF drive becomes $NI_3 - I_2$, turning the transistor OFF more rapidly. The quiescent OFF base-emitter voltage is, however, limited to $V_{D3} + V_{D2} - V_{D1}$, regardless of the number of OFF inputs, so the initial condition for turn ON is virtually independent of prior drive.

Diodes D_2 provide a forward voltage drop to force most of I_3 through D_1 during turn OFF. The situation at the instant all inputs to one AND gate go positive is illustrated in Figure 5.47, in which the OFF voltage during the transient at the transistor input is V_F. Some forward current will flow through D_2 and D_3, and it must be accounted for in the design. Because forward currents are generally much larger than diode reverse leakage currents, we can approximate the forward characteristics by

$$I_{D1} \cong I_{R1} e^{V_{D1}/E_T},$$
$$I_{D2} \cong I_{R2} e^{V_{D2}/E_T},$$
$$I_{D3} \cong I_{R3} e^{V_{D3}/E_T}, \tag{5.92}$$

with the added conditions

$$mI_{D3} = I_{D2},$$
$$-V_{ces} + V_{D3} + V_{D2} - V_{D1} - V_F = 0. \tag{5.93}$$

The voltage across D_3 can be found from (5.92) and (5.93), in terms of the

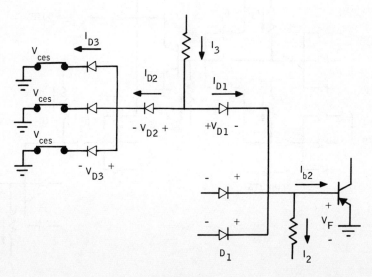

Figure 5.47

voltage across D_2 and diode parameters, as

$$V_{D3} = V_{D2} + E_T \ln \frac{I_{R2}}{mI_{R3}}. \qquad (5.94)$$

For similar diodes and usual values of m, this voltage will be considerably smaller than V_{D2}, and it is on the conservative side to neglect $V_{D3} \ll V_{D2}$. With $I_3 = I_{D1} + I_{D2}$, the portion of the OFF drive current that flows through D_1 is

$$I_{D1} = \frac{I_3}{1 + p_D}, \qquad (5.95)$$

in which a parameter related to the division of current is defined by

$$p_D \equiv \frac{I_{R2}}{I_{R1}} e^{(V_F + V_{ces})/E_T}. \qquad (5.96)$$

The relative diode parameter p_D should be as small as possible, thus imposing a constraint on the choice of diode types such that $I_{R2} \ll I_{R1}$.

The OR diodes connected to negative inputs in Figure 5.47 were reverse-biased before the beginning of the OFF transient and remain so after the transient. However, the voltage across them must change by $V_N + V_F$ before turn-OFF can begin. The charge that must be transferred is $Q = (N - 1)C_D(V_N + V_F) \equiv NQ_{NF}$, and it is supplied by $I_{D1} - I_2$ before I_{b2} begins to flow. The turn-ON transient is entirely dependent on I_2. Diodes D_1 may all have been forward-biased prior to the transient and must eventually recover. Their recovery current, supplied by the collector load of the preceding stage, is in the turn-ON direction, so that slow recovery does not oppose transistor switching. It is conservative to neglect this effect. Capacitance charging may similarly be neglected as far as its effect on turn-ON is concerned.

The collector load current I_1 must absorb I_3 for each driven gate, plus mI_R, as shown in Figure 5.48. The forward current prior to recovery is on the order of I_3/m. If the diode recovery and transistor turn-ON times are taken equal, $T_R = T_N$, then

$$I_R \cong \frac{I_3}{km} e^{-T_N/\tau_R}. \qquad (5.97)$$

Using the same approximations as in the preceding *DTL* analysis and combining the above ON, OFF, and LOAD conditions, we may obtain an explicit expression for the performance measure of this circuit; thus

$$\frac{n}{\beta} \leq \frac{(1 - e^{-T_N/\tau_a})(1 - e^{-T_F/\tau_a})}{(1 + p_D)[1 + (1/k) e^{-T_N/\tau_R}][1 + (\tau_a/T_F)Nq(1 - e^{-T_N/\tau_a})(1 - e^{-T_F/\tau_a})]},$$

$$(5.98)$$

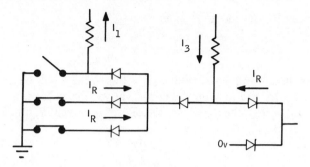

Figure 5.48

in which $q \equiv Q_{NF}/\tau_N I_1$, and $N \gg 1$ is assumed for simplicity. We see that to a first order the undesirable ON overdrive effect on performance is avoided. Except for the degrading diode current-division ratio p_D, diode recovery, and fan-in diode charging, the frequency response resembles that of the one-input saturating RTL circuit. Function (5.98) is plotted in Figure 5.26 for comparison with the parameters $p_D = \tau_R = 0$ and $Nq = 1.0$.

Direct-coupled transistor logic

The duality between diode logic and direct-coupled transistor logic, $DCTL$, was noted in Chapter 4. The amplifying property of transistors, however, means that $DCTL$ networks must receive special design treatment.

The two worst-case conditions at a current-summing point are illustrated in Figure 5.49. In (*a*) one input transistor is turned ON by I_b, driving the voltage at the summing point to V_{ces}, the maximum collector saturation voltage. Because V_{ces} is the difference between the emitter- and collector-junction voltages, provided the transistor is well into saturation, the voltage at the summing point is less than V_{beN}, the voltage required to turn a transistor ON. The n load transistors are therefore nearly OFF (or only slightly ON). In order to achieve this state of affairs it is necessary that βI_b be greater than I_1 less the leakage currents I_L through the remaining $N - 1$ input transistors that are nearly OFF, and the small base ON currents of the n nearly OFF load transistors. In (*b*) all input transistors are nearly OFF, and I_1 supplies the ON drive to the n load transistors. Assuming all devices are identical, the circuit constraint is

$$I_1 \geq NI_L + nI_b. \tag{5.99}$$

The design problem consists of finding a relationship between I_L and I_b as follows. It is conservative in (*a*) to assume that the one ON input transistor

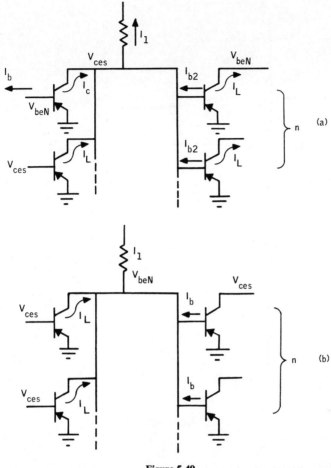

Figure 5.49

must supply I_1 and to neglect the possible load reduction due to $(N - 1)I_L + nI_{b2}$. Then the base-emitter voltage V_{beN} and collector saturation voltage V_{ces} of a transistor driven by I_b with collector current I_1 can be calculated or determined from curves such as that of Figure 5.50a. This value of V_{ces} is the applied base voltage of the nearly OFF transistors, and V_{beN} is their maximum collector voltage. Because $|V_{beN}| > |V_{ces}|$, the collector junctions of these transistors are reverse-biased, and they are therefore, by definition, in the active region. It is a very nonlinear portion of the active region, however, and must be treated accordingly. When the collector is reverse-biased, the collector current is primarily a function of base-emitter

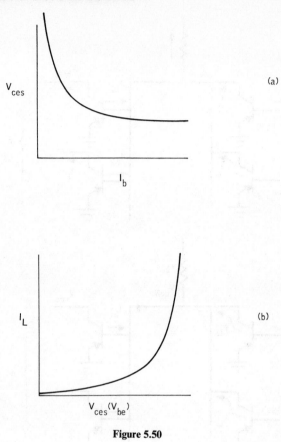

Figure 5.50

voltage, in this case the V_{ces} of the saturated input. With V_{ces} known, I_L can be calculated or determined from curves such as that of Figure 5.50*b*. Therefore I_L is a dependent function of I_b, $I_L = f(I_b)$, and Figure 5.50*a* and *b* can be combined into a single curve such as that of Figure 5.51. A graphical solution is now possible, because (5.99) can be written in the linear form

$$I_L = \frac{I_1}{N} - \frac{n}{N}I_b. \tag{5.100}$$

A design is possible if (5.100) and $f(I_b)$ intersect; cases illustrating different values of fan-out are shown in Figure 5.51.

An analytic solution that sheds some light on the device parameters of importance is also possible, based upon the ideal-transistor equivalent

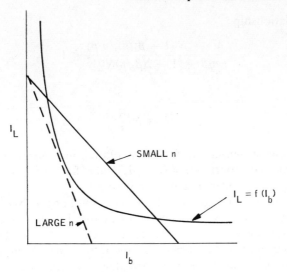

Figure 5.51

circuit. Conservatively assuming the ON transistor driven by I_b carries the entire load current I_1, we can give the magnitude of the collector voltage, if the transistor is well into saturation, by

$$e^{V_{ces}/E_T} \cong \frac{1}{\alpha_I}\left(\frac{1 + I_1/\beta_I I_b}{1 - I_1/\beta I_b}\right), \tag{5.101}$$

in which the approximations $\beta I_b - I_1 \gg I_{co}$ and $I_1 \gg I_{eo}$ have been used, and $\beta_I \equiv 1/(1 - \alpha_I)$ is defined for convenience. Providing the collector saturation voltage applied to the bases of the succeeding nearly OFF transistors is sufficient to forward-bias the base-emitter junction, $e^{V_{be}/E_T} \gg 1$. With the collector junctions of these transistors reverse-biased, an approximate expression for the magnitude of leakage collector current is

$$I_L \cong \frac{I_{co}}{1 - \alpha_N \alpha_I}(1 + \alpha_I e^{V_{be}/E_T}). \tag{5.102}$$

Because V_{be} in (5.102) equals V_{ces} in (5.101), the design constraint (5.99) can be put in the form

$$I_1 \geq \frac{NI_{co}}{1 - \alpha_N \alpha_I}\left(1 + \frac{1 + I_1/\beta_I I_b}{1 - I_1/\beta I_b}\right) + nI_b. \tag{5.103}$$

The resulting quadratic possesses a solution only when fan-in and fan-out obey the relationship

$$\frac{4[1 + N'(1 - \beta_I/\beta)](n/\beta)}{\{n/\beta + [1 - 2(\beta_I/\beta)N']\}^2} \leq 1, \qquad (5.104)$$

in which

$$N' \equiv N\frac{\beta I_{co}}{I_1}$$

This limitation on fan-in and fan-out is plotted in Figure 5.52 for several values of β_I/β. The importance of $I_1 \gg \beta I_{co}$ and of forward current gain much greater than reverse current gain is emphasized.

One of the greatest dangers in *DCTL* is current hogging by one transistor at the expense of others. This occurs when the driven transistors, whose bases are all at the same potential, do not exhibit identical base-current–base-voltage characteristics. Generally, the ratio of maximum base current to minimum base current is approximately constant over a reasonable range of base voltage. This is illustrated in Figure 5.53 for a typical range of characteristics and can be represented by $\overline{I_b} \cong k\underline{I_b}$. The worst case occurs when $n - 1$ driven transistors are drawing $\overline{I_b}$ and the remaining one is drawing $\underline{I_b}$. The current sink must then supply the drive

$$I_1 \geq NI_L + (n - 1)\overline{I_b} + \underline{I_b}. \qquad (5.105)$$

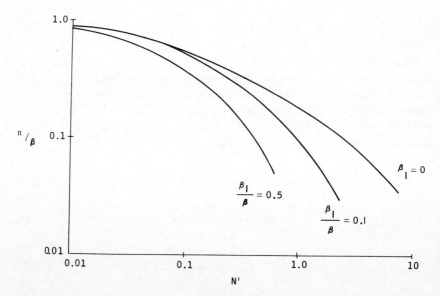

Figure 5.52

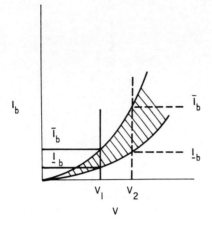

Figure 5.53

Leakage current I_L is in the worst case a function of I_b. It is conservative to assume all n (rather than $n - 1$) loads at maximum current, in which case n is simply replaced by kn in the performance constraint (5.104) and the performance depicted in Figure 5.52 is divided by the factor k. Because this maximum-to-minimum ratio can be large, a substantial degradation may occur.

Some degree of equalization can be obtained with the hybrid *RTL-DCTL* arrangement of Figure 5.54, at the expense of increasing voltage swings over those of pure *DCTL*. Variations in V_{be} of the driven transistors have less effect upon base drive as R is increased. For a fixed value of E, however, increasing R also eventually reduces the available drive. An optimum value exists, as can be shown by a simplified analysis. Let the unknown voltage at the current-summing point be V. The worst possible load condition is for all driven transistors to have small V_{be}. Therefore, let $V_{be} \to 0$, in which case, for $I_L \to 0$,

$$\frac{E - V}{R_1} \geq n\frac{V}{R}. \tag{5.106}$$

The most lightly driven transistor, however, may have a large value of ON base voltage $\overline{V}_{be}$. For $I_L \to 0$,

$$\frac{V - \overline{V}_{be}}{R} \geq \frac{1}{\beta}\frac{E}{R_1}. \tag{5.107}$$

The value of R that maximizes fan-out within the constraints of (5.106) and

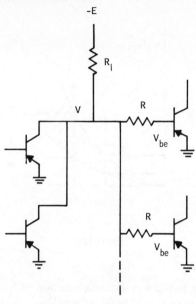

Figure 5.54

(5.107) is

$$\frac{R}{\beta R_1} = \left(\frac{\overline{V}_{be}}{E}\right)^{1/2}\left[1 - \left(\frac{\overline{V}_{be}}{E}\right)^{1/2}\right]. \tag{5.108}$$

In practice, a somewhat smaller value of R is desirable to provide the overdrive for reliable circuit operation. The relationship between performance and relative resistor values is plotted in Figure 5.55.

Switching in networks of pure $DCTL$ elements occurs in a highly nonlinear region of operation and has only been analyzed on a crude basis. Because of certain similarities to the behavior of $RCTL$ elements, however, the results shed some light on the underlying principles common to both classes of circuit, and can also be extended to some degree to flip-flop switching criteria. The hybrid RTL-$DCTL$ circuit of Figure 5.54 reduces essentially to RTL behavior if voltage swings are sufficiently large, and it has therefore been described in previous sections. Note that the turn-OFF transient of the hybrid circuit relies entirely upon the switching time constant of the transistor, because no reverse OFF overdrive is provided.

Switching behavior of pure DTL circuits can be based on assumed ideal transistors in the system of Figure 5.56. For $t < 0$, the switch is closed, T_1 and other transistors of the first stage are OFF, and T_2 and other transistors

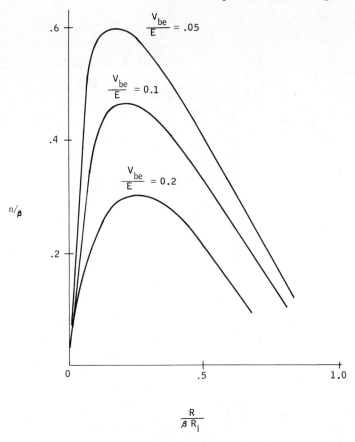

$$\frac{V_{be}}{E} = .05$$

$$\frac{V_{be}}{E} = 0.1$$

$$\frac{V_{be}}{E} = 0.2$$

n/β

$$\frac{R}{\beta R_l}$$

Figure 5.55

of the second stage are ON. Fan-out n is taken to be the same for all stages. At $t = 0$, the switch opens, and T_1 begins to turn ON. Its collector current reduces the drive available to second-stage transistors, and they eventually turn OFF. The reason for considering two stages simultaneously is as follows. For the ideal transistor, as long as the device is in the active region (although in the process of turning OFF), the base-emitter junction is forward-biased. If the base-emitter voltage of the driven transistor (T_2) is greater in magnitude than that of the driving transistor (T_1), the collector junction of the latter is reverse-biased, and it remains in the active region. The situation is depicted in simplified form in Figure 5.57. To a first order, the emitter junction voltage is directly related to the stored base charge and collector current, because it is the logarithm of the ratio of minority-carrier densities across the junction.

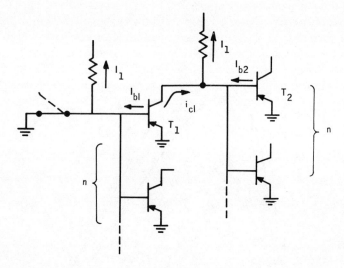

Figure 5.56

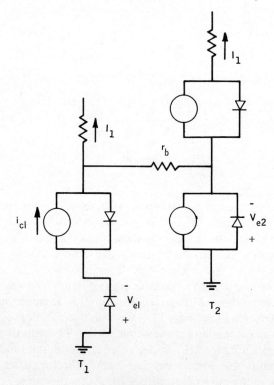

Figure 5.57

242

It is therefore convenient to examine the time histories of base charges rather than those of emitter voltages. These are shown in Figure 5.58 for a constant collector current I_1 in T_2. With T_1 in the active region, the base charge builds up in accordance with the single-pole model, under the drive I_1 divided among the n sharing elements ($I_{b1} = I_1/n$ in Figure 5.56). Transistors T_2 begin in the saturation region, and base charge decreases under the reduced drive. At some point in time the carrier densities at the two junctions become equal, the collector junction of T_1 changes from reverse to forward bias, and

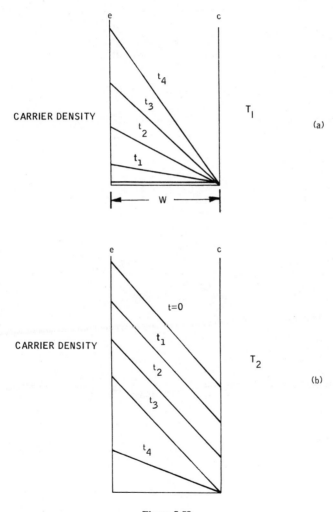

Figure 5.58

the remainder of the transient occurs with T_1 in saturation. The two junction voltages are sketched in Figure 5.59. Although some charge continues to be removed from the base of T_2 after this time, it is conservative to consider the remainder of the OFF transient to be entirely the result of charge recombination (τ_a of T_2). The time at which reverse drive ceases can be estimated from the charge diagrams of Figure 5.58. The first-order carrier concentration at the emitter of T_1 (assuming instantaneous charge redistribution) is, in terms of base charge,

$$p_1(t) = \frac{2}{qW}Q_1(t), \qquad (5.109)$$

in which $Q_1(t)$ is the total base charge in transistor T_1.

The concentration at the emitter of T_2, which is in saturation, is

$$p_2(t) = \frac{1}{qW}[Q_2(t) + \tau_N I_1], \qquad (5.110)$$

in which $Q_2(t)$ is the total base charge in transistor T_2.

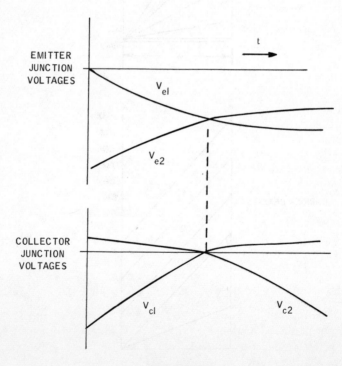

Figure 5.59

The change in switching rate occurs when $p_1(t) \cong p_2(t)$, or when

$$\frac{2Q_1(t)}{\tau_N I_1} = \frac{Q_2(t)}{\tau_N I_1} + 1. \tag{5.111}$$

The time dependence of the two base charges is also available from the first-order switching model. For T_1,

$$\frac{Q_1(t)}{\tau_N I_1} = \frac{\beta}{n}(1 - e^{-t/\tau_a}) \tag{5.112}$$

in the active region, with the ON drive I_1/n. T_2 starts from an initial base-charge value of $\tau_a I_1/n$ and receives an incremental reverse OFF drive i_{c1}/n (see Figure 5.56). The collector current of T_1 is in turn a function of β and I_1/n. The complex-frequency form of the change in base charge in T_2 is thus

$$\frac{\Delta Q_2(s)}{\tau_N I_1} = \left(\frac{\beta}{n}\right)^2 \frac{1}{s(\tau_a s + 1)^2}. \tag{5.113}$$

Inserting initial conditions, and converting to the time domain, we obtain

$$\frac{Q_2(t)}{\tau_N I_1} = \frac{\beta}{n}\left\{1 - \frac{\beta}{n}\left[1 - \left(1 + \frac{t}{\tau_a}\right)e^{-t/\tau_a}\right]\right\}. \tag{5.114}$$

Functions (5.112) and (5.114) are shown in Figure 5.60 with $Q_2(0) = Q_1(\infty) = (\beta/n)\tau_N I_1$. With heavy overdrive ($\beta \gg n$), less than about half a transistor time constant is occupied by the first portion of this transient, and some charge remains in T_2 after T_1 goes into saturation. Keeping in mind that switching after T_1 reaches saturation is relatively independent of fan-out, we can obtain a reasonable measure of performance by asking when Q_2 would reach zero or cutoff (rather than a finite fraction of its initial value) if the transient were to continue as before. Setting $Q_2(t) = 0$ in (5.114) yields

$$\frac{n}{\beta} = 1 - \left(1 + \frac{T}{\tau_a}\right)e^{-T/\tau_a}. \tag{5.115}$$

This is plotted in Figure 5.61. At high frequencies, (5.115) is closely approximated by $(T/2\tau_a)^2$. Note the high-frequency drop-off of 12 dB/octave characteristic of the saturating RTL NAND circuit ($N = m$).

Resistor-capacitor-transistor logic

Resistor-capacitor-transistor logic, $RCTL$, is another of the several approaches that have been taken to improve frequency response by avoiding the saturation region. The principle may be described as one of arranging the circuit so that large drives can occur during switching, but only the small drives required for dc stability can occur at other times.

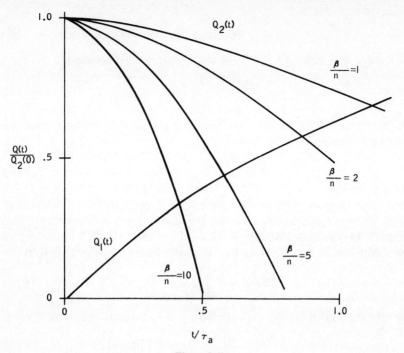

Figure 5.60

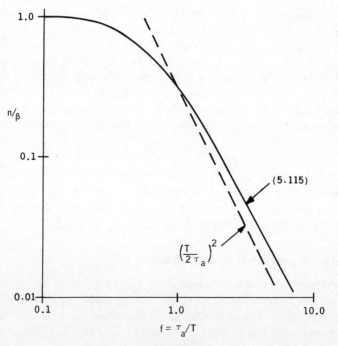

Figure 5.61

The basic one-input *RCTL* inverter is depicted in Figure 5.62. Stability conditions (dc) are, of course, identical to those of the one-input *RTL* element and need not be considered further here. The only variable to determine is the value of the capacitor C for proper transient behavior. The role of C may be described as one of preventing the input voltage $e(t)$ from exceeding the range $0 < |e(t)| < |V|$ during the time the transistor is going from OFF to ON or the reverse, that is, $e(t)$ represents the collector voltage of a prior stage, and we wish to prevent that stage from reaching either collector clamp or saturation during the switching transient. Ideally, the input voltage would just reach the clamp, V at the instant the driven transistor reaches the edge of saturation, or would just reach 0 V at the instant the driven transistor reaches cutoff.

An approximate value for C can be obtained from a first-order analysis of switching behavior. For this purpose, let I_{co} be sufficiently small such that with ideal dc inputs, 0 and $-V$, OFF bias I_B is negligible. Further, let V_{be} during switching be negligibly small compared to the input voltage, so that the base-emitter voltage may be regarded as zero. This idealization amounts to assuming either that the base resistance is very small or that switching times are not extremely short. If $e(t)$ approaches a step input, base resistance dominates charge transfer and must be taken into account. This case is treated in Chapter 6 in connection with flip-flop triggering, where the trigger pulse normally comes from an external, low-impedance

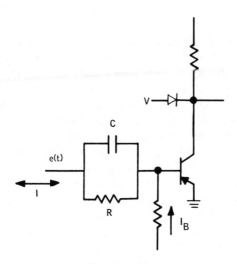

Figure 5.62

source. In decision elements in which $e(t)$ is the collector voltage of a previous stage, this high-current case does not usually arise.

Considering the ac circuit only with the above idealizations and a single-pole function for the transistor, we see that the change in base charge in response to a change in drive current I of arbitrary waveform is

$$\Delta Q(s)\frac{\tau_a I(s)}{\tau_a s + 1}. \tag{5.116}$$

Similarly, the change in input voltage is

$$\Delta E(s) = I(s)Z(s) = \frac{RI(s)}{\tau s + 1}, \tag{5.117}$$

in which $\tau \equiv RC$. Suppose we let $\tau = \tau_a$. Then the ratio of (5.116) to (5.117) is a constant regardless of drive waveform, and

$$\frac{\Delta Q(t)}{\Delta e(t)} = \frac{\tau_a}{R}. \tag{5.118}$$

This is compatible with the desired behavior. That is, because the ON base drive under dc conditions is V/R, the change in base charge in going between OFF and ON is $\Delta Q = \tau_a V/R$, while the change in input voltage between the same two conditions is V. Therefore C is ideally determined by the dc design and the value of $\tau = \tau_a$. In practice, the value of C is normally selected to be somewhat larger than this limit as a result of nonideal transistor behavior.

Systems of RTL decision elements can be converted on a one-to-one basis to $RCTL$, as indicated in Figure 5.4. This type of arrangement is seldom used, however, because even under dc conditions more than one input to a transistor can be ON, in which case the advantage of nonsaturating operation is partially lost. For high-speed operation, hybrid $RCTL$ and voltage-switching DTL of the type indicated in Figure 5.5a offer improved performance. A several-level system of this type, incorporating a single stage of diode gating between inverters, is shown in Figure 5.63. Operation is as follows. If at least one input to a diode gate goes positive, the voltage at R_2 goes positive, and reverse drive through capacitor C turns the transistor OFF. After the input signal reaches its final value (0 V), I_B maintains OFF drive for dc stability. If all inputs to a diode gate go negative, diodes are reverse-biased, and I_2 turns the transistor ON. After the voltage at R_2 reaches its final value $(-V)$, the collector clamp prevents further excursion, and current $I_g \equiv V/R_g$ maintains ON drive for dc stability. Note that for finite switching times, $I_2 > I_g$, and the collector clamp diode current is not insignificant.

For a first-order analysis of this system, consider all current sources to be ideal, perfect diodes, $V_{be} = 0$, and $I_B = I_{co}$ negligible compared to switching

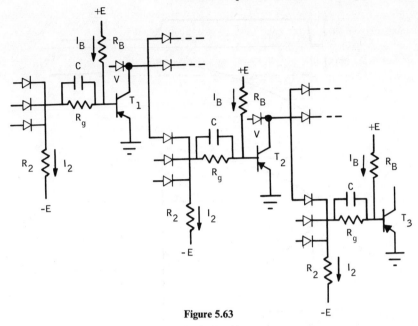

Figure 5.63

currents. For a system fan-out of n, the collector current at the edge of saturation must be $I_s \geq nI_2$, and R_g is immediately determined by the ON requirement that $I_g \geq I_s/\beta$. Let all of the inputs to the first stage go negative at $t = 0$. Because turn-ON of T_1 affects the turn-OFF of T_2, the system switching time will be taken as that necessary to reduce the collector current in T_2 to zero. This is similar to the $DCTL$ case and might correspond, for example, to the flip-flop situation in which the output of T_2 could be an input to T_1. Current and stored base-charge waveforms are illustrated in Figure 5.64. The worst-case value of maximum collector clamp current requires some consideration. The clamp diode of a given circuit (e.g., T_1) may have a lower forward drop than the clamp diodes of other inputs to the gates it is driving. It may therefore be slightly reverse-biasing the other inputs and must consequently supply all of $I_2 - I_g$ for each of n loads. This possibility is of course extreme, because the corresponding diode inputs to the load gates would be forward-biased, tending to cause other diodes to conduct and equalize the load. It is on the conservative side, however, to assume this worst case, because if T_1 is the only positive-going input to each of its loads, a slight rise in the collector voltage of T_1 will force it to assume the entire burden. After the collector comes off clamp, the net OFF drive to each of the n loads becomes $i_{c1}/n - I_2$. Since the ON drive to T_1 is I_2, the OFF drive to loads T_2 would reach an eventual value $\beta I_2/n - I_2$ if the transient continued

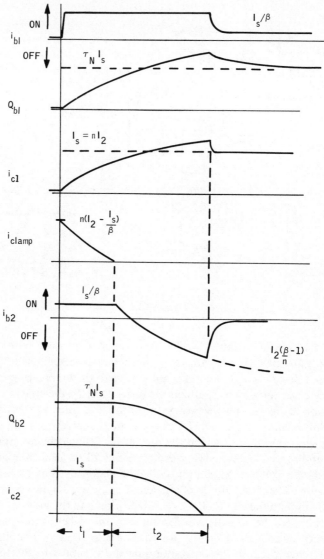

Figure 5.64

indefinitely. Under this OFF drive, the base-charge and collector currents decrease to zero, terminating the active-region portion of the transient. With load T_2 OFF a high impedance is presented to T_1, and because at this instant i_{c1} exceeds nI_2 by ni_{b2}, T_1 goes temporarily into saturation. If the input capacitor to T_1 is selected so that the input voltage reaches the clamp,

V, at the instant T_2 turns OFF, the ON drive is instantaneously reduced to I_s/β, and the excess charge in the base of T_1 decays to $\tau_N I_s$ with the time constant τ_a.

The first period t_1 is determined by the collector current rise in T_1. Inserting the above conditions in the standard rise-time expression, we obtain

$$\frac{t_1}{\tau_a} = \ln \frac{1}{1 - n/\beta + (n/\beta)^2}. \tag{5.119}$$

After t_1, the change in base drive to T_2 is

$$\Delta I_{b2}(s) = I_2 \left(\frac{\beta}{n} + \frac{n}{\beta} - 1 \right) \frac{1}{s(\tau_a s + 1)}. \tag{5.120}$$

The change in base charge is given by (5.116). Inserting initial conditions and converting to the time domain (with time starting at t_1), we obtain

$$\frac{Q_{b2}(t)}{\tau_N I_s} = 1 - \left[\left(\frac{\beta}{n} \right)^2 - \frac{\beta}{n} + 1 \right] \left[1 - \left(1 + \frac{t}{\tau_a} \right) e^{-t/\tau_a} \right]. \tag{5.121}$$

Setting (5.121) equal to zero determines t_2. At high switching frequencies (requiring $n \ll \beta$) solutions to (5.119) and (5.121) are closely approximated by $t_1 \cong t_2 \cong \tau_a n/\beta$. Because $T = t_1 + t_2$, under these conditions,

$$\frac{n}{\beta} \cong \frac{T}{2\tau_a}. \tag{5.122}$$

We confirm, therefore, that the ideal system possesses the 6 db/octave limiting-frequency response characteristic of nonsaturating circuits. The relationship between performance and switching time given by (5.119) and (5.121) is plotted in Figure 5.65 (solid curve).

Finite repetition rates affect system performance, because the ON base charge following a switching transient may not have decayed to its quiescent value $\tau_N I_s$ before a new transient is initiated. Therefore, the initial value may exceed the quiescent value used in (5.121), and t_2 will in fact be longer for a given system fan-out. Referring to Figure 5.64, we have

$$\frac{Q_{b1}(T)}{\tau_N I_s} = \frac{\beta}{n} [1 - e^{-T/\tau_a}]. \tag{5.123}$$

For the high switching rates at which (5.122) is valid, the value of $Q_b(T)$ can therefore exceed its quiescent value by a factor of 2. Taking finite T_p into

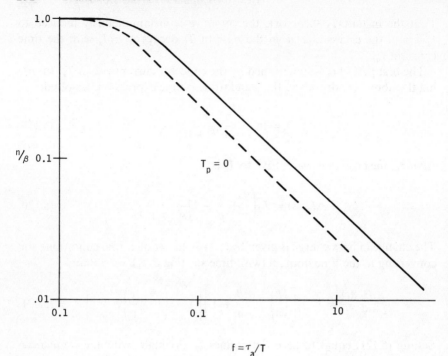

$$f = \tau_a/T$$

Figure 5.65

account, the correct expression for the OFF base charge is

$$\frac{Q_{b2}(t)}{\tau_N I_s} = 1 + \left[\frac{\beta}{n}(1 - e^{-T/\tau_a}) - 1\right]e^{-T_p/\tau_a}$$

$$- \left[\left(\frac{\beta}{n}\right)^2 - \frac{\beta}{n} + 1\right]\left[1 - \left(1 + \frac{t}{\tau_a}\right)e^{-t/\tau_a}\right]. \qquad (5.124)$$

Setting (5.124) equal to zero defines t_2. Performance versus total switching time $T = t_1 + t_2$ is plotted in Figure 5.65 (dotted curve) for the extreme case $T_p = 0$. The form of the function can be verified by noting that at the high frequencies at which $t_2 \cong T/2$, the solution to (5.124) for $T_p = 0$ is approximated by $n/\beta \le T/4\tau_a$, representing a factor-of-2 degradation in performance over (5.122).

Two-level diode gating, identical to that discussed in connection with *DTL* elements (Figure 5.32), is also widely used with *RCTL* elements. In this case the requirements on the coupling capacitor value can be obtained in a particularly straightforward manner. Referring to Figure 5.66, in which

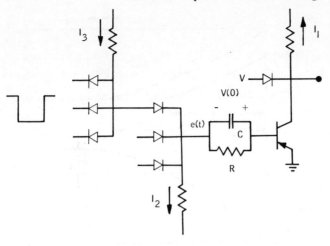

Figure 5.66

dc OFF bias is omitted for simplicity, turn-ON occurs with drive I_2, and turn-OFF occurs with drive $I_3 - I_2$. Capacitor constraints arise from the following conditions:

1. If the voltage across the capacitor input to an OFF element has not discharged to zero at the instant of turn-ON, $V(0) > 0$, the capacitor must be large enough so the input voltage $e(t)$ does not change by more than $V - V(0)$ during the turn-ON transient (see solid-curve waveform of Figure 5.67a).

2. If, however, the capacitor voltage has discharged to zero, with the same ON drive current an additional time is required for $e(t)$ to reach clamp voltage V. The transistor is therefore driven into saturation (see Figure 5.67b, dotted curve).

3. In the worst turn-OFF case the base charge in an ON transistor has not decayed to its quiescent value, and the capacitor must be large enough so that $e(t)$ does not reach 0 V before the base charge $Q(T_F)$ reaches zero (see the dotted waveform of Figure 5.67c).

4. If, however, the base charge has had time to return to its quiescent value before the beginning of the transient, with the same OFF drive it will reach zero in a shorter time, and a finite voltage $V(0)$ will remain across the capacitor (see the solid waveforms of Figure 5.67c and d).

To a first order, ON and OFF event requirements are compatible. For example, assume switching times are short compared to device and network time constants, and neglect charges that must be assigned to junction and

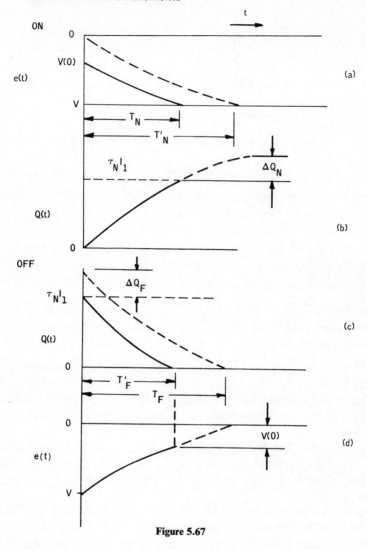

Figure 5.67

resistor losses. Then the ON conditions are

$$C[V - V(0)] \geq Q,$$

$$\Delta Q_N = CV(0),$$

(5.125)

in which Q is the quiescent base charge at the edge of saturation. The OFF conditions are

$$CV \geq Q + \Delta Q_F,$$

$$V(0) = V - \frac{Q}{C}.$$

(5.126)

For any finite period between transients, $\Delta Q_N > \Delta Q_F$, and the above conditions are compatible. Required ON current drive is unchanged by the presence of a residual capacitor charge, but OFF current drive must be increased to remove the excess base charge. For a selected value of this excess, the performance measure is given to a first order by the proportionality

$$\frac{n}{\beta} \sim \frac{(1 - e^{-T_N/\tau_a})(1 - e^{-T_F/\tau_a})}{1 - e^{-(T_N + T_F)/\tau_a} + q\, e^{-T_F/\tau_a}(1 - e^{-T_N/\tau_a})},$$

(5.127)

in which $q \equiv \Delta Q_F/Q$. This function is plotted in Figure 5.68 for $T_N = T_F = T$. Note that the response is relatively insensitive to q and closely approximates the 6 dB/octave high-frequency behavior characteristic of nonsaturating circuits.

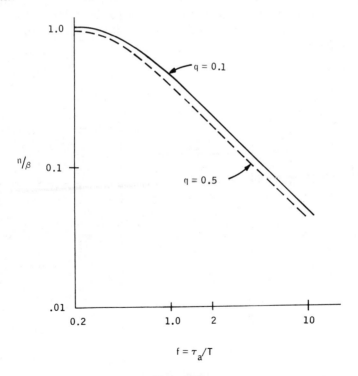

Figure 5.68

Current-Mode Logic Elements

The desirability of nonsaturating circuits and the high-frequency characteristics of the emitter follower have led to a number of decision element forms incorporating large negative feedback by means of an emitter resistor. These circuit forms, usually termed *current-mode logic*, are often particularly suitable for integrated device fabrication techniques. The price paid for improved high-frequency performance is a substantial increase in circuit power dissipation.

The circuit of Figure 5.7*a* can be taken as typical of this class of device, in which the diode D_1 may or may not be replaced by a transistor, as in Figure 5.7*b*. For first-order analysis, consider the emitter resistor R_E and the interstage coupling bias I_B to represent ideal current sources. Let the external base-input resistors be eliminated, so that only internal transistor base resistance r_b need be accounted for. The worst-case dc ON state of one stage of elements is depicted in Figure 5.69. One of N input transistors is ON, and its collector current must be sufficient to drive the voltage e to a value that will cut OFF all n driven loads. Driven transistors are held OFF by the difference between e and the diode voltage V_D due to current I_E flowing through the diode when all driven transistors of a given stage are OFF. Some minimum current I_z must be provided to maintain the voltage drop V_z across the zener diode.

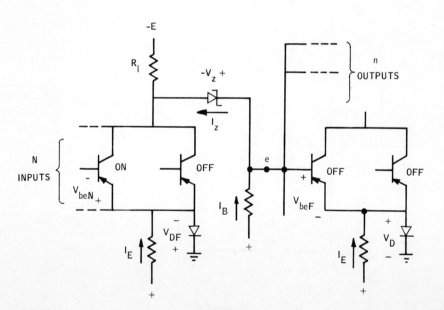

Figure 5.69

Therefore the bias current I_B is given by

$$I_B \geq I_z + nI_{co}. \tag{5.128}$$

Conservatively, let $r_b I_{co}$ be negligible, and let the base-emitter voltage required to insure that a transistor is OFF be V_{beF}. This voltage may in fact be slightly negative according to the convention of Figure 5.69, because collector current is usually negligible if V_{beF} is small compared to the necessary ON base-emitter voltage. A positive OFF bias is assumed here for reliable circuit action and noise suppression. Conservatively neglecting the leakage of OFF driving transistors, the collector current of the single ON source must satisfy

$$I_c \geq \frac{E(1 - \Delta_1)}{R_1} - I_z, \tag{5.129}$$

in which

$$\Delta_1 = \frac{V_z - V_D - V_{beF}}{E}.$$

The worst-case OFF state for the same set of elements is depicted in Figure 5.70. Here the current through R_1 must be sufficient to drive n loads to the ON state. It is assumed that the design arranges for the output voltage e to be sufficiently negative so that a small reverse voltage V_{DF} appears across the diodes in the driven stages for noise suppression. Each driven transistor must be capable of passing current I_c. Including leakage of the input transistors, the required current in R_1 is

$$\frac{E(1 - \Delta_2)}{R_1} \geq n\frac{I_c}{\beta} + I_B + NI_{co}, \tag{5.130}$$

in which

$$\Delta_2 = \frac{V_z + V_{DF} + V_{beN} + r_b I_c/\beta}{E}.$$

Combining (5.128), (5.129), and (5.130), we obtain the limiting dc performance measure of this system. Thus,

$$\frac{n}{\beta} > \frac{1 - \Delta_2 - (NI_{co}/I_c)(1 - \Delta_1) - (I_z/I_c)(\Delta_2 - \Delta_1)}{(1 - \Delta_1)(1 + \beta I_{co}/I_c)}. \tag{5.131}$$

Note that V_z must be selected to maintain the collector junction voltage of the driving stage in Figure 5.69 negative, thereby avoiding saturation. Thus

$$V_z > V_D + V_{DF} + V_{beN} + V_{beF} + \frac{r_b I_c}{\beta}. \tag{5.132}$$

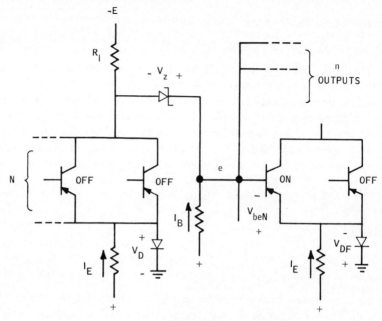

Figure 5.70

Observe that the function (5.131) closely resembles the form of limiting behavior for other types of elements [see (5.16) or (5.46)]. The importance of similar design choices (e.g., $I_c \gg \beta I_{co}$ and $E \gg V_z$) is again emphasized.

In order to provide some additional noise immunity an alternative design approach is sometimes taken in which a total ON-to-OFF signal swing of the source voltage e is specified. For this purpose, external base-input resistors R_g may be added to the circuit for a total base circuit resistance of $R = R_g + r_b$. The effect of specifying signal swing, $\Delta e = V$, is equivalent to requiring V to satisfy (5.132) with r_b replaced by R. This specification determines R_g and causes Δ_2 in (5.130) and (5.131) to be replaced by

$$\Delta_2 = \frac{V + V_z - V_D - V_{beF}}{E} \tag{5.133}$$

An additional constraint is sometimes imposed by requiring the signal swing to be symmetrical with respect to ground (or other reference). This has the effect of replacing Δ_1 and Δ_2 by $(V_z + V/2)/E$ and requiring the zener diode to satisfy $V_z > V/2 + V_{DF} + V_{beN}$.

It is worth noting that the behavior of the described circuit is sensitive to variable loading conditions. For example, suppose the circuit is designed for a fan-out of *n*, but that its use in a system requires only a minimum fan-out (i.e., *n* = 1). It can be seen that the signal *e* in Figure 5.70 will have a value more negative than that just necessary to turn the driven stage ON. The additional input signal swing imposes difficult requirements on the current source I_E and increases transistor dissipation. These problems can be avoided by providing clamping circuits that standardize the signal swing.

Expensive and power-consuming zener coupling diodes are eliminated in some systems by alternating *n-p-n* and *p-n-p* decision element stages. Such a scheme is shown in single fan-in form in Figure 5.71, in which the *p-n-p* stages are referenced to ground, and the *n-p-n* stages are referenced to − *V*. Required circuit behavior can be stated as follows. An ON *p-n-p* stage must supply a collector current sufficient to drive its collector voltage to a value more positive than − *V* by $V_{beF} + V_{beN} + r_b I_c/\beta$, but negative with respect to ground, in order to avoid saturation. At the same time it must supply base current I_c/β to *n* stages of ON *n-p-n* circuits. An ON *n-p-n* stage must draw a collector current sufficient to bring its collector voltage negative with respect to ground by the same series of voltage drops, but not as negative as − *V*, while absorbing ON base currents from *n* stages of *p-n-p* circuits. The performance is similar to that of diode-coupled elements.

Transient behavior of *CML* elements is critically dependent upon the highly nonlinear switching characteristics of junction voltages and base resistances. Fortunately, the large emitter degeneration produces fast

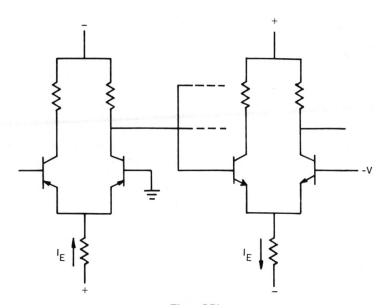

Figure 5.71

response, and first-order analysis can be based on simple charge-transfer models that imply that switching times are short compared to the transistor recombination rate. For this purpose, and to simplify the discussion, consider a network with zener diode level shifting, signal clamps, and transistor current-shunting paths. Let the signal swing be symmetrical about ground, $\pm V$, include all base resistors in r_b, and assume transistors to have identical characteristics. The turn-ON situation is depicted in Figure 5.72, in which the input signal goes from $+V$ to $-V$ at $t = 0$, and T_1 begins to turn from OFF to ON. The charge that is inserted into the base of T_1 is effectively the same charge as is removed from the base of T_2; that is, I_{b1} may be regarded as flowing through the path r_{b1}, T_1 base, T_2 base, and r_{b2}. Thus

$$I_{b1} \cong \frac{V - (V_{be1} - V_{be2})}{2r_b}. \tag{5.134}$$

The junction voltages V_{be1} and V_{be2} are primarily functions of emitter currents I_{e1} and I_{e2}, with $I_{e1} + I_{e2} = I_E$. As I_{e1} begins to flow, V_{be1} increases rapidly at first and thereafter more slowly. As I_{e2} decreases from I_E to zero, V_{be2} changes slowly at first and then rapidly as $I_{e2} \rightarrow 0$. Waveforms are sketched in Figure 5.73, with I_{b1} given by (5.134) and e_E by

$$e_E = \frac{-V + (V_{be1} + V_{be2})}{2}. \tag{5.135}$$

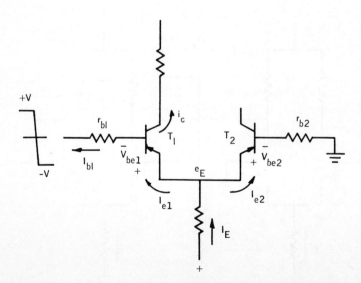

Figure 5.72

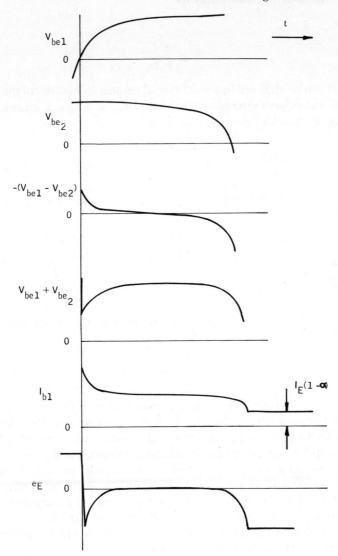

Figure 5.73

During most of the transient, $V_{be1} \cong V_{be2}$, and $I_{b1} \cong V/2r_b$. The charge accumulated in the base of T_1 is $Q = I_{b1}t$, and using a single-pole transistor frequency characteristic the collector current is $i_c = Q/\tau_N$. With switching complete when i_c reaches a specified value I_c, the turn-ON time of the circuit is therefore

$$T_N \cong \frac{\tau_N I_c}{V/2r_b}. \tag{5.136}$$

This result can be obtained in a more formal manner by analysis of the circuit with both transistors in the active region and $V_{be1} = V_{be2} = 0$ assumed. The governing relationships are

$$I_{b1}(s) = \frac{V(s) + E_E(s)}{r_b},$$

$$I_{b2}(s) = \frac{E_E(s)}{r_b},$$

$$\Delta I_{ei}(s) = \left[\frac{1}{1 - \alpha(s)}\right]\Delta I_{bi}(s), \qquad \text{with } i = 1, 2,$$

$$I_{e1} + I_{e2} = I_E,$$

$$I_c(s) = \alpha(s)I_{e1}(s). \tag{5.137}$$

Combining (5.137), inserting initial conditions, and converting to the time domain, we obtain

$$i_c(t) = \frac{\beta V}{2r_b}(1 - e^{-t/\tau_a}). \tag{5.139}$$

For $T_N \ll \tau_a$, (5.139) can be linearly approximated, with the result (5.136).

The OFF transient follows the same sequence of events, with the roles of T_1 and T_2 interchanged. Assuming the initial collector current (base charge) prior to turn-OFF is the same as that required to complete the ON transient, (5.136) is also valid for T_F. These approximate results can also be used if transistor response does not constitute the only time constant of interest. For example, including both collector junction and load capacitances has the effect of replacing τ_N by $\tau_N + \tau_c + \tau_L/\beta$, with $\tau_c = R_L C_c$ and $\tau_L = R_L C_L$. In the case considered here, with nearly constant-current sources, R_L approaches $2r_b$ during the switching period.

The LOAD transient conditions require some consideration of the driving source. For simplicity, it is assumed here that the source is a zero-impedance switch to a voltage $V - V_z$, which is opened at $t = 0$ to initiate the ON transient or closed to initiate the OFF transient. The circuit is illustrated in Figure 5.74. When the switch opens the output voltage falls to $-V$ and ON current I_{b1} flows in the driven circuits. During the transient it is necessary that $I_1 \geq nI_{b1}$. After the transient is complete, ON drive falls to $I_e(1 - \alpha)$, and the excess current flows through the clamp diode. In fact, of course, the

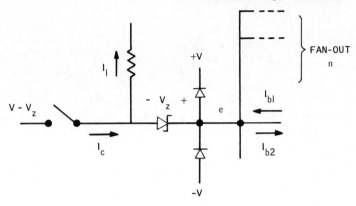

Figure 5.74

switch represents the collector current of a stage in the process of turning OFF. It is therefore conservative to take the total turn-ON time as $T_F + T_N$. Similarly, when the switch closes, the output voltage rises to $+V$, and OFF current I_{b2} flows in the loads. During the transient, the switch must therefore pass a current $I_c = I_1 + nI_{b2}$. After the transient is complete, OFF drive drops to zero (I_{co}), and the excess drive is taken by the clamp diode. Again, I_c is the collector current of a stage turning ON, and because the output voltage may not rise until nearly the end of the switching period it is conservative to take total turn-OFF time as $T_N + T_F$. Eliminating I_1 above, taking $T_N = T_F = T$, and using (5.136), we obtain the limiting fan-out. Thus

$$n \leq \tfrac{1}{2} \frac{T}{\tau_N}. \tag{5.138}$$

This result suggests the limiting 6 dB/octave frequency-response characteristic of nonsaturating circuits. It is illusory, however, because at high switching rates the single-pole frequency model yields optimistic results.

The two-pole model provides the following criteria for the required ON and OFF drive currents:

$$I_{b1} \geq \frac{I_c}{\beta} \left[\frac{1}{f(T_N)} \right], \tag{5.139}$$

in which

$$f(T_N) = 1 - \frac{e^{-T_N/\tau_a}}{1 - 1/\beta} + \frac{e^{-T_N/\tau_N}}{\beta - 1},$$

and

$$I_{b2} \geq \frac{I_c(0)}{\beta}\left[\frac{1}{f(T_F)} - 1\right],$$

in which $I_c(0)$ is the collector current just prior to the OFF transient. At the high frequencies for which the exponentials can be approximated by $e^x \cong 1 + x + x^2/2$, collector current and ON drive are related by

$$\frac{I_c}{\beta I_{b1}} \cong \frac{T^2}{2\tau_a\tau_N}. \tag{5.140}$$

We might expect, therefore, that the frequency response of the system has a characteristic drop-off of 12 dB/octave at higher rates. Note that the proportionality is $T^2/2\tau_a\tau_N$ for *CML* devices, rather than $(T/\tau_a)^2$ as in the usual saturating decision element, representing a high-frequency improvement by a factor of $\beta/2$. Taking $I_c(0) = I_c = I_1 + nI_{b2}$ as before, the performance

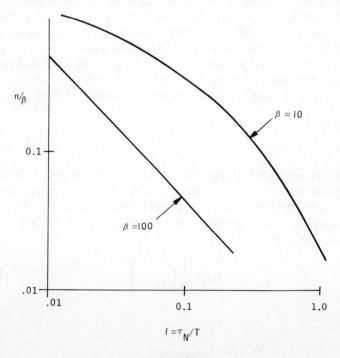

Figure 5.75

measure of the ideal circuit can be written as

$$\frac{n}{\beta} \le \frac{1}{1/f(T_N) + 1/f(T_F) - 1}. \tag{5.141}$$

At high frequencies, the performance is limited by

$$\frac{n}{\beta} \le \frac{T^2}{4\tau_a\tau_N}, \tag{5.142}$$

with $T = T_N = T_F$. Function (5.141) is plotted in Figure 5.75 for equal ON and OFF times, and two values of β. Because $f(T)$ is a function of β, net fan-out, rather than the ratio n/β is a more appropriate performance measure for this circuit. It is plotted in Figure 5.76 for the same conditions. Although low-gain systems are flat to higher frequencies, the fan-out in all cases is proportional in the limit to $(T/\tau_N)^2$.

Degrading factors similar to those considered in connection with other types of decision elements reduce performance below the ideal. The main factors to consider include diode recovery times, junction and stray capacitance charging, nonideal current sources, and tolerances.

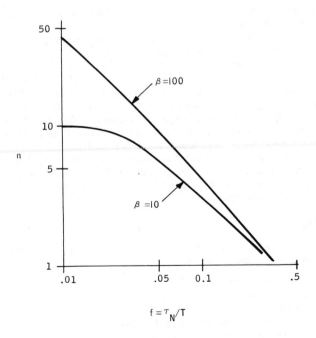

Figure 5.76

REFERENCES

[1] S. C. Chao, "A Generalized Resistor-Transistor Logic Circuit and Some Applications," *IRE Trans. Electron. Computers*, **EC-8**, 1, 8–12 (March 1959).

[2] R. C. Minnick, "Linear-input Logic," *IRE Trans. Electron. Computers*, **EC-10**, 6–16 (March 1961).

[3] W. D. Rowe and G. H. Roger, "Transistor NOR Circuit Design," *AIEE Commun. Electron.*, 31, 263–267 (July 1957).

[4] M. W. Marcovitz and E. Seif, "Analytical Design of Resistor-coupled Transistor Logic Circuits," *IRE Trans. Electron. Computers*, **EC-7**, 2, 109–119 (June 1958).

[5] W. J. Wray, Jr., "DC Design of Resistance Coupled Transistor Logic Circuits," *IRE Trans.*, **Vol. CT-6**, 3, 304–310 (September 1959).

[6] J. W. Easley, "Transistor Characteristics for Direct-coupled Transistor Logic Circuits," *IRE Trans. Electron. Computers*, **EC-7**, 1, 6–16 (March 1958).

[7] P. W. Becker, "Static Design of Transistor Diode Logic," *IRE Trans. Circuit Theory*, **CT-8**, 4, 461–467 (December 1961).

[8] W. Dunnet, E. Auger, and A. Scott, "Analysis of *TRL* Circuit Propagation Delay," *Proc. 1958 EJCC, Philadelphia*, 99–108 (December 1958).

[9] L. M. Spandorfer and J. B. Schwarz, "Microelectronic Logic Circuits," *Solid State Design*, 5, 7, 50–56 (July 1964).

[10] P. M. Ansbro, "Application of the Transistor Charge Control Model to Predict *RTL* and *DTL* Transient Response," *Solid State Design*, 5, 6, 19–25 (June 1964).

[11] R. Bohn, I. Feinberg, and T. Longo, "50 Mc Monolithic Integrated Circuits for Digital Application," *Solid State Design*, 4, 7, 25–31 (July 1963).

[12] W. D. Roehr (ed.), *Switching Transistor Handbook*, Motorola Semiconductor Products, Inc., Phoenix, Ariz., 1963.

[13] W. J. Roth and J. D. Standeven, "Designing for Efficiency with Diode-Transistor Logic," *Solid State Design*, 4, 9, 23–27 (September 1963).

[14] G. H. Goldstick, "Comparison of Saturated and Nonsaturated Switching Circuit Techniques," *IRE Trans. Electron. Computers*, **EC-9**, 2, 161–175 (June 1960).

[15] G. H. Goldstick and D. G. Mackie, "Design of Computer Circuits Using Linear Programming Techniques," *IRE Trans. Electron. Computers*, **EC-11**, 4, 518–530 (August 1962).

[16] J. B. Atkins, "Worst-Case Circuit Design," *IEEE Spectrum*, **2**, 3, 152–161 (March 1965).

[17] W. H. Hailey, "Diode-Transistor Minority Carrier Lifetime Relationship in Integrated *DT* Logic Circuits," *Solid State Design*, **6**, 6, 21–26 (June 1965).

[18] W. Bongenaar and N. deTroye, "Worst-Case Considerations in Designing Logical Circuits," *IEEE Trans. Electron. Computers*, **EC-14**, 4, 590–599 (August 1965).

[19] J. A. Narud and C. S. Meyer, "Characterization of Integrated Logic Circuits," *Proc. IEEE*, **52**, 12, 1551–1564 (December 1964).

EXERCISES

NOTE: For simplicity in all of the following exercises, assume circuit parameters are as listed below, unless otherwise specified:

Minimum transistor current gain $\beta \geq 50$,
Active-region time constant $\tau_a \leq 1.0\ \mu s$,
Saturation-region time constant $\tau_s \leq 1.0\ \mu s$,
Maximum permissible collector current $I_s \leq 10\ mA$,
Nominal power supply values $E = \pm 12\ V$,
Nominal binary signal levels = 0 and $-4\ V$.

5.1 A system containing 1,000 of the standard three-input circuits of Figure 5.31 is to be constructed. It must have a probability of at least 99 percent of surviving for 10 h. The failure rates of resistors and diodes are

Component	Failure rate (fraction per hour)
Resistors	8×10^{-8}
Diodes	10×10^{-8}

What is the upper limit on transistor failure rate to meet system specifications?

5.2 A system is to be constructed of the same standard inverter circuits as above, consisting of four diodes, three resistors, and one transistor each. It is necessary that the system operate for 10 h with a probability of survival of at least 90 percent. It has been determined that the failure rates of resistors and diodes remain approximately constant over the range of interest, but that transistor failure rate increases in proportion to $1 + I_s/20$, with I_s in milliamperes. The failure rates for low I_s ($I_s \ll 20$) are

Component	Failure rate (fraction per hour)
Resistor	1×10^{-6}
Diode	1×10^{-6}
Transistor	2×10^{-6}

Find the maximum value of collector current allowable if the number of circuits required in the system is 1,000.

5.3 A system is to be constructed using 10,000 decision elements of the above type. The mean time to failure of the system must be at least 100 h, and the failure rates of resistors and diodes are known to be as follows:

Component	Failure rate (fraction per hour)
Resistors	8×10^{-8}
Diodes	10×10^{-8}

A curve of failure rate versus power dissipation for transistors is given in Figure P.5.1. It has been decided that the turn-ON time (T_r) and turn-OFF time (T_F) of decision elements should be one-fourth of the period between clock pulses (T_p). The binary signal values are to be 0 and -4 V. Find the limiting value of collector saturation current to be used in the design.

5.4 Consider an RTL circuit similar to that of Figure 5.15 implementing the two-input NAND function. A fan-out of $n \geq 15$ is required under dc conditions. Supply voltages are ± 10 V, and signal levels are 0 and -4 V. What is the maximum tolerance on resistors that can be used in this circuit? (Assume for simplicity zero-tolerance power supplies, a perfect transistor, $V_{be} = V_{ces} \cong 0$ V, and a perfect diode, $V_D \cong 0$ V).

5.5 The generalized RTL circuit of Figure 5.15 is restricted to three inputs and designed to mechanize the logic function $\bar{x} + \bar{y}\bar{z}$.
(a) Find R_1, R_B, and the values of the three input resistors that minimize power dissipation while providing proper dc operation.
(b) The given decision element must drive the x inputs to n other elements, the y inputs to n others, and the z inputs to $2n$ others. What is the maximum value of n?

5.6 It is decided to change the logic of the decision element of Problem 5.5 to the three-input NAND function, $\overline{x \cdot y \cdot z}$ by making all resistors of equal value R.
(a) Find R_1, R_B, and R if the required turn-ON and turn-OFF times are 0.5 μs.
(b) What is the maximum fan-out of the circuit?

5.7 Consider an RTL circuit like that of Figure 5.15 designed to implement the complement of the three-out-of-five majority logic function ($m = 3$, $N = 5$). Let power supply voltages

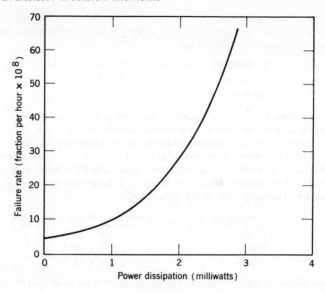

Figure P.5.1

be nominally ± 12 V, and let signal voltages be 0 and -4 V. Assume diode effects are negligible, but transistor parameters must be accounted for. They are $I_{co} = 20\,\mu\text{A}$ and $I_{eo} = 16\mu\text{A}$. Find R_1, R_B, and the values of input resistors that minimize power dissipation while providing correct circuit operation if the tolerance to be anticipated on both resistors and power supplies is ± 2 percent. Include effects due to finite base-emitter and collector-emitter drops (V_{be} and V_{ces}).

5.8 An *RTL* circuit implements the logic function, $f = x_3(\bar{x}_2 + x_1)$ using ± 5 percent resistors. The ± 12 V power supplies are well regulated (negligible compared to ± 5 percent), as are the 0 and -3 V logic levels. Find design values for the resistors if the transistors can be regarded as approximated by ideal equations with $I_{co} = 10\,\mu\text{A}$, $\alpha_N = 0.98$ and $\alpha_I = 0.80$.

5.9 The standard voltage-switching *DTL* circuit of Figure 5.31 is to have the following properties: fan-in $N = 10$ and $R_1 = 1\,\text{k}\Omega$. Find the values of R_B and R_g for the following successively less idealistic conditions:

(a) Assume perfect transistor and diodes and zero tolerances.

(b) Assume perfect transistor and diodes and ± 5 percent tolerances on all resistors and power supplies.

(c) Ideal transistor and diode properties as follows, ± 5 percent tolerances on all resistors and power supplies:

Transistor	Diodes
$I_{co} = 20\,\mu\text{A}$	$I_R = 20\,\mu\text{A}$
$I_{eo} = 16\,\mu\text{A}$	$V_D = 1.0$ V

What is the worst-case maximum current load on an input signal under each of the above assumptions?

5.10 Consider two-level diode voltage-switching transistor logic of the type illustrated in Figure 5.32. Repeat Problem 5.9, finding R_B, R_g, and R_2 based upon each of the successive approximations (a), (b), and (c) if the fan-in at each level is ten; that is, $N_1 = N_2 = 10$.

5.11 Find R_B, R_g, R_2, and maximum fan-out n for the two-level diode voltage-switching transistor logic above (Problem 5.10) if maximum turn-ON and turn-OFF times $T_r = T_F = 0.5\ \mu s$ are specified, and both resistors and power supplies are subject to ± 5 percent tolerance variations. Assume successively less idealistic conditions as in Problem 5.9.

5.12 What is the dc value of e_c in the antisaturation circuit of Figure 5.28 if $I_B = 0$, $I_1 = 10\ mA$, $R_2 = R_3 = 2\ k\Omega$, and the input signal is at $-4\ V$, and if transistor and diode parameters are as in Problem 5.9(c)?

5.13 The *DCTL* decision elements shown in Figure 5.6 must yield a fan-out of 5. What is the maximum fan-in permitted if the transistors all possess the characteristics given in Problem 5.9(c) and $I_1 = 10\ mA$?

5.14 For $t < 0$, both inputs to the circuit of Figure P.4.2 are at 0 V. At $t = 0$, one input goes to $-4\ V$. What is the time for the collector of the second transistor T_2 to reach $-4\ V$ if the effective load on its output at f_2, is a 1-kΩ resistor to ground?

5.15 What is the rise time T_r and fan-out n of the circuit of Figure 5.28 if circuit values are as follows:

$$I_1 = 10\ mA,$$

$$I_B = 0.4\ mA,$$

$$R_2 = R_3 = 2\ k\Omega,$$

$$V = -4\ V.$$

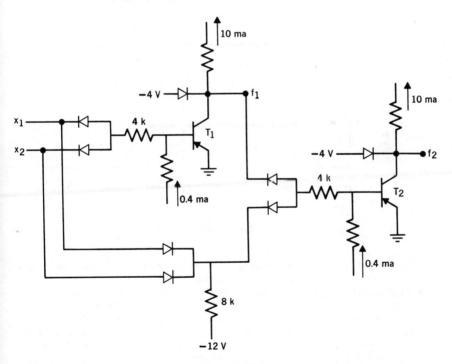

Figure P.5.2

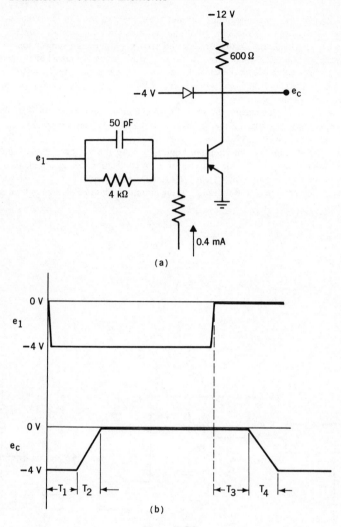

(a)

(b)

Figure P.5.3

Assume that the collector is connected directly to the inputs of n other identical circuits. Let maximum forward diode and base-emitter drops be $V_D = V_{be} = 0.5$ V.

5.16 What is the time for the collector of T_2 in the circuit of Figure P.5.2 to reach clamp (-4 V) if for $t < 0$, $x_1 = x_2 = 0$ V, and for $t > 0$, $x_1 = 0$ V, $x_2 = -4$ V? Solve for each of the successively less idealistic conditions specified in Problem 5.9(a), (b), and (c).

5.17 Prior to $t = 0$, the switch in Figure 5.56 is closed, the first-stage transistors (T_1) are OFF, and the second-stage transistors (T_2) are ON. At $t = 0$, the switch opens. At what time do the second-stage transistors turn OFF ($I_c \rightarrow 0$) if the fan-out $n = 4$ is the same at each stage? Transistor parameters are given in Problem 5.9(c), and $I_1 = 10$ mA.

5.18 If the circuit of Figure 5.15 implements the three-input NAND function $\overline{x \cdot y \cdot z}$, find values for R_1, R_B, and R_g that minimize power dissipation and maximize fan-out, while meeting the requirement that the maximum turn-ON and turn-OFF times are less than 0.5 μs. Assume maximum base-emitter forward drop is 1.0 V. What is the maximum fan-out n of the circuit?

5.19 Find maximum values of R_1, R_B, R_g, and n for the circuit of Figure 5.31c if maximum turn-ON and turn-OFF times are to be less than 0.5 μs. Assume transistor and diode properties as in Problem 5.9(c), resistor tolerances of ± 5 percent, power supply tolerances of ± 1 percent, and fan-in $N = 8$.

5.20 Find maximum values of R_B, R_2, and N for the circuit of Figure 5.36 if it is required that maximum turn-ON and turn-OFF times be less than 0.5 μs. Assume successively less idealistic conditions, as in Problem 5.9. Let both first- and second-level inputs be $N = m = 5$.

5.21 Find maximum values of R_B, R_g, R_2, and n, and minimum value of C, in the circuit of Figure 5.63, if maximum ON and OFF times are $T_N = T_F \leq 1.0$ μs. Assume successive conditions as in Problem 5.9, and maximum fan-in N of 8.

5.22 The input and output waveforms for the circuit of Figure P.5.3a are shown in Figure P.5.3b. Find T_1, T_2, T_3, and T_4 if collector-base capacitance $C_c = 20$ pF.

Chapter 6

Transistor Memory Elements

The preceding chapter dealt exclusively with the synthesis of combinatorial decision elements—circuits without memory of prior events other than that inherent in circuit delays. It has thus been assumed that no feedback paths exist in the logic networks so far considered. This is a very serious constraint. By far the most interesting networks—and all computer systems of any significance—are ones in which feedback is not only tolerated, but employed to great advantage to mechanize long-term memory in the form of active bistable devices.

In this chapter attention is confined to only four types:

RS: reset-set flip-flop. Sets to state 1 on receipt of an S input; resets to 0 on receipt of an R input. $R = S = 1$ not allowed. Maintains previous state if $R = S = 0$.

T: trigger. Maintains previous state if $T = 0$. Switches to opposite state if $T = 1$.

JK: same as RS with $J \sim R$, $K \sim S$, except that $J = K = 1$ is allowed, and produces the effect of a trigger.

RST: reset-set-trigger. As the name implies, an RS flip-flop with an additional trigger input.

From these verbal definitions, straightforward design procedures lead directly to truth tables for the various devices, as listed in Chapter 1.

Memory Element Form

RS. Let R = reset input signal,
S = set input signal,
q = present state of circuit, or $Q(t)$,
Q = next state of circuit, or $Q(t + 1)$.

When $R = S = 0$, $Q = q$; when $S = 1$ and $R = 0$, $Q = 1$; when $R = 1$ and $S = 0$, $Q = 0$; when $R = S = 1$, Q is a don't care, because this combination is not allowed. In fact, of course, the circuit will do something if $R = S = 1$ occurs, and this behavior can be specified in such a way as to simplify the circuit design as much as possible. For purposes of network synthesis it is desirable to have both Q and $\bar{Q}$ equations in a similar form, because this results in symmetry of the network. Because Q is unspecified when $R = S = 1$, we are free to fill in the truth table in such a way that Q and $\bar{Q}$ are in conflict. Let

$$Q = \bar{R}(S + q),$$

$$\bar{Q} = \bar{S}(R + \bar{q}), \tag{6.1}$$

with the corresponding tables.

R	S	q	Q	$\bar{Q}$
0	0	0	0	1
0	0	1	1	0
0	1	0	1	0
0	1	1	1	0
1	0	0	0	1
1	0	1	0	1
1	1	0	0	0
1	1	1	0	0

The other piece of information required for synthesis is the logic function performed by the decision elements to be employed in the design. One of the simplest decision elements previously considered is the two-input RTL circuit. With the bias properly selected and $V \equiv 1, 0\,V \equiv 0$, the logic function of the circuit for inputs x and y is

$$f = \bar{x}\bar{y}. \tag{6.2}$$

To generate the network, put equations (6.1) in the form of (6.2). For example,

$$Q = R(\overline{\bar{S}\bar{q}}) = \bar{R}f_1, \qquad f_1 = \bar{S}\bar{q},$$

$$\bar{Q} = S[\overline{R(\bar{q})}] = \bar{S}f_2, \qquad f_2 = \overline{R(\bar{q})}. \tag{6.3}$$

Decision elements may now be substituted directly for terms in the above expressions. Starting with Q, the result is the circuit of Figure 6.1. The delay (or hold) function mechanizes the fact that the correct value of Q appears at $t + 1$, in response to inputs at time t. Part or all of this delay may be inherent in the decision elements themselves. The dotted connection mechanizes the fact that $q = Q(t)$. Ignoring for the moment the specific mechanization of the delay element, we may redraw the dc circuit of the RS flip-flop in the conventional manner, as in Figure 6.2.

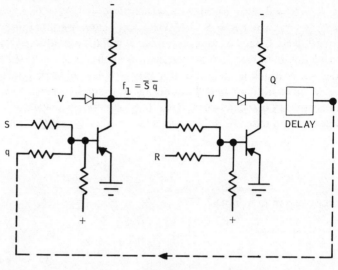

Figure 6.1

It is apparent that the techniques developed for decision element design can be applied directly to the synthesis of memory elements. Thus further design of the decision elements themselves is not required. It remains only to design delay circuits whose properties match the specifications on over-all circuit behavior.

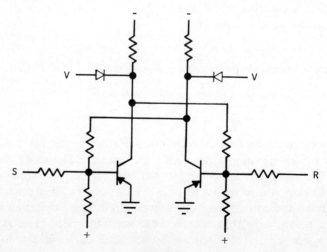

Figure 6.2

Alternative forms of *RS* flip-flops can be obtained by starting with other forms of basic decision element. For example, using DCTL elements, the network of Figure 6.3 mechanizes the logic (6.3) for *Q* with negative inputs defined as binary 1. Redrawn in conventional form, this network becomes the simple flip-flop of Figure 6.4. In the case of CML, two different types of decision elements are commonly combined to generate the flip-flop. One is the two-input NOR gate of Figure 6.5a. Another is the standard inverter with complementary outputs of Figure 6.5b. In this illustration two separate inputs are shown, although as an inverter one would be returned to a fixed supply, because it is necessary that $x \neq y$. This requirement can be met by cross coupling, as in Figure 6.5c. When these decision elements are combined to mechanize the logic of (6.3), the result is shown in Figure 6.6. Again redrawn in conventional form, the flip-flop circuit is that of Figure 6.7.

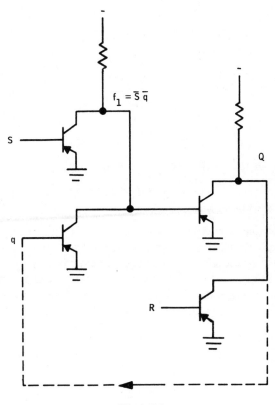

$$f_1 = \overline{S}\,\overline{q}$$

Figure 6.3

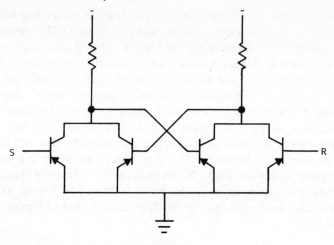

Figure 6.4

Trigger. The specification of the trigger flip-flop is as follows: when $T = 0$, $Q = q$; when $T = 1$, $Q = \bar{q}$. The truth table is

T	q	Q
0	0	0
0	1	1
1	0	1
1	1	0

and Q may be written as the EXCLUSIVE-OR of T and q,

$$Q = T \oplus q. \tag{6.4}$$

To mechanize (6.4) in a two-stage cascade, it is necessary to go to a more complex basic decision element for synthesis. It can be shown that the decision element of Figure 6.8 is the only three-input type capable of generating the EXCLUSIVE-OR logic in a two-stage cascade. In this figure the relative values of the input conductances are noted; thus, when z is negative, two units of base current are drawn, while either x or y draw only one unit. With the bias properly adjusted, the function becomes

$$f = \bar{z}(\bar{x} + \bar{y}). \tag{6.5}$$

After some manipulation, it is possible to write (6.4) in the form

$$Q = \{\bar{q}[\overline{T} + \overline{(\bar{q})}]\}(\overline{T} + \bar{q})$$

$$= \bar{f}_1(\overline{T} + \bar{q}), \tag{6.6}$$

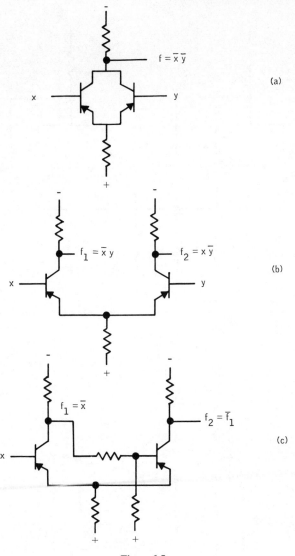

$f = \bar{x}\,\bar{y}$

(a)

$f_1 = \bar{x}\,y$ $f_2 = x\,\bar{y}$

(b)

$f_1 = \bar{x}$ $f_2 = \bar{f}_1$

(c)

Figure 6.5

in which

$$f_1 = \bar{q}[\,\bar{T} + \overline{(\bar{q})}\,].$$

Decision elements with the logic of (6.5) may therefore be substituted directly for the terms in (6.6), with the result shown in Figure 6.9. It is drawn in this

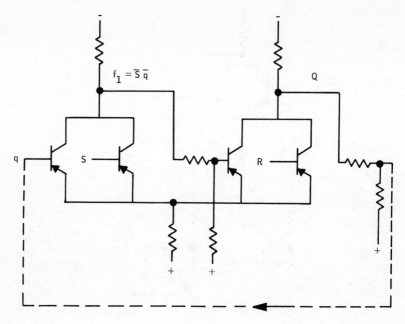

Figure 6.6

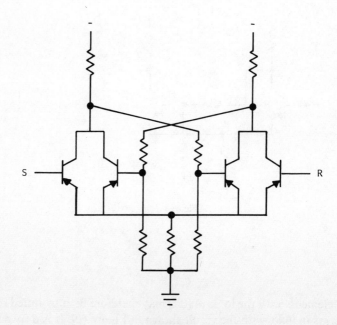

Figure 6.7

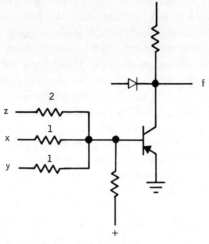

Figure 6.8

particular manner to emphasize the combinatorial logic aspects of the circuit. Starting from $\bar{Q}$, a similar result is obtained, and the symmetry remains. Therefore, the point labeled f_1 in Figure 6.9 must correspond to the quantity $\bar{Q}$, and delay units can be inserted as indicated. Redrawing the dc portion

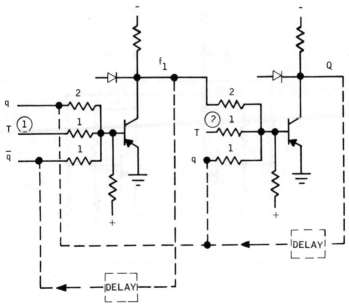

Figure 6.9

of the circuit results in the form of trigger flip-flop shown in Figure 6.10. As it stands, this circuit will not operate correctly, precisely because the delay circuits have been removed. Nevertheless, it illustrates at least one interesting aspect of trigger flip-flop design. Note the feedback resistors from the collector to the base of each transistor. These have the effect of making it easier to turn ON the OFF transistor (because negative logic is assumed, $T = 1$ corresponds to a turn-ON signal). Thus the trigger input T is "assisted" by the state of the circuit itself. This is a universal characteristic of trigger flip-flops and is sometimes termed *pulse steering* or *pulse gating*.

The self-gating function in the trigger flip-flop is more clearly brought out in a form based on *DTL* decision elements. A highly simplified *DTL* element is used in Figure 6.11 in a two-level network which generates, in somewhat redundant but symmetrical form, the function of (6.4). Redrawn, again without incorporating the required delay circuits, the flip-flop appears as in Figure 6.12. Here pulse steering can be easily identified. For example, suppose T_1 is OFF and T_2 is ON. When T goes negative, the left-hand negative AND gate is activated, and I_2 can begin to turn T_1 ON. The right-hand AND gate is held at about 0 V by the collector of T_2, and therefore T_2 receives no additional ON drive.

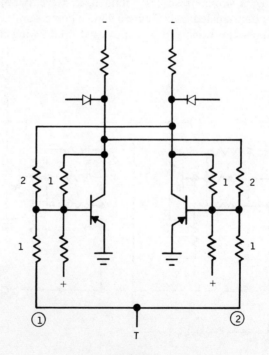

Figure 6.10

JK. The specification of the JK flip-flop is as follows: J sets Q to 1; K sets Q to 0; when $J = K = 0$, $Q = q$; and when $J = K = 1$, $Q = \bar{q}$. The resulting truth table is

J	K	q	Q
0	0	0	0
0	0	1	1
0	1	0	0
0	1	1	0
1	0	0	1
1	0	1	1
1	1	0	1
1	1	1	0

The logic may be written as

$$Q = J\bar{q} + \bar{K}q. \tag{6.7}$$

The strong similarities of this table to both RS and T logic would indicate that the resulting circuit might also be similar. This is indeed the case. If the connection between points 1 and 2 in Figure 6.9 is broken and J is applied to 1 and K is applied to 2, this circuit mechanizes (6.7). Equivalent pairs of points are also marked in Figures 6.10, 6.11, and 6.12. This result confirms the idea implicit in the truth table that a JK flip-flop acts like a trigger when $J = K$ and like a simple RS flip-flop when $J \neq K$.

RST. The specification of the RST flip-flop is as follows: when $R = S = 0$, the flip-flop is a normal trigger device ($Q = T \oplus q$); otherwise, it is a normal RS device ($Q = S + Rq$); $R = S = 1$ and $T(R + S) = 1$ are not allowed. The resulting table is

R	S	T	q	Q	Q (completed)
0	0	0	0	0	0
0	0	0	1	1	1
0	0	1	0	1	1
0	0	1	1	0	0
0	1	0	0	1	1
0	1	0	1	1	1
0	1	1	0	X	1
0	1	1	1	X	0
1	0	0	0	0	0
1	0	0	1	0	0
1	0	1	0	X	0
1	0	1	1	X	0
1	1	0	0	X	0
1	1	0	1	X	0
1	1	1	0	X	0
1	1	1	1	X	0

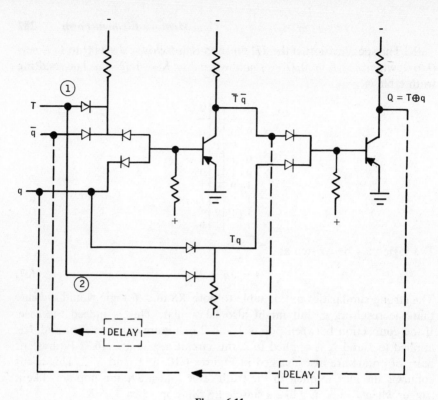

Figure 6.11

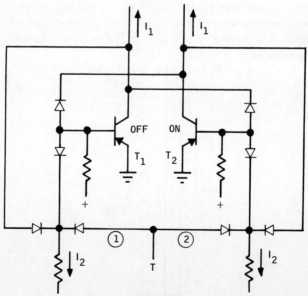

Figure 6.12

The don't care entries in the table may be filled in any desired manner, the only constraint being one of circuit simplicity. One possible way is shown in the last column of the table, and yields the logic

$$Q = \bar{R}[S(\bar{q} + \bar{T}) + (\bar{T}q + T\bar{q})].\tag{6.8}$$

It can be verified that Q is obtained by use of the function

$$f = \bar{z}\bar{w}(\bar{x} + \bar{y})\tag{6.9}$$

in cascade. An RTL decision element for this function is given in Figure 6.13. The corresponding flip-flop circuit is given in Figure 6.14. It can be seen by inspection that when $R = S = 0$, the corresponding resistors are effectively out of the circuit, and it reduces to a trigger. When $T = 0$, it becomes an RS with negative feedback around each stage, making triggering more sensitive. The RST flip-flop can also be generated from that of Figure 6.12 simply by adding R and S input gates to the transistor bases. With crossover coupling performed by resistors, the circuit appears as in Figure 6.15.

Flip-Flop Memory and Time-Dependent Components

The requirement for appropriate delay circuits, indicated specifically in Figures 6.1 and 6.9, has so far been neglected. In order to achieve proper flip-flop action specific attention must be given to this portion of the circuit. In some very restricted cases no additional delay is necessary. For example, it is intuitively clear that the RS circuit of Figure 6.2 is not time-dependent; that is, it is possible to drive the R or S inputs with signals whose duration is

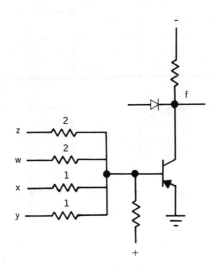

Figure 6.13

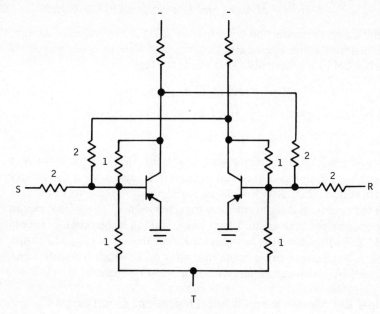

Figure 6.14

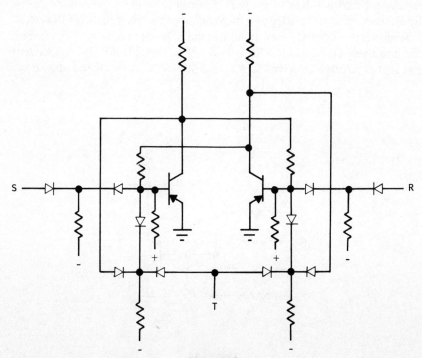

Figure 6.15

284

indefinitely long. On the other hand, the trigger circuit of Figure 6.10 or 6.12 requires a consideration of time, because in effect the same drive signal is being applied to both inputs. Some portion of the circuit must therefore be reserved for remembering its previous state. Investigation of this problem begins with an analysis of the time-dependent behavior of the simplest circuits.

RS switching and logic. Consider the switching behavior of the simple *RS* flip-flop of Figure 6.2. The circuit is reproduced, with currents and voltages labeled, in Figure 6.16. Assume that the action of the input is to turn ON the initially OFF transistor T_1 with a negative step of V at S. The sequence of events is illustrated in Figure 6.17 and is as follows:

1. The collector current i_{c1} rises from zero toward βi_{b1} with the transistor's active-region time constant τ_a.

2. With current in the diode clamp on the collector of T_1 reduced to zero, e_{c1} rises. For the nearly constant-current load shown, this rise would be almost instantaneous.

3. The step change at e_{c1} causes i_{b2} to reverse direction and T_2 to begin turning OFF. This portion of the transient terminates with T_2 coming out of saturation and entering the active region.

4. With reverse base current still flowing, i_{c2} decreases toward $-\beta I_2$ with an active-region time constant τ_a.

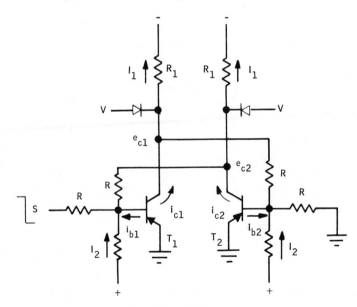

Figure 6.16

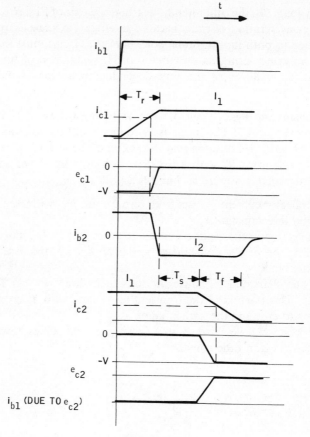

Figure 6.17

5. The collector current i_{c2} continues to decrease until $i_{c2} = 0$. During this time no further voltage changes occur in the circuit. When i_{c2} reaches zero, T_2 is completely cut off, the base input impedance becomes large, and a small positive reverse bias appears at the base of T_2.

An important question at this point is the following: what are the limits on the duration of the input signal S? There is certainly a minimum duration. If S is removed (returned to 0 V) too soon, T_2 will not come out of saturation, and regeneration cannot occur. There is, however, no limit on the maximum duration of S; that is, no harm is done if S remains negative indefinitely, provided it is returned to its normal value sufficiently far ahead of subsequent inputs to permit the circuit to recover.

The above discussion might indicate that the circuit of Figure 6.16 is ready to use in a general logic system. A very simple example will show that this is not the case. Suppose RS flip-flops are to be employed in mechanizing a counter. The logic for the bits, $1, 2, 4, \ldots$, of the counter might be as follows:

$$S_1 = \bar{Q}_1 C,$$

$$R_1 = Q_1 C,$$

$$S_2 = \bar{Q}_2 Q_1 C,$$

$$R_2 = Q_2 Q_1 C, \text{etc.,}$$

in which C is a common timing signal. Suppose the negative AND terms are mechanized by means of diode gates. Then the complete circuit for, say, the second flip-flop in the counter chain, is given in Figure 6.18. However, because $Q_1 C$ is the same input to both sides, the circuit reduces to that of a trigger. This circuit cannot trigger properly, however, because at some time during the switching transient, the two halves of the flip-flop are in an identical condition with identical drive. For example, when $e_{c1} = e_{c2}$, collector and base currents are equal on the two sides, and the input signal $Q_1 C$ is also the same on both sides. Thus the subsequent behavior of the circuit will be indeterminate. We see, therefore, the necessity for storing information about the flip-flop's previous state during the time of transition.

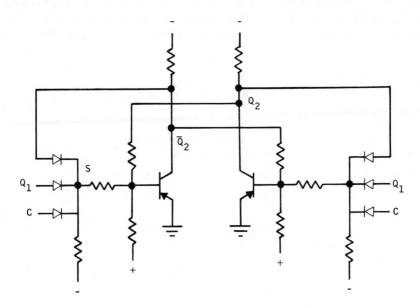

Figure 6.18

A widely used technique for converting the circuit of Figure 6.18 to a trigger is to add capacitor storage to some point or points in the circuit. The capacitor-coupled circuit of Figure 6.19 is perhaps the most common form. In this circuit, the cross-coupling capacitors, initially charged to different voltages, maintain sufficient charge during the switching transient to prevent the two halves of the circuit from ever reaching identical states. Although the circuit of Figure 6.19 will indeed trigger properly, it will be shown later that additional forms of storage are required for correct circuit operation in a general logic system. Nevertheless, it furnishes a good starting point from which to analyze the behavior of elements containing time-dependent coupling components.

Capacitor-coupled decision elements. The introduction of capacitor storage into the trigger structure makes it resemble a cascade network composed of *RCTL* decision elements. Thus, the circuits of Figures 6.18 and 6.19 can be converted to the open-loop hybrid form shown in Figure 6.20. An investigation of the switching characteristics of this circuit sheds some light on trigger action. Note that Figure 6.20 is equivalent to Figure 6.18 with C in parallel with the crossover resistors and the diode gates making up $Q_1 C$ replaced by direct trigger (T) drive of the input resistor.

As a first approach, consider both T_1 and T_2 loaded by n similar circuits with T_1 initially OFF and T_2 initially ON. The time to turn T_2 OFF is the quantity of interest, because at that time its output is satisfactory for driving the input to T_1, as indicated by the dotted connection, and the trigger signal T may be removed. Assume I_1 and I_2 are constant-current sources, and that the dc

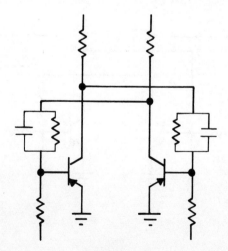

Figure 6.19

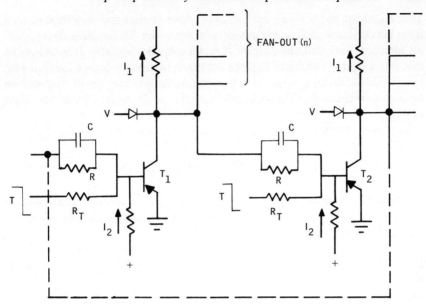

Figure 6.20

current in a coupling resistor R is sufficient to keep a transistor turned ON. The current I_2 must be sufficient to keep a transistor turned OFF under dc conditions, and, to a first approximation, is simply the worst-case transistor I_{c0}. Because this current is usually small compared to other quantities, it can be neglected in this approximate development. Thus,

$$I_3 \equiv \frac{V}{R} \cong \frac{I_1}{\beta}. \tag{6.10}$$

With T_1 heavily loaded, little current will be flowing through its clamp diode, and the time to come off clamp can also be neglected. This implies

$$I_1 \cong nI_3. \tag{6.11}$$

The load condition at the output of T_2 may be unknown, but if T_2 is turned completely OFF there will be sufficient drive available to its outputs, by (6.11). When Q_2, the base charge in T_2, reaches zero, therefore, switching may be considered completed. At $t = 0$, the trigger signal goes negative. Let the current drawn by R_T with the base clamped to near 0 V by the base-emitter junction be I_T. Then the base current to T_1 becomes I_T, and including the time-varying collector current I_{c1} from T_1 the current to T_2 becomes

$$I_{b2} = \frac{I_1 - I_{c1}}{n} + I_T. \tag{6.12}$$

This equation implies two assumptions. One is that the impedances seen from the collector of T_1, corresponding to n parallel RC coupling circuits, are all identical and that the current therefore divides equally. It also implies that the collector voltage of T_1 does not reach saturation before the transient is over. This, in turn, imposes a minimum requirement on C that will be treated subsequently. The collector current of T_1 is related to the base current by $\beta(s)$.

The base charge of T_2 is related to its base current by

$$Q_2(s) = Q_2(0)\frac{1}{s} + \Delta I_{b2}(s)\frac{\tau_a}{\tau_a s + 1}, \tag{6.13}$$

in which ΔI_{b2} is the change in base drive to T_2. Because the initial value of I_{b2} is I_1/n, $\Delta I_{b2} = -(I_{c1}/n) + I_T$. The initial value of base charge in T_2, due to dc conditions, is

$$Q_2(0) = \tau_a I_{b2}(0) = \frac{\tau_a I_1}{n}. \tag{6.14}$$

Using a single-pole approximation for $\beta(s)$ and assuming both transistors are identical, we obtain the form of the base charge in T_2. Thus

$$\frac{Q_2(s)}{\tau_a I_T} = \frac{I_1/I_T}{n}\frac{1}{s} + \frac{1}{s(\tau_a s + 1)} - \frac{\beta}{n}\left[\frac{1}{s(\tau_a s + 1)^2}\right]. \tag{6.15}$$

The corresponding time function is

$$\frac{Q_2(t)}{\tau_a I_T} = \frac{I_1/I_T}{n} + (1 - e^{-t/\tau_a}) - \frac{\beta}{n}\left[1 - \left(1 + \frac{t}{\tau_a}\right)e^{-t/\tau_a}\right]. \tag{6.16}$$

Thus Q_2 consists of a positive term related to its dc value, an additional positive term related to the trigger input to T_2, and a negative term due to the cross-coupling drive from T_1. Some idea of the switching time can be obtained by setting (6.16) equal to zero and approximating the exponentials by their first-order expansions, $e^x \cong 1 + x$. Solving the resulting quadratic for T, the trigger switching time, we obtain

$$T \cong \tau_a\frac{n}{2\beta}\left\{1 + \left[1 + 4\frac{I_1/\beta}{I_T}\left(\frac{\beta}{n}\right)^2\right]^{1/2}\right\}. \tag{6.17}$$

The case of greatest interest occurs when the fan-out n is small. This would be the case, for example, if buffer output amplifiers were used in conjunction with the basic flip-flop. The trigger would then be required to drive only a fan-out of 2, one being the output amplifier, the other being the complementary side of the flip-flop. With n small in (6.17), the switching time approaches

$$T \cong \tau_N \left(\frac{\beta I_1}{I_T} \right)^{1/2}. \qquad (6.18)$$

The desirability of a large impulse of trigger drive current I_T is thereby apparent. This suggests capacitively coupling the trigger input directly to the base, in which case the drive current is limited only by the base resistance of the transistor. Such systems are indeed widespread, having the advantages of both high rate and controlled quantity of charge transfer.

Regeneration. The foregoing analysis was based on the rather pessimistic assumption that the loop gain in the trigger never reaches unity. In a properly designed circuit, however, some operating range is normally reached within which the loop gain of the flip-flop exceeds unity, and through which the switching process proceeds under regenerative conditions. Some idea of the most optimistic minimum switching time can be obtained by investigating transient behavior under these conditions. Fastest switching occurs when the fan-out is effectively $n = 1$, corresponding only to the cross-coupling network loading each collector. Thus the circuit reduces to that of Figure 6.21, in which the impedance Z is the parallel RC coupling network. Again, assume that for dc OFF stability, $I_2 \cong I_{co}$, which is relatively small and can be neglected for first estimates. For dc ON stability, the current through Z must be sufficient to maintain a transistor at the edge of saturation. Thus the dc ON base current is approximately I_1/β. At $t = 0$, the trigger input T goes negative, driving both bases with a turn-ON current I_T. With T_2 initially ON, its starting base charge is

$$Q_2(0) \cong \frac{\tau_a I_1}{\beta}. \qquad (6.19)$$

That in T_1 is $Q_1(0) \cong 0$. The change in OFF drive to T_2 is

$$\Delta I_{b2}(s) \cong I_{c1}(s) - I_T(s). \qquad (6.20)$$

This neglects the time for the collector of T_1 to come off clamp, because we are concerned for the moment only with regenerative behavior. It thus neglects the corresponding rise of I_{c1} from 0 to $I_1(1 - 1/\beta)$ required to get the collector of T_1 off clamp; this amounts to assuming that $I_T \gg I_1/\beta$. The collector current of T_1 is still given by $\beta(s)I_{b1}(s)$, but with regeneration present, the base current of T_1 becomes

$$I_{b1}(s) = I_1(s) - I_{c2}(s) + I_T(s). \qquad (6.21)$$

This assumes that T_2 is initially at the edge of saturation and therefore comes instantaneously into the active region. Of course, the same condition as

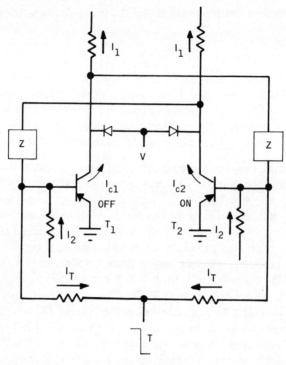

Figure 6.21

before applies to the cross-coupling impedance, namely, that the voltage across the impedance does not change by more than V during the transient, so that T_1 does not reach saturation and the collector of T_2 does not reach the clamp before the transient is over. Again consider the transient to be complete when Q_2 reaches zero. The relationship between Q_2 and I_{c2} will be assumed to be instantaneous; that is,

$$I_{c2}(s) = \frac{Q_2(s)}{\tau_N}, \qquad (6.22)$$

which effectively neglects the transit-time delay τ_N relative to the recombination time constant τ_a. Combining the above conditions and noting that $\beta = \tau_a/\tau_N$, we see that the complex-frequency form for Q_2 becomes

$$\frac{Q_2(s)}{\tau_a(I_1/\beta)} = \left[\frac{1}{1 - \beta^2/(\tau_a s + 1)^2}\right]$$

$$\times \left[\frac{1}{s} + \beta\frac{I_T}{I_1}\frac{1}{s(\tau_a s + 1)} - \beta^2\left(1 + \frac{I_T}{I_1}\right)\frac{1}{s(\tau_a s + 1)^2}\right]. \qquad (6.23)$$

The denominator of the first bracket in (6.23) can be factored into

$$(\tau_a s + 1 - \beta)(\tau_a s + 1 + \beta) \cong \beta^2(\tau_N s - 1)(\tau_N s + 1), \tag{6.24}$$

in which the approximation $\beta \gg 1$ is used. The result is

$$\frac{Q_2(s)}{\tau_a(I_1/\beta)} = -\left(1 + \frac{I_T}{I_1}\right)\frac{1}{s(\tau_N s + 1)(\tau_N s - 1)} + \frac{1}{\beta}\frac{I_T}{I_1}\frac{\tau_a s + 1}{s(\tau_N s + 1)(\tau_N s - 1)}$$

$$+ \frac{1}{\beta^2}\frac{(\tau_a s + 1)^2}{s(\tau_N s + 1)(\tau_N s - 1)}. \tag{6.25}$$

The corresponding time function is

$$\frac{Q_2(t)}{\tau_a(I_1/\beta)} = -\left(1 + \frac{I_T}{I_1}\right)\left(-1 + \tfrac{1}{2}e^{-t/\tau_N} + \tfrac{1}{2}e^{+t/\tau_N}\right)$$

$$+ \frac{1}{\beta}\frac{I_T}{I_1}\left[-1 + \tfrac{1}{2}\left(1 - \frac{\tau_a}{\tau_N}\right)e^{-t/\tau_N} + \tfrac{1}{2}\left(1 + \frac{\tau_a}{\tau_N}\right)e^{+t/\tau_N}\right]$$

$$+ \frac{1}{\beta^2}\left[-1 + \tfrac{1}{2}\left(\frac{\tau_a}{\tau_N} - 1\right)^2 e^{-t/\tau_N} + \tfrac{1}{2}\left(\frac{\tau_a}{\tau_N} + 1\right)^2 e^{+t/\tau_N}\right]. \tag{6.26}$$

Collecting terms and again using the approximation $(\tau_a/\tau_N) = \beta \gg 1$, we obtain

$$\frac{Q_2(t)}{\tau_a(I_1/\beta)} \cong \left(1 + \frac{I_T}{I_1}\right) - \tfrac{1}{2}\left(1 + \frac{I_T}{I_1}\right)e^{-t/\tau_N} + \tfrac{1}{2}\left(1 - \frac{I_T}{I_1}\right)e^{t/\tau_N}. \tag{6.27}$$

Because of the idealizing assumptions, this is a time function that *can* go in the direction of turning T_2 more heavily ON. This is because both T_1 and T_2 are assumed to be active. However, if T_2 goes ON, it will in fact saturate, whereas the equations imply it will turn T_1 further OFF. The condition to ensure that Q_2 goes to 0 is $I_T > I_1$, in which case the coefficient of the term involving the positive exponent in (6.27) is negative. After any reasonable number of time constants ($t \gg \tau_N$), the term involving the negative exponent in (6.27) becomes negligible. Therefore, in terms of magnitudes, for $Q_2(T) = 0$,

$$T \cong \tau_N \ln 2\left(\frac{I_T/I_1 + 1}{I_T/I_1 - 1}\right), \qquad I_T > I_1. \tag{6.28}$$

Although the above result cannot be taken too literally, it does mean that if both transistors are in the active region, switching will be completed within a few time constants τ_N. Compare this result with that of (6.18), in which (with $I_T > I_1$) the switching time becomes something less than $\sqrt{\beta}$ times τ_N. These functions are compared for one value of β in Figure 6.22.

Another conclusion to draw from this analysis is that flip-flop switching time can be dominated by any switching that occurs with either or both transistor outputs clamped. The clamp can be either to the collector clamp voltage V via the clamp diode, or to ground via the collector junction (i.e., in saturation). In either case the value of the corresponding crossover impedance Z is simply the value of coupling resistor. Switching times under these conditions can therefore be determined as in decision element analyses in which base drive is a constant during the switching period. This effect is particularly important if the fan-out n is greater than 1. For the circuit of Figure 6.21, with $I_T \gg I_1/\beta$, the time for the OFF transistor to come off clamp is approximately $T_r \cong \tau_N(I_1/I_T)$. This function is also plotted in Figure 6.22. If, however, the circuit of Figure 6.21 must drive other loads and each requires a drive of I_1 for rapid switching, under the most unfavorable condition the time to come off clamp can approach nT_r.

The presence of a growing exponential in (6.27) means that the complex-frequency response function of the circuit must possess a pole in the right half plane. This viewpoint has been taken by Pederson [1], and by Lynn and Pederson [3] in analyzing the switching transient in multivibrators and flip-flops. In the former case the circuit can take on a particularly simple

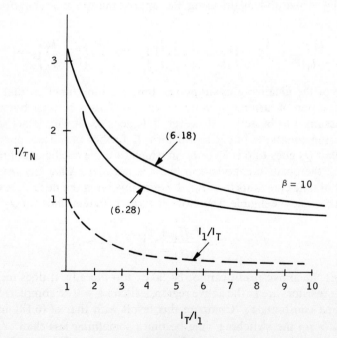

Figure 6.22

form, and constant-current drives such as shown in Figure 6.21 need no longer be assumed. In principle, to produce a free-running multivibrator, it is only necessary to let the trigger input remain true at all times. Then, by definition, the circuit changes state each time period. Under these conditions, however, dc stability is no longer required. Thus the resistor R in the cross-coupling impedance Z and the OFF bias I_2 can be removed.

The result is the circuit of Figure 6.23. Circuit action is as follows. Assume T_1 is OFF and its collector voltage at $-E$. T_2 is ON, and the voltage across C_1 is therefore approximately E. R must be sufficient to keep T_2 at the edge of saturation, or $R = \beta R_1$, for equal supply voltages. If, prior to this time, C_2 is also charged to about E, with T_2 ON the base voltage of T_1 must initially be approximately $+E$. This voltage decays with the time constant RC_2 toward $-E$ until it reaches about 0 V, or enough to begin turning T_1 ON. With T_1 ON, collector current is fed back to the base of T_2 in the turn-OFF direction, and the circuit becomes regenerative. Assuming only small changes in voltage across the capacitors during the switching period, we see that at the end of this time T_1 is ON, and T_2 is OFF with a positive base voltage of approximately E. The capacitor C_2 then charges to the voltage E with a time constant R_1C_2, and the cycle is repeated.

The regenerative portion of this cycle can be explored by using the single-pole transfer characteristic of the transistor, and by considering the effect of the coupling network. Because both transistor bases remain clamped to

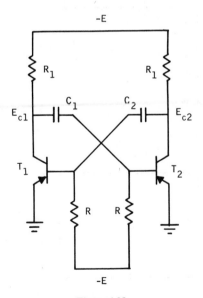

Figure 6.23

approximately 0 V via the base-emitter forward-biased junction during this period, the coupling network appears as in Figure 6.24a. The ac equivalent is thus as shown in Figure 6.24b, with the transfer function

$$\Delta I_b(s) = \Delta I_c(s)\frac{\tau_1 s}{\tau_1 s + 1},$$ (6.29)

in which $\tau_1 \equiv R_1 C$. The initial value of charge in the base of the ON transistor at the instant regeneration begins is $Q(0) = \tau_a I_T = \tau_a(E/R)$. The ON drive to the previously OFF side (e.g., T_1) is

$$I_{b1}(s) = I_T\left(\frac{1}{s}\right) - \Delta I_{c2}(s)\frac{\tau_1 s}{\tau_1 s + 1}.$$ (6.30)

The change in drive (in the OFF direction) to the opposite side is given by

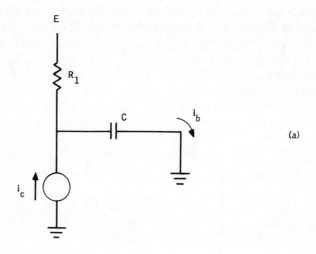

(a)

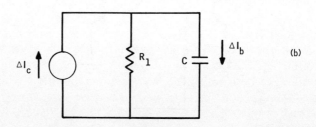

(b)

Figure 6.24

(6.29). Noting that

$$\Delta I_{c2}(s) = \frac{1}{\tau_N}\left[Q_2(0)\left(\frac{1}{s}\right) - Q_2(s)\right], \tag{6.31}$$

we see that the above conditions result in a transition function of the form

$$\frac{Q_2(s)}{\tau_a I_T} = \left[\frac{1}{1 - \beta^2\tau_1{}^2s^2/(\tau_1 s + 1)^2(\tau_a s + 1)^2}\right]$$

$$\times \left[\frac{1}{s} - \frac{\tau_1\tau_a\beta}{(\tau_1 s + 1)(\tau_a s + 1)^2} - \frac{\beta^2\tau_1{}^2s}{(\tau_1 s + 1)^2(\tau_a s + 1)^2}\right]. \tag{6.32}$$

The characteristic function of this expression is determined by the denominator in the first bracket. Clearing of fractions and factoring, the roots of (6.32) are thus the solutions to

$$[(\tau_1 s + 1)(\tau_a s + 1) - \beta\tau_1 s][(\tau_1 s + 1)(\tau_a s + 1) + \beta\tau_1 s] = 0. \tag{6.33}$$

Clearly, only the first factor can produce positive roots. With $\tau_a/\tau_N = \beta \gg 1$, the sign of the possible positive root of (6.33) is determined by the sign of

$$\frac{1}{\tau_N} - \frac{1}{\tau_1}. \tag{6.34}$$

Therefore for regeneration to occur the necessary condition is

$$\tau_N < \tau_1. \tag{6.35}$$

It is also of interest to examine the constraints if the effect of collector junction capacitance C_c is included. This can be done by replacing τ_a by $\tau_a + \beta\tau_c$ in (6.33), with $\tau_c \equiv R_1 C_c$. In this case the constraint reduces to

$$\tau_N + \tau_c < \tau_1\left(1 + \frac{\tau_c}{\tau_N}\right). \tag{6.36}$$

General initial conditions. It has been shown that for the flip-flop of Figure 6.21 proper regeneration occurs provided the trigger pulse is greater than the collector current $I_T > I_1$. For the multivibrator of Figure 6.23 regeneration occurs provided the passive-circuit time constant is greater than that of the amplifier, $\tau_1 > \tau_N$. It is of importance to inquire into the initial conditions in both collector current and circuit voltages (across the capacitors) that, in general, will produce regenerative behavior. It will then become the task of the trigger pulse to establish these initial conditions.

A simplified analysis of the general case has been carried out by Lynn and Pederson [3]. Consider the simplified general circuit of Figure 6.25. The

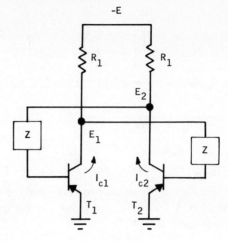

Figure 6.25

impedances Z are parallel RC networks. Thus

$$Z(s) = \frac{R}{\tau s + 1},\qquad(6.37)$$

with $\tau = RC$. During the period of interest, both transistors are in the active region, and the base voltages are assumed clamped to near 0 V by the forward-biased emitter junction. Initial conditions are the initial charges in the bases of the two transistors, $Q_i(0)$, and the initial collector voltages $E_i(0)$, which are equivalent to the initial charges on the capacitors, $Q = CE$. The charge-transfer function of the transistor is still assumed to be of the form

$$Q(s) = Q(0)\frac{\tau_a}{\tau_a s + 1} + I_b(s)\frac{\tau_a}{\tau_a s + 1},\qquad(6.38)$$

and collector current is assumed to be given by $I_c(s) = Q(s)/\tau_N$. The collector current of each transistor drives an impedance similar to that of Figure 6.24b. In this case, however, the impedance becomes

$$Z_1(s) = \frac{R_p}{\tau_1 s + 1},\qquad(6.39)$$

in which

$$R_p = \frac{R_1 R}{R_1 + R},$$

$$\tau_1 = R_p C.$$

The collector voltage can be written as the sum of two terms. One is the voltage that would occur with the transistor OFF; the other is a result of I_c driving the load impedance. Thus, in terms of magnitudes,

$$E_i(s) = E_i(0)\frac{\tau_1}{\tau_1 s + 1} + E_i(f)\frac{\tau_1}{s(\tau_1 s + 1)} - I_{ci}(s)Z_1(s), \qquad (6.40)$$

in which $E_i(f)$ is the final collector voltage with the transistor OFF and is of the magnitude

$$E_i(f) = \frac{R}{R_1 + R}E. \qquad (6.41)$$

Base drive is, of course, given by $I_{bi}(s) = E_j(s)/Z(s)$, with $i \neq j$. Note that this analysis assumes identical impedance values and transistor characteristics on both sides. In this case $E_1(f) = E_2(f) = E(f)$. Further, note that if β is at all large, dc stability permits $R \gg R_1$, and therefore $\tau_1 \ll \tau$, and $R_p \cong R_1$. In particular, saturation-mode operation requires that $R \leq \beta R_1$. Combining the above conditions, we can put an expression for the charge in one of the transistor bases in the form,

$$Q_i(s) = \left\{ \frac{1}{1 - [\beta(R_p/R)(\tau s + 1)/(\tau_1 s + 1)(\tau_a s + 1)]^2} \right\}$$
$$\times F[Q_i(0), Q_j(0), E_i(0), E_j(0), E(f)], \qquad (6.42)$$

in which $F[\ \ ]$ is a complex-frequency function of the initial and final conditions. The roots of the denominator in the first term of (6.42) will determine the form of the transient. Factoring, letting $R \cong \beta R_1$, and assuming $\beta \gg 1$, we see that two of these roots occur at approximately zero frequency. The signs of the other two are determined by

$$p \cong \frac{1}{\tau_N} - \frac{1}{\tau_1},$$

$$q \cong -\frac{1}{\tau_N} - \frac{1}{\tau_1}. \qquad (6.43)$$

In order to obtain a positive pole in the frequency function we again encounter the requirement that

$$\tau_1 > \tau_N, \qquad (6.44)$$

in which case, $p > 0$. For $\tau_1 \gg \tau_N$, the switching speed is thus dominated by τ_N, the alpha time constant of the transistor. When (6.42) is converted to a time function and terms collected, the form of the switching transient becomes

$$Q_i(t) = a_0 + a_1 e^{pt} + a_2 e^{qt} + a_3 t + a_4 t^2. \tag{6.45}$$

Again, after any reasonable number of time constants, $1/p$, the positive exponential term will dominate the transient. Therefore the coefficient a_1 will determine the final direction toward which the circuit goes. It is given approximately by

$$a_1(i) \cong \frac{b}{2}\{[Q_i(0) - Q_j(0)] + C[E_i(0) - E_j(0)]\}, \tag{6.46}$$

in which $b \cong 1$. This important result substantially defines the conditions for proper trigger action. It means essentially that whichever transistor has the most charge stored both in its base and in its input impedance will be the one that is eventually turned ON. Note that in (6.46) voltages are given as initial magnitudes across the coupling network, and as far as turning ON a transistor is concerned, the most energy is stored in its input impedance when the voltage across the capacitor is near zero.

A number of particular cases of interest can be identified from (6.46). For example, suppose $C \to 0$, reducing the flip-flop to a simple resistor-coupled network. Then the trigger pulse must establish more charge in the base of the transistor that is to be turned ON. Assume this is the previously OFF transistor, and that recombination does not substantially reduce the amount of charge in the previously ON transistor during the short trigger pulse. Then the amount of charge transferred must at least exceed that in the base of the ON transistor. If the latter is at the edge of saturation, the supplied trigger charge must be $Q_T > \tau_N I_s$. Another case of interest is that of turning previously ON transistors OFF by means of the trigger pulse. Again the same amount of charge must be transferred during the trigger pulse (in this case in a direction to remove base charge), and the condition at the end of the trigger pulse is approximately $Q_1(0) = Q_2(0) = 0$. Therefore the stored charge remaining on the coupling capacitors entirely determines the subsequent direction of flipping. One of the disadvantages of triggering by inserting a pulse via the collector node is also made apparent by (6.46). For example, if it is desired to turn the ON transistor OFF, a positive current pulse can be inserted at the collector of the OFF transistor, say T_2. This pulse, passing through the coupling impedance Z, has the effect of changing both $E_2(0)$ and $Q_1(0)$. However, the change in $E_2(0)$ is in the wrong direction, and C must be large enough so that $CE_2(0) > Q_1(0) = Q_T$. Otherwise $E_2(0) = E_1(0) = 0$, $Q_1(0) = Q_2(0) = 0$, and (6.46) is indeterminate. This same problem shows up when turning ON the OFF transistor, it being necessary that the voltage across the coupling impedance not change by the difference between the two voltage states.

In addition, the trigger pulse must also supply most of the collector current of the ON stage for the pulse duration.

The effect of the triggering pulse on the charge stored in the coupling network can be eliminated by replacing the coupling resistor with a zener diode, such that a constant voltage always appears across the coupling capacitor. A circuit of this type that has been analyzed by Linvill [2] is illustrated in Figure 6.26. It possesses the added advantage of providing an antisaturation circuit. Analysis yields essentially the same criterion as (6.44) for proper regenerative behavior. In this case $\tau_1 = R_1 C$ is the time constant of the emitter network, the collector resistors acting as small dc bias sources.

Minimum and maximum C, and recovery. The foregoing discussion implies a minimum constraint on the value of the coupling capacitor, C, since resistor values in the remainder of the network are established by dc stability and power-level constraints. Thus $R_1 C = \tau_1 \geq \tau_N$ fixes the minimum C. Note that in the limiting case of $R \cong \beta R_1$, this is equivalent to requiring that

$$RC = \tau \geq \tau_a. \tag{6.47}$$

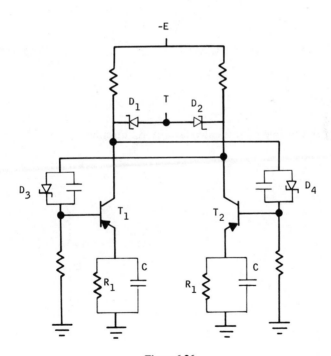

Figure 6.26

This same result was obtained in the preceding chapter, without regeneration present, for the case of cascade *RC* coupled decision elements.

Another view of this constraint is also illuminating. In order for rapid switching to occur—with or without regeneration—it is necessary that the voltage driving the coupling impedance not reach any clamp level that might be present. The circuit of interest is shown in Figure 6.27, in which V represents a voltage clamp level, and the diode to ground may be regarded as equivalent to the collector junction of the driving transistor. In the turn-ON case, it is necessary that $e(t)$, starting from about 0 V, not reach $-V$ before $Q(t)$, the base charge of the transistor, reaches some prescribed level such as saturation $\tau_N I_s$. If this constraint is not met, the turn-ON drive reduces to V/R, the dc value, and the remainder of the turn-ON transient may occur at a very slow rate. A similar argument applies to the turn-OFF case, when $e(t)$ must not reach 0 V before $Q(t)$ reaches zero. The small turn-OFF bias required for dc stability is not shown in Figure 6.27. As a first approximation, assume that the transistor base impedance is negligible. Then if the base charge obeys (6.38) and the time constant of the coupling circuit is chosen equal to the transistor active-region time constant ($\tau = \tau_a$),

$$\frac{\Delta Q(t)}{\Delta e(t)} = \frac{\tau_a}{R}, \tag{6.48}$$

regardless of the driving waveform. Let $\Delta Q(T) = \tau_N I_s$ and $\Delta e(T) = V$. Then

$$\frac{I_s}{\beta} = \frac{V}{R}, \tag{6.49}$$

which is compatible with the dc stability requirement.

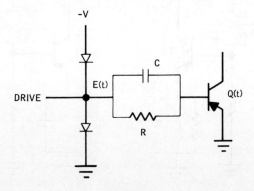

Figure 6.27

Still another way of estimating C is to assume the transient is so short that the charge transferred through R can be neglected. Then, with the voltage swing V available.

$$C \geq \frac{\Delta Q}{V} = \frac{\tau_N I_s}{V}. \tag{6.50}$$

With R determined by (6.49) for dc stability, $RC = \tau \geq \tau_a$.

The maximum value of C is determined by the repetition rate, or, putting it in the more usual way, C and tolerances impose one constraint upon the repetition rate. This is simply another way of stating that the period of a free-running multivibrator is established by the timing capacitor. Consider the circuit of Figure 6.23 with T_1 OFF, T_2 ON, $|e_{c1}| = |E|$, and the base voltage of T_1 just arriving at 0 V from a previous turn-OFF value. At this point regeneration occurs. If this period is short, the change in charge across C_1 is approximately that required to turn T_2 OFF, or $\Delta V_{c1} \cong Q_2(0)/C_1$. When T_2 turns OFF, its base impedance becomes large, and T_1 rapidly saturates. Waveforms are as shown in Figure 6.28. The period T_p is seen to be determined by the time constant of $C_1 = C_2 = C$ and R. Provided T_p is sufficiently long so that the collector voltage has reached stability $(-E)$ prior to the switching transient, the initial voltage appearing at the base of the OFF transistor following the switching transient is $E - Q(0)/C$. The period then becomes

$$T_p = \tau \ln \left[2 - \frac{Q(0)}{EC} \right], \tag{6.51}$$

in which $\tau = RC$. If stability has been reached, $Q(0)/EC \cong \tau_a/\tau$, and (6.51) can be written as

$$T_p \cong \tau \ln \left(2 - \frac{\tau_a}{\tau} \right), \tag{6.52}$$

in which it is required that $\tau > \tau_a$. With this constraint (6.52) is closely approximated by $T_p \cong 0.7\tau > 0.7\tau_a$. For any reasonable value of β, this means that $T_p \gg \tau_N$. Such a result also satisfies the initial condition that the transient at the OFF collector be completed before regeneration begins again.

In the usual multivibrator case $T_p \gg \tau_N$, and the above conditions can be easily satisfied. In the case of flip-flops, however, it is commonly desirable to achieve as high a repetition rate as the basic components will permit. Under these conditions, both the transient at the base of the recently turned-OFF transistor, and the transient base charge of the recently turned-ON transistor must be examined. Consider the circuit of Figure 6.29 and the following situation. At the instant the trigger pulse arrives, the voltage

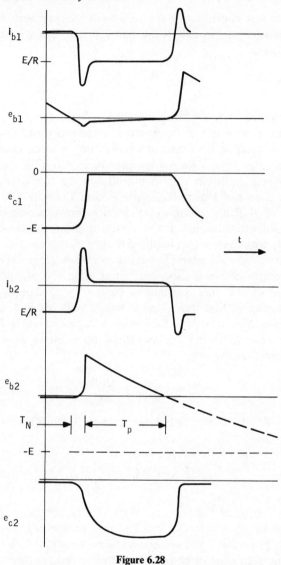

Figure 6.28

across the capacitor to the base of the OFF transistor has not yet quite reached its quiescent OFF value. Assume the latter is approximately 0 V. Let the charge on the coupling capacitor be $Q_o \equiv VC$ when exactly V appears across it, and let a change in voltage be regarded as a fractional charge on the capacitor, $\Delta_c Q_o$. During the same time that the voltage at the OFF base has been decaying

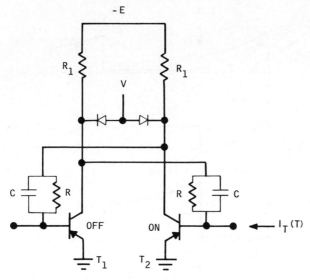

Figure 6.29

toward zero, the excess charge stored in the base of the ON transistor has also been decaying toward its quiescent value, which is that required to just maintain saturation. Waveforms are shown in Figure 6.30, in which Q_{ci} are the charges stored on the coupling capacitors, and Q_{bi} are the charges stored in the transistor bases. When the trigger pulse arrives, it is uncertain whether the excess base charge has reached zero, and to ensure that the ON transistor is completely turned OFF, the trigger pulse can be made larger than nominal. Let the maximum possible excess base charge be $\Delta_b Q_s$, with $Q_s \equiv \tau_N I_s$ being the saturation base charge. Then

$$Q_T = Q_s(1 + \Delta_b), \qquad (6.53)$$

with $Q_T = I_T T$ being the charge inserted during the trigger pulse to ensure that the OFF condition exists. As far as the following transient is concerned, however, the worst case arises when $Q_b = Q_s$. In this case, once Q_s has been removed from the base, most of the excess trigger charge will be deposited on the coupling capacitor attached to the previously ON base. It therefore takes on a possible total charge $Q_o + \Delta_b Q_s$. Switching will now occur, and because no further charge can be removed from the base of T_2, T_1 will go rapidly into saturation. During this time the charge on the coupling capacitor does not change appreciably, but a positive OFF voltage appears at the base of T_2. During the subsequent recovery time this charge will decay exponentially toward approximately 0 V with a time constant $\tau = RC$.

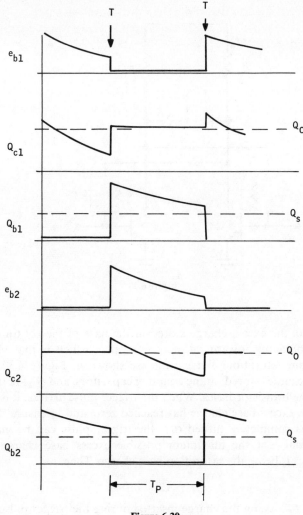

Figure 6.30

At the end of the recovery time T_p the net charge remaining must not exceed the limiting value $\Delta_c Q_o$. Thus

$$\Delta_c Q_o \geq (Q_o + \Delta_b Q_s)\, e^{-T_p/\tau}. \tag{6.54}$$

During the switching action, most of the charge on the capacitor to the base of T_1 is deposited in the base of T_1. The worst recovery problem will occur when this charge is a maximum, or when $Q_{c1} = Q_o$. However, to be sure that the OFF transistor turns completely ON during switching, the nominal value of Q_o may be selected so that more than the saturation charge is transferred. In particular, because the available capacitor charge may be less than Q_o by the amount $\Delta_c Q_o$,

$$Q_o \geq Q_s + \Delta_c Q_o. \tag{6.55}$$

This charge, placed rapidly in the base of T_1, will decay toward its saturation value Q_s with the time constant τ_a. Therefore, at the end of the recovery period, the limit on the excess base charge is given by

$$\Delta_b Q_s \geq (Q_o - Q_s) e^{-T_p/\tau_a}$$
$$= (\Delta_c Q_o) e^{-T_p/\tau_a}. \tag{6.56}$$

Together, (6.54) and (6.56) impose the allowable constraints on T_p. Note that if $I_s \cong \beta V/R$, the ratio of capacitor to base saturation charge is approximately $Q_o/Q_s \cong \tau/\tau_a$. Then, eliminating Δ_c between (6.54) and (6.56), we obtain

$$e^{-T_p(1/\tau_a + 1/\tau)} < \frac{\Delta_b}{\tau/\tau_a + \Delta_b} \cong \Delta_b \frac{\tau_a}{\tau}. \tag{6.57}$$

Similarly eliminating Δ_b, we obtain

$$e^{-T_p(1/\tau_a + 1/\tau)} + \frac{1}{\Delta_c} e^{-T_p/\tau_a} < 1. \tag{6.58}$$

With $\tau \cong \tau_a$, the second term is dominant in (6.58), implying that

$$e^{-T_p/\tau_a} < \Delta_c. \tag{6.59}$$

It is usually desirable to allow the circuit to approach dc conditions before retriggering occurs; that is, normal operation specifies Δ_b and $\Delta_c \ll 1$. Then with $\tau \cong \tau_a$, both (6.57) and (6.59) require $T_p > \tau_a$. Note that the previous analysis of the regenerative switching period T_N meant, in effect, that T_N will be several times τ_N. For good stability, the above results mean that T_p must be several times τ_a. Thus the ratio of the repetition period to the switching time is on the order of the current gain; $T_P/T_N \sim \beta$.

In practice, the ultimate limit on repetition rate is imposed by circuit component tolerances. The important effect of tolerances can be seen in a very simple way by considering only the contributions of R, C, τ_N, τ_a, and I_s to the charge remaining on the coupling capacitor at the completion of a switching transient. It has already been noted that $\tau \geq \tau_a$. Including tolerance effects, $\underline{RC} \geq \tau_a$, from which the nominal value of C must be

$$C \geq \frac{\tau_a}{R} \frac{1 + \Delta_\tau}{(1 - \Delta_r)(1 - \Delta_c)}, \tag{6.60}$$

in which Δ_i is the fractional tolerance on component or quantity i. It is also necessary that the capacitor charge transferred to the base during turn ON be some specified amount, for example, that corresponding to saturation, or $\underline{VC} \geq \tau_N I_s$. However, in fact, the amount of charge required in the base at the beginning of the transient may be at its lower limit, $\underline{Q} = \tau_N I_s$, and the transistor will saturate after this amount of charge is delivered. Therefore some charge Q_r remains on the capacitor and will also be delivered to the base, causing the transistor to be driven beyond saturation. This quantity is

$$Q_r = \overline{V}\overline{C} - \underline{Q}. \tag{6.61}$$

From the foregoing inequalities it can be found that the maximum value of Q_r is

$$\frac{Q_r}{Q_s} = (1 - \Delta_\tau)(1 - \Delta_v)\left(\frac{1 + \Delta_v}{1 - \Delta_v}\frac{1 + \Delta_c}{1 - \Delta_c}\frac{1 + \Delta_\tau}{1 - \Delta_\tau}\frac{1 + \Delta_r}{1 - \Delta_r} - 1\right). \tag{6.62}$$

For all tolerances small and equal, (6.62) is approximately $Q_r/Q_s \cong 8\Delta$, demonstrating the important influence even of modest tolerance values.

Additional storage. It has been shown that valid flip-flop action demands some minimum energy storage in the network external to the transistors. In the next section flip-flops in complete logic systems will be considered. In these systems additional storage, beyond that necessary for regenerative behavior is, if not mandatory, at least highly desirable. This additional storage has been provided for in a great variety of ways, sometimes so disguised that identification is difficult. By far the most popular techniques have been those involving capacitor storage and transistor base-charge storage. Other methods have, for example, included inductor storage and diode charge storage. Only the first two will be treated in detail.

To obtain a qualitative picture of the need for external storage, return to the logic problem which originally led to the requirement for storage internal to the flip-flop. This was the general counter logic problem. In particular, consider the second bit (or any higher-order bit) with the *RS* equations,

$$R_2 = Q_2 Q_1 T,$$

$$S_2 = \overline{Q}_2 Q_1 T.$$

A highly simplified flip-flop circuit, including negative AND diode gates for mechanizing the logic, is shown in Figure 6.31. The small dc OFF base bias is again omitted (or could be supplied by a normal, slightly positive value of *T*). Note that *T* could equally well be inserted via a diode at the gates, with constant I_T gate drive, as in Figure 6.18. It is drawn in this manner for simplicity. Assume that Q_1 is true (negative), T_1 is OFF, T_2 is ON, and T

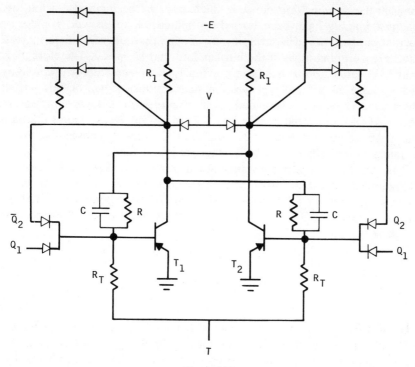

Figure 6.31

suddenly goes negative. Because both $\bar{Q}_2$ and Q_1 are negative, the base drive to T_1 increases from zero (OFF) to I_T through R_T. Because Q_2 is positive, however, the base drive to T_2 does not change (for ideal components). Now the collector current of T_1 begins to increase and to reduce the clamp current through its collector diode to V. However, the collector of T_1 is also driving n other gates. In the worst case, T_1 may be the only input to each of these gates that is going positive. Thus it must supply the full load current, which is nI_T. We can imagine that the collector of T_1 is slightly positive with respect to other inputs to its load gates, and that all of this current may be flowing through the clamp diode. Therefore i_{c1} must rise to about nI_T before e_{c1} comes off clamp and regeneration can begin (by duality, with positive AND gates, R_1 must be sufficiently small to draw nI_T, but T_1 may happen to be lightly loaded, in which case all of this current flows through the clamp diode). When T_1 comes off clamp, e_{c1} begins to rise. This will eventually be troublesome because there is feedback from the collector of T_1 to its gate input. However, if C is sufficiently large, the rise will be slow, and other events will occur first. In particular, the excess collector current will flow through the coupling impedance to the base of T_2, beginning to turn it OFF. T_2 must initially have been turned ON sufficiently to provide a collector current of about nI_T. However, it may happen that other inputs to its load gates are assuming most of the burden, and that $i_{c2} \ll nI_T$. Therefore T_2 is heavily in saturation. It has been previously demonstrated, for example by (6.18), that even without regeneration such a transient will occupy only a short time. Therefore T_2 soon comes out of saturation, and regeneration can occur. At this point, the trigger input T can be removed because the base charge in T_1 exceeds that in T_2, and with C properly chosen the stored coupling charge is favorable to T_1, thus satisfying (6.46).

The waveform sequence corresponding to the above events is sketched in Figure 6.32. The duration of the time period is given by the usual constant-current rise time for t_1, by (6.17) with $I_1 = nI_T$ corresponding to the base charge originally in T_2 for t_2, and by something on the order of several time constants τ_N for t_3. These are

$$t_1 = n\tau_N$$

$$t_2 = \tau_N\sqrt{n\beta},$$

$$t_3 = k\tau_N, \qquad k > 1. \tag{6.63}$$

Although these values are only qualitative, they do indicate the dependence of switching time on fan-out. Now suppose flip-flop Q_1 is not so adversely loaded. In particular, its equivalent transistor T_1 may be only lightly loaded, in which case its collector comes off clamp almost immediately. Similarly, its equivalent transistor T_2 may be heavily loaded, in which case it is at the edge of saturation, and regeneration occurs almost immediately. Therefore, the collector carrying the signal Q_1 will rise to 0 V at a time much less than $t_1 + t_2$ as given above. However, this has the effect of removing the drive signal I_T to the base of T_1 ($\bar{Q}_2$) before regeneration occurs—possibly before the collector of T_1 comes off clamp. This is equivalent to removing the trigger input too soon (see the dotted T waveform of Figure 6.32), and Q_2 will not flip, resulting in a failure.

This intolerable situation leads to the rule that, in general, a fast flip-flop cannot serve as an input to a slow flip-flop unless additional storage is provided to retain the information contained in the state of the gating network prior to switching time.

A great improvement is possible if the charge necessary for triggering can be stored, under gate control, over a relatively long time between trigger signals. In this way fan-out can be substantially improved—at the expense of bandwidth. In addition, it is usually possible to decouple the energy delivered for triggering from the gate control of the external storage. This is variously called *decoupling, pulse gating,* or *pulse steering.* It has the

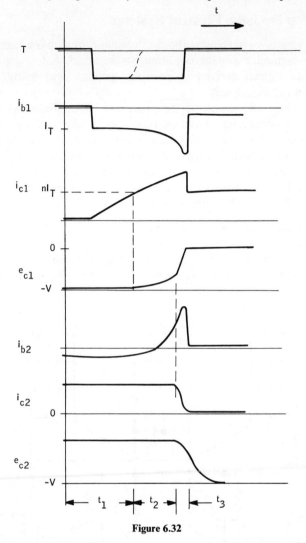

Figure 6.32

general effect of making flip-flops insensitive to events elsewhere in the network during the switching transient. It may also be described as converting flip-flop decision element networks into two-phase systems. During one phase (trigger input), no external energy is stored, but previously stored energy is delivered to produce selective flip-flop switching. During the second phase, no flip-flop switching occurs, while energy is selectively stored in the remainder of the network.

Flip-Flop Decision Element Systems

A huge variety of particular flip-flop configurations have been designed, each with particular system requirements in mind. A few representative examples have been derived in previous sections, and many others are described in the References.

Although it is impossible to do justice to all of the many variations, certain major categories of circuit type appear to cover the majority of particular cases. The remaining sections are devoted to three types of network, in simplified form, which are illustrative. They are (a) the *RCTL* flip-flop with capacitor storage of gating information, (b) the mixed *DCTL* and *RTL* flip-flop with transistor storage of gating information, and (c) the current-switching flip-flop and gating network.

RCTL flip-flop and gated capacitor. A highly simplified flip-flop and gating system is illustrated in Figure 6.33. Operation is as follows. Assume T_2 is ON and T_1 is OFF. A common timing signal or clock, T, is present on each gate input and is normally negative. If the other inputs to the right-hand

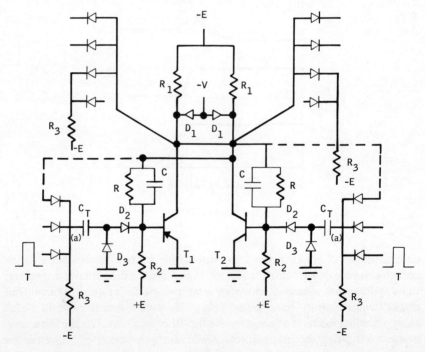

Figure 6.33

negative AND gate are also negative, C_T on the input to T_2 will charge to $-V$ through D_3 and R_3 to $-E$. If at least one input to the left-hand gate is positive, C_T on the input to T_1 will remain charged to about 0 V. When the trigger signal T_1 goes positive, points a are driven to 0 V. This does not affect the input to T_1, because the storage capacitor there is already at 0 V. However, because the charge on C_T at the input to T_2 cannot change instantaneously, D_3 becomes reverse-biased, D_2 becomes forward-biased, and most of the charge on C_T is delivered to the base of T_2, turning it OFF. When regeneration occurs, D_2 at the input to T_2 becomes reverse-biased. T may remain positive indefinitely without affecting circuit performance, but it may be returned to $-V$ (or a more negative value) at any time that flip-flop switching has been completed in the system. Note that during the time T is at 0 V, point a is effectively decoupled from the other input signals that may undergo various transitions but can only be more negative than a. When flip-flop switching is completed, T returns to $-V$, and gate storage begins. Again, if all other inputs are negative, the corresponding C_T will charge. Otherwise, it remains at about 0 V.

A variety of additional niceties are omitted from this circuit for simplicity. For example, D_3 may be returned to a small negative voltage rather than to ground (or the emitters may be returned to a positive voltage) for two reasons. One is to avoid shunting the base of an ON transistor via the diodes D_2 and D_3. The other is to provide a measure of noise immunity, so that small positive signals on the inputs will not cause D_2 to conduct. For this purpose, a small forward bias current may be desirable to insure that D_3 is clamped to its anode voltage. Base clamp diodes or other means for improving recovery time are also omitted for simplicity. It is assumed that signal levels are determined by collector saturation and the collector clamp to $-V$. Although the clamping function could be equally well performed in the gating structure, no substantial difference in circuit performance is obtained.

Note that the circuit as described is a simple *RS* flip-flop. If inputs to both storage capacitors C_T are negative simultaneously, incorrect operation may ensue. This situation can be avoided by connecting one of the gate inputs to the collector of the opposite transistor, as shown by the dotted lines. Then, even if all other inputs are negative on both sides, only that storage capacitor at the input to the ON transistor is allowed to charge. This has the effect of converting the circuit to a trigger (if T is the only other input, or if the third input to each side is tied to the same signal) or to a *JK* flip-flop (if the third input is denoted J on one side and K on the other). Thus the nomenclature of the flip-flop is rather arbitrary. If the dotted connection or its equivalent is made internally, the device is a *JK*; if the third input diodes are tied together internally, it is a trigger; if an additional diode is added to the gates and both of the above connections are made internally, it becomes a *RST* or *JKT*; and if none of these connections are made, it is a simple *RS*.

It is worth noting that with the addition of the gate storage capacitor C_T, there is no longer an absolute requirement for memory in the coupling impedance C. However, if C is eliminated, it is no longer possible to trigger by turning the ON transistor OFF with a minimum charge. If this is done, the base charges in the two transistors become $Q_1(0) = Q_2(0) = 0$, and with $C = 0$, (6.46) is indeterminate. Thus C_T must be large enough so that the ON transistor is not only turned OFF, but is also held OFF long enough for the previously OFF transistor to begin turning ON. As soon as a reliable difference in base charges exists, the trigger signal may be removed. The trigger circuit of a *DTL* flip-flop incorporating no internal storage is illustrated in Figure 6.34.

The most important quantity to determine in the circuit of Figure 6.33 is the minimum value of C_T required for reliable triggering. Other circuit parameters are determined by dc stability factors, including tolerance effects, and C is chosen to meet the regeneration criterion. For example, the maximum R_1 must be capable of keeping a transistor in saturation; thus $(E - V)/R_1 \gtrless I_s/\beta$. Maximum R must also fulfill this requirement, with a drive voltage V. R_2 must supply I_{c0}. For a selected saturation current level, fan-out is limited by $I_s \geq nI_3$, neglecting $I_1 = E/R_1$, with $I_3 = E/R_3$. R_1 is, in general, chosen to be much less than its maximum value to improve

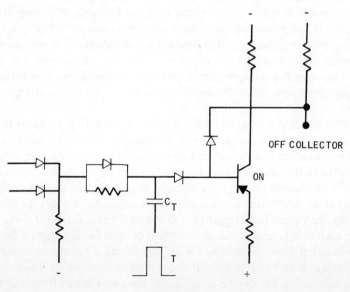

Figure 6.34

switching time, but this does not substantially affect the design or gain bandwidth limitations. With C_T determined, the repetition rate establishes I_3 and the consequent fan-out.

Because the coupling diodes between T and the base of the ON transistor are forward-biased when the trigger pulse goes positive, the base-emitter impedance of the transistor assumes importance. The equivalent circuit can be represented as in Figure 6.35a, in which R_D is the diode forward resistance, R_b is the total base resistance, R_e is the emitter bulk resistance, and C_d is the equivalent diffusion capacitance of the base. As long as the transistor is in the saturation or active region, the emitter junction voltage V_e is negative. However, its excursion is relatively small, and if C_d is large, it may be considered a constant during the switching period. In addition, usually $R_b \gg R_e$ in high-speed transistors. Normally one can select diodes so that $R_b \gg R_D$. An alternative way to handle this problem is to assume that the diodes

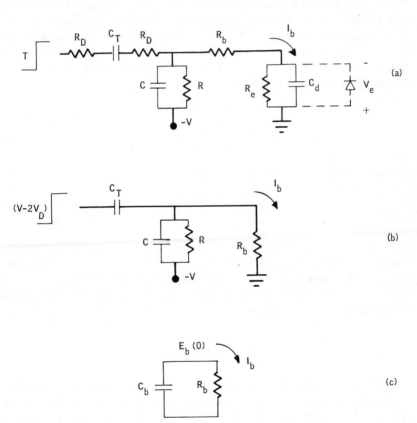

Figure 6.35

possess a fixed maximum forward drop V_D that subtracts from the available amplitude of the trigger pulse. If the trigger amplitude is V, the net available voltage is $V - 2V_D$, and the diodes may be otherwise neglected. The small charge required for the recovery of D_3 is neglected in the following first-order analysis, but it may be subtracted from that finally calculated or used to reduce the available voltage drive by Q_3/C_T, in which Q_3 is the charge necessary for diode recovery. In a detailed analysis, similar charges across the gate diodes must be accounted for. With these simplifications, the input circuit reduces to that of Figure 6.35*b*. Note that it is assumed that the collector of T_1 remains on clamp (at $- V$) during this transient. In fact, it does not matter whether or not this occurs. If T_1 comes unclamped, a pulse appears on its output (doing no damage because all gates are already being driven by T), and the trigger pulse sees a somewhat higher impedance shunting the base. Therefore more charge is delivered to the base. If it does not come off clamp, some charge is shunted to C. However, this charge is in the correct direction for further turn OFF, thereby improving the regeneration criterion (6.46). Note that both T_1 and T_2 are lightly loaded at this time because all gates are being driven by T; consequently, all output diodes are reverse-biased. Observing that generally $R \gg R_b$, the ac equivalent circuit reduces to that of Figure 6.35*c*, in which

$$C_b = C_T + C,$$

$$E_b(0) = \frac{C_T}{C_T + C}(V - 2V_D). \tag{6.64}$$

The base current as a function of these quantities is

$$I_b(s) = \frac{E_b(0)}{R_b} \frac{\tau_b}{\tau_b s + 1}, \tag{6.65}$$

in which $\tau_b \equiv R_b C_b$. During this time, provided T_1 remains on clamp, a continuous turn-ON current is also being supplied through R. Therefore, if regeneration never occurs, the base charge Q_2 will eventually return to its quiescent ON condition. Using a single-pole expression for base charge before regeneration, we obtain

$$Q_2(t) = Q_2(0) - I_b(0)\frac{\tau_a \tau_b}{\tau_a - \tau_b}(e^{-t/\tau_a} - e^{-t/\tau_b}), \tag{6.66}$$

in which $I_b(0) = E_b(0)/R_b$. This function is sketched in Figure 6.36 for values of $I_b(0)$ that do and do not produce cutoff. This figure may also be taken to represent I_{c2}, based on a single-pole approximation, or I_{c2} before a delay of τ_N, based on a double-pole frequency function.

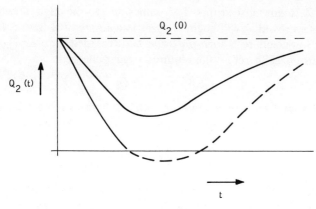

Figure 6.36

In the usual case, $\tau_a \gg \tau_b$, and $Q(t)$ reaches a minimum at a time

$$t_m \cong \tau_b \ln \frac{\tau_a}{\tau_b}. \tag{6.67}$$

At this time, again approximating under the condition $\tau_a \gg \tau_b$, the value of remaining charge is

$$Q(t_m) \cong Q(0) - \tau_b I_b(0). \tag{6.68}$$

This is equivalent to assuming that all of the charge transferred by a current waveform

$$I = I_b(0)\, e^{-t/\tau_b}$$

is retained for infinite time. On the other hand, an overly conservative estimate is based on the assumption that only the charge transferred during the first time constant of τ_b is effective for switching, or $0.63\, \tau_b I_b(0)$. Clearly, the range of possible values is not large.

The question remains as to the required value of $\tau_b I_b(0)$ for triggering action. If (6.46) applies and all components are ideal; that is, $Q_2(0) = Q_s = Q_o$, an infinitesimal trigger will theoretically initiate proper regeneration. Because of variations in loading, however, this is no longer valid. Regardless of the instantaneous action of T on the load gates, T_2 may previously have been heavily loaded. Therefore its base charge must be sufficient to supply I_s. If, in fact, it is lightly loaded (in the limit, no external loads, and $I_{c2} = I_1 = E/R_1$), it is heavily in saturation. If the trigger pulse does not bring it out of saturation, regeneration cannot occur because T_1 will never experience a change in drive. Suppose $E \gg V$, so that I_1 is approximately constant, and let $I = V/R$. It can be seen that if T_2 comes just to the edge of cutoff when OFF drive is removed, it will begin to turn ON again with a drive I. However, T_1 will be subjected to a turn-ON drive I_1. Provided $I_1 > I$, T_1 will then correctly turn ON. Therefore the required trigger charge is

$$\tau_b I_b(0) \cong Q_T \geq Q_s. \tag{6.69}$$

Using (6.64) and the approximation $Q_s/Q_o \cong \tau_a/\tau$, we see that the ideal limiting value of C_T becomes

$$C_T \geq \frac{Q_s}{V(1 - \Delta_D)} = \frac{\tau_a}{\tau} \frac{C}{1 - \Delta_D}, \tag{6.70}$$

in which $\Delta_D \equiv 2V_D/V$. The repetition rate can be derived from this result by observing that during the time the trigger signal is negative, C_T must charge to $-V$ if all gate inputs are also negative. For the ideal case, assume $E \gg V$ so that I_3 is approximately constant. Then

$$T_P = \frac{Q_T}{I_3} \geq \frac{Q_s}{I_3}. \tag{6.71}$$

But fan-out is limited by $n = I_s/I_3$, and $I_s = Q_s/\tau_N$. Therefore

$$T_P \geq n\tau_N. \tag{6.72}$$

This represents the ultimate limit on performance in the ideal case and for a complete cycle ($2T_P$) corresponds to a gain bandwidth product of

$$GB = \frac{n}{2T_P} = \frac{\omega_N}{2}. \tag{6.73}$$

At low frequencies, of course, fan-out is limited by β. Therefore, the frequency-response function of n/β appears as in Figure 6.37. This figure assumes that provision is made for rapid recovery, as previously discussed, so that flip-flop recovery time is not the limiting factor.

Transistor charge storage. In the last example gate information is stored temporarily in a capacitor, and no input or output amplification assists the basic flip-flop. Capacitors are awkward to manufacture in integrated circuit form, however, and additional amplification helps to overcome some of the tolerance problems associated with devices in that form. There is thus considerable motivation for constructing all-transistor and -resistor circuits in which the necessary memory function is performed by base-charge storage.

Control of the delivery of the triggering charge can be achieved by means of either a series or shunt arrangement. A simplified shunt trigger system is shown in Figure 6.38 for turning the OFF transistor ON. Operation is as

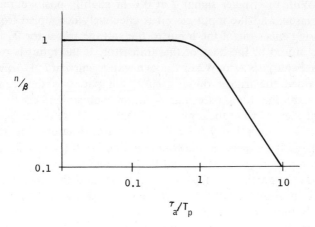

Figure 6.37

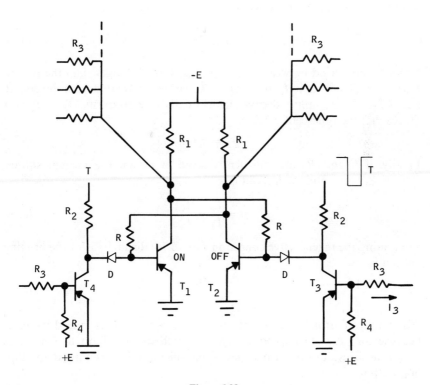

Figure 6.38

follows. With the trigger signal T at $0\,V$ or slightly positive, diodes D are reverse-biased, and the input circuit is effectively decoupled from the flip-flop. During this time, if the input to the storage transistor T_3 is negative, charge is stored in the base of this transistor. If the input is positive, the stored base charge is zero. When T goes negative, current I_T is drawn through R_2. Provided the trigger pulse is short compared to the recombination time constant, the stored base charge cannot change significantly during this time, regardless of what the input signal may do. Therefore, if Q_3, the stored charge in T_3, is such that $Q_3 \geq \tau_N I_T$, T_3 remains in saturation, and only the small voltage V_{ces} appears at the coupling diode D. If $Q_3 < \tau_N I_T$, the current that T_3 can pass is $Q_3/\tau_N < I_T$, and the remainder is available as a turn-ON drive to T_2. In particular, if $Q_3 = 0$, T_3 is OFF, and the charge delivered to T_2 is $I_T T_T$. If at the end of the trigger pulse the charge stored in the base of T_2 exceeds that stored in T_1, the regenerative criterion (6.46) is satisfied for proper reversal. This condition simply means that $I_T T_T = Q_T > Q_s$, the base charge originally stored in T_1.

Insight into the limiting behavior of this type of circuit can be obtained from a simple first-order analysis. The limiting trigger current that can be shunted by T_3 without producing a noise pulse input to T_2 is

$$I_T \leq \frac{Q_3}{\tau_N}. \tag{6.74}$$

Suppose the period between trigger pulses is much longer than the recombination time constant $T_P \gg \tau_a$. Then R_4 can be eliminated from the circuit, and if $I_3 = 0$, Q_3 simply decays exponentially to zero during T_P. If $I_3 > 0$, Q_3 reaches the quiescent value

$$Q_3 = \tau_a I_3. \tag{6.75}$$

Finally, if E and R_1 are sufficiently large to resemble a current source, $I_1 = E/R_1 = I_s$, and the fan-out is

$$n = \frac{I_1}{I_3} = \frac{I_s}{I_3}. \tag{6.76}$$

Combining the regenerative condition with (6.74) through (6.76), the limiting fan-out is

$$\frac{n}{\beta} \leq \frac{T_T}{\tau_N}. \tag{6.77}$$

This implies that n/β can exceed unity, if $T_T > \tau_N$. This is indeed the case, because for very slow-speed switching the theoretical limit to fan-out is $n < \beta^2$ as a result of the two stages of amplification between input and flip-flop output.

If, on the other hand, the repetition rate is such that $T_P \ll \tau_a$, R_4 must be present to remove charge remaining in T_3 when the input is zero. If the maximum charge which might be present is Q_3, the current I_4 passed by R_4 must, to a first order, be

$$I_4 \geq \frac{Q_3}{T_P}. \tag{6.78}$$

With the input negative and initially no charge stored, the input must be able to store sufficient charge for shunting the trigger, or

$$I_3 - I_4 \geq \frac{Q_3}{T_P}. \tag{6.79}$$

With I_1 again constant, combining (6.78) and (6.79) with the regenerative condition yields

$$n \leq \frac{T_P T_T}{2\tau_N^{2}}. \tag{6.80}$$

As might be expected for saturating circuits of this type, (6.80) indicates that fan-out drops off at the rate of 12 dB/octave at high frequencies; that is, (6.80) may be written in the form

$$n \leq \frac{T^2}{2\tau_N^{2}}\left(1 - \frac{T_P}{T}\right)\frac{T_P}{T}, \tag{6.81}$$

in which $T \equiv T_P + T_T$. The maximum fan-out therefore occurs for the symmetrical triggering waveform $T_P = T_T = T/2$, in which case $n \leq T^2/8\tau_N^{2}$.

The above simplified analysis does not take into account the relatively slow switching rate associated with the flip-flop itself when the cross-coupling capacitors are omitted. For the RTL circuit shown in Figure 6.38 without turn-OFF bias, turning OFF the previously ON transistor requires a time on the order of τ_a (simple charge decay toward zero). On the other hand, it is desirable to operate with $T_T < \tau_a$, because otherwise trigger charge decays significantly during the trigger pulse, and a much larger value of initial charge must be stored in order to deliver enough to the flip-flop. In view of these considerations it is usual practice to operate systems with $T_P > T_T$. Because the switching behavior of flip-flops in isolation has been considered in previous sections, attention is focused in the following discussions only on those basic limitations imposed by the input trigger-coupling circuits.

A series stored-charge trigger arrangement is shown in simplified form in Figure 6.39. In this case T_3 either blocks or passes the trigger pulse. Allowing for recombination, we write the current passed by T_3 during the trigger pulse as

$$I_T(t) = \frac{Q_3(0)\,e^{-t/\tau_a}}{\tau_N}. \tag{6.82}$$

Provided $T_T \ll \tau_a$, and allowing for recombination in both T_3 and T_2, we can write the amount of charge delivered during the pulse (approximately) as

$$Q_T \cong Q_3(0)\frac{T_T}{\tau_N}\left(1 - \frac{T_T}{2\tau_a}\right)^2, \tag{6.83}$$

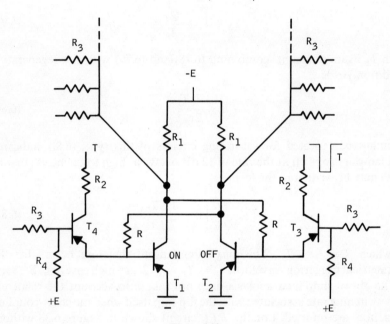

Figure 6.39

corresponding to a small correction to (6.74). The initial charge $Q_3(0)$ is established as before by the input signals prior to the trigger pulse.

Two disadvantages are apparent in the simplified systems of Figures 6.38 and 6.39. One is the saturating input *RTL* circuitry leading to the response limitations of (6.80) and (6.81). Another is that input signals are not decoupled from the charge storage device during triggering. Changes in input signals during triggering, if occuring sufficiently early, may cause noise pulses to appear at the flip-flop inputs. A more elaborate input system to overcome these disadvantages has been described by Davies, Seeds, and

Shou [19]. A simplified schematic diagram of the control portion of this circuit is given in Figure 6.40, in which a series triggering arrangement via the charge-storage transistor T_3 is used. The input to the storage transistor is a *DCTL* gate consisting of T_4, T_5, T_6, and other input transistors in parallel. Operation is as follows. If any one of the *DCTL* inputs is turned ON, the base of T_3 is driven to V_{ces}, and T_3 is turned OFF. When the trigger pulse arrives at the collector of T_3, it is blocked, and no change of state occurs. If all inputs to the *DCTL* gate are OFF, current I_3 through R_3 stores charge in the base of T_3. When the trigger pulse arrives it is passed to the base of T_2, and a change of state occurs. Note that the circuit can be arranged so that

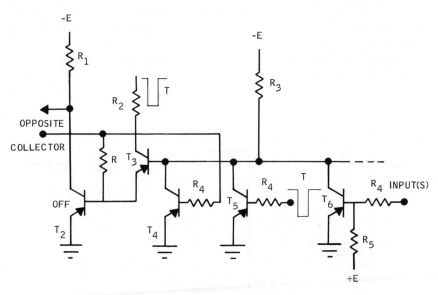

Figure 6.40

current does not flow in the emitter of T_3 during charging, for example, by making the trigger signal slightly positive during this period. The transistor T_4, connected to the opposite collector of the flip-flop, converts the circuit to a *JK* (or trigger) logic. This input behaves otherwise just as any other input signal. The transistor T_5 performs two functions: (a) it removes all input signals from the base of T_3 during the trigger pulse, thus decoupling the storage device from external signals, and (b) it removes charge from the base of T_3, so that recovery is not limited by the recombination time constant.

The location of the storage function is not restricted to occur at the input to the flip-flop; it can equally well be placed at the output, or at both input and output. A highly simplified arrangement for output storage is shown in Figure 6.41. Triggering is by turning the OFF transistor ON. When the trigger pulse goes negative, if all other inputs to the diode gate are also negative, current flows in R_3 from the base of the transistor. If at least one input is positive, the gate output is held positive, and no change in input occurs. The dotted connections convert the circuit from an *RS* to a *JK* or trigger.

Qualitative requirements for proper operation are as follows. The trigger pulse must last long enough to establish regenerative conditions in the slowest flip-flop. For the initial conditions shown in Figure 6.41, this amounts to requiring that $I_3 \cong E/R_3$ must, under worst-case conditions, supply during T_T more charge to the base of T_2 than is originally stored in T_1. During T_T, inputs to the gates must not change; otherwise, the trigger may be removed from the input to the previously OFF transistor, and/or the previously ON transistor may receive a noise pulse. This amounts to requiring that the

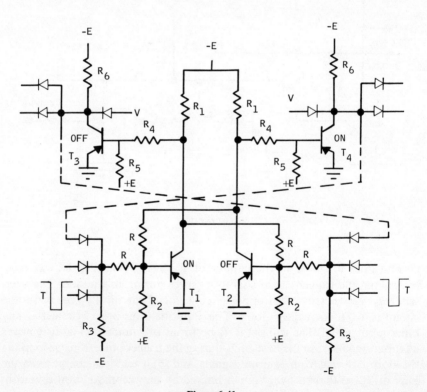

Figure 6.41

fastest flip-flop, which may be regarded as changing state instantaneously, not change the output of T_3 or of T_4 for a time at least exceeding T_T. In the case of the OFF output inverter T_3, this means that the turn-ON drive (through R_4) must not bring the collector of T_3 off clamp, even under zero-load conditions, before T_T is over. In the case of the ON output inverter T_4, this means that the turn-OFF drive (through R_5) must not bring T_4 out of saturation, even under maximum-load conditions, before T_T is over. These conditions are clearly antagonistic, and together they impose the ultimate limits on fan-out and switching time, as before.

Current-mode flip-flops. The high switching rates of current-mode decision elements suggest their use in high-speed flip-flops. Indeed, a number of similar versions of this form of memory element have been developed, operating at nanosecond rates. Typical examples are those described by Yourke [15], Walsh [20], Jarvis *et al.* [18], and Rapp and Robinson [17]. All follow essentially the same synthesis route as previously discussed, namely that of coupling two negative-gain decision elements in a bistable loop.

Synthesis can be readily visualized by considering the cascade current-mode decision element circuit of Figure 6.42*a*. The two decision elements operate on small voltage swings around 0 V. If the base is more negative than zero, the transistor conducts, and the shunt diode at the emitter is reverse-biased. If the base is more positive than zero, the shunt diode conducts the current through R_E to ground, the base-emitter junction is reverse-biased, and the transistor is OFF. Because only small voltage swings on the order of two diode (or junction) drops are required, the energy going into charging and discharging stray capacitance is minimized, thus increasing speed.

A feedback connection which converts the cascade into a memory element is shown dotted in Figure 6.42*a*. In addition, it is common practice to recognize that the shunt diodes and base-emitter junctions of the two circuits must play complementary roles, one of each pair being in conduction and the other cut off. Therefore the base-emitter junction of one decision element can provide the functions of the shunt diode for the other. Furthermore, the ground voltage reference for the shunts is arbitrary and can be placed at any point. The result of these considerations is the basic common-mode flip-flop of Figure 6.42*b*.

Some provision must of course be made for coupling trigger signals to this basic circuit. For this purpose, synthesis follows a line of reasoning similar to the above. In particular, cascade direct-coupled current-mode decision elements of the type of Figure 6.43*a* are feedback-coupled in the manner shown in Figure 6.43*b*. The input transistors are normally both OFF,

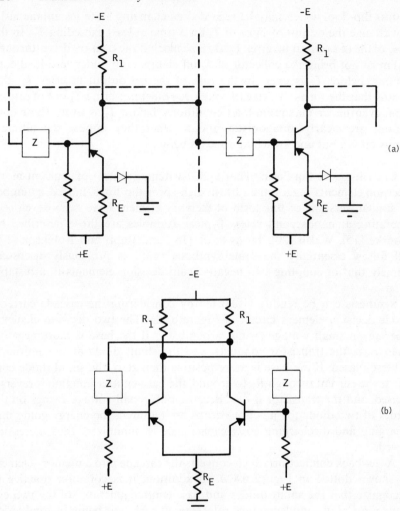

Figure 6.42

and one of the two cross-coupled transistors is ON. If one of the input transistors is turned ON, the other three are turned OFF. It is then the job of the input transistor to establish the required initial conditions for proper regeneration to occur when it is turned OFF again. In particular, if the input transistor is ON for a relatively long time and instantaneously goes OFF, the situation in the regenerative pair at the initial instant is that the stored base charge in each is zero. The regenerative direction is therefore determined by the stored charge in the cross-coupling impedance Z.

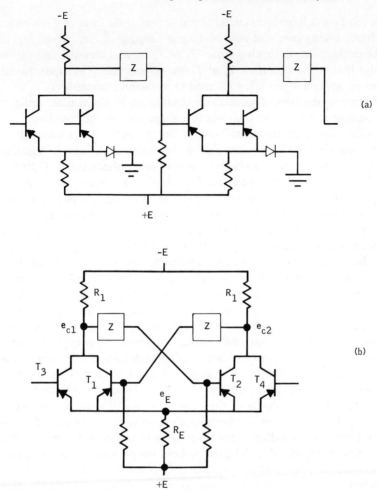

Figure 6.43

To be specific, suppose T_3 has been ON for some time and that its input signal begins to go positive. If signal swings are small, E/R_E is approximately a constant current, and the voltage e_{c1} remains nearly constant and relatively positive. Because both T_2 and T_4 are OFF, the voltage e_{c2} is also constant and relatively negative, being determined by the passive coupling network. Therefore the base voltage at T_2 is more positive than that at T_1; the base voltage at T_4 can be assumed to be sufficiently positive to keep T_4 OFF throughout the process. As the input to T_3 goes positive, e_E rises with it (small, approximately fixed drop across the base-emitter junction). As e_E

rises, the first voltage level encountered is that at the base of T_1, which then conducts, taking over the role of T_3 and keeping T_2 OFF. Thus, regardless of the previous state, turning either T_3 or T_4 ON and then OFF sets (or resets) the flip-flop to the condition T_1 or T_2 ON, respectively. The input transistors in this arrangement are often referred to as *pullover* transistors.

Because of the large degeneration introduced by the emitter resistor R_E, it is desirable to avoid further loss in voltage gain in the coupling network. For this reason, attenuationless coupling networks are commonly used, zener diodes being a convenient method for obtaining a fixed voltage change. Both series and parallel level-shifting networks, illustrated in Figure 6.44*a* and *b*, respectively, have been used. In both cases the constant current I_z is usually obtained from bias resistors to fixed voltage supplies, the network approximating a constant-current source for small signal swings.

More sophisticated arrangements than that of Figure 6.43*b* are necessary if feedback around the flip-flop is permitted, that is, if a trigger or *JK* flip-flop is to be mechanized. In this case additional external storage of the type previously discussed must be provided to furnish an unambiguous memory of the previous state of the flip-flop during switching. Because of the high switching speeds of current-mode circuits, only a relatively modest time delay or storage is required, and it is frequently hidden in the natural time delays of associated input and output amplifiers. A simplified version of Rapp and Robinson's circuit representing such a case is shown in Figure 6.45. Operation is as follows. The trigger signal T is normally negative, thus turning OFF the pullover transistors T_3 and T_4. When T goes positive, if one of the inputs is also positive, the corresponding gate turns OFF, which in turn turns ON either T_3 or T_4. Switching action is then as before. If, however, one or both of the flip-flop outputs serves as one of its own inputs, sufficient delay must be provided in the gates to hold the previous state. For example, if the dotted connections are made and T_2 is initially ON, T_3 will be turned ON when T goes positive. The condition for proper triggering is that T_2 be turned OFF, but that the feedback signal from its output to the set input be delayed sufficiently so that T_3 can remain ON while this occurs.

Some insight into both proper triggering behavior and switching speed can be obtained by examining the regenerative conditions, as before. The analysis is slightly more complicated, because of the common-mode coupling in addition to cross coupling. However, a particularly simple case occurs when the cross-coupling circuit is of the fixed level-shifting variety. Under this condition, no differential energy is stored in the cross-coupling impedances, and the direction of regeneration depends only on initial stored charge in the bases (and stray capacitances). Consider the circuit of Figure 6.46, with a fixed voltage V across Z. Let the base charges in the two transistors be given by (6.38), as before. Assume both transistors in the active region,

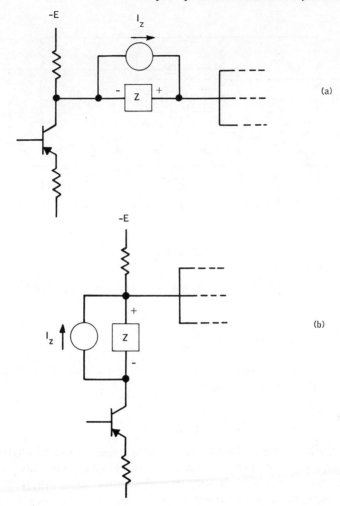

(a)

(b)

Figure 6.44

with initial stored charges of $Q_1(0)$ and $Q_2(0)$, respectively. Let the base current be given by the voltage across $R_b = r_b + R$, the total base resistance, not shown in Figure 6.46. If both base-emitter junctions are forward-biased (active region), the change in voltage across them will be small and can be incorporated into the fixed coupling voltage V. With this simplification, in terms of magnitudes,

$$I_{bi}(s) = \frac{1}{R_b}\left[E_{cj}(s) - E_E(s) - V\right], \tag{6.84}$$

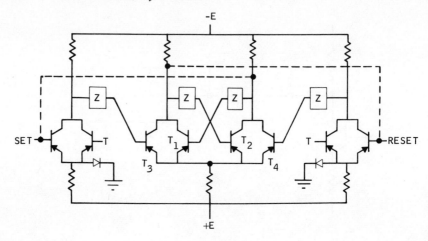

Figure 6.45

with $i \neq j$. The current in R_E is the sum of the emitter currents I_{e1} and I_{e2}, or

$$E_E(s) = \frac{R_E}{1 - \alpha_N(s)}[I_{b1}(s) + I_{b2}(s)].\qquad(6.85)$$

The net base current is the difference between the current through R_1 and the collector current into the same node. Thus

$$I_{bi}(s) = \frac{E - E_{cj}(s)}{R_1} - I_{cj}(s),\qquad(6.86)$$

in terms of magnitudes, with $i \neq j$. In combining these conditions with (6.38) it is convenient to eliminate the common-mode signal. For this purpose, we define $\Delta Q \equiv Q_1 - Q_2$. Using the first-order relationship between collector current and base charge, $I_c(s) = Q(s)/\tau_N$, we can reduce the above conditions to

$$\Delta Q(s)\left[1 - \frac{\beta}{(1 + R_b/R_1)(\tau_a s + 1)}\right] = \Delta Q(0)\frac{\tau_a}{\tau_a s + 1}.\qquad(6.87)$$

In the usual case, $\beta \gg 1$ and $R_b \ll R_1$. Using these approximations, we can reduce (6.87) to the simple time function

$$\Delta Q(t) = \Delta Q(0)\, e^{t/\tau_N}.\qquad(6.88)$$

This simply means that switching goes in the direction of the initial difference in stored energy, $\Delta Q(0)$, and at a positive exponential rate dominated by the α cutoff frequency.

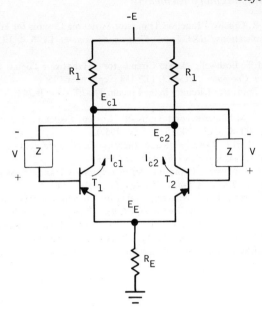

Figure 6.46

REFERENCES

[1] D. O. Pederson, "Regeneration Analysis of Junction Transistor Multivibrators," *IRE Trans. Circuit Theory*, **CT-2**, 171–178 (June 1955).

[2] J. G. Linvill, "Nonsaturating Pulse Circuits Using Two Junction Transistors," *Proc. IRE*, **43**, 826–834 (July 1955).

[3] D. Lynn and D. Pederson, "Switching and Memory Criteria in Transistor Flip-Flops," *IRE Trans. Circuit Theory*, **CT-7** (Supplement), 92–99 (August 1960).

[4] L. Hill, D. Pederson, and R. Pepper, "Synthesis of Electronic Bistable Circuits," *IEEE Trans. Circuit Theory*, **CT-10**, 1, 25–35 (March 1963).

[5] R. Henle and J. Walsh, "The Application of Transistors to Computers," *Proc. IRE*, **46**, 6, 1240–1254 (June 1958).

[6] T. R. Finch, "Transistor Resistor Logic Circuits for Digital Data Systems," *Proc. WJCC, May, 1958*, **T-107**, 17–22 (March 1959).

[7] J. B. Angell, "Direct-coupled Logic Circuitry," *Proc. WJCC, May, 1958*, **T-107**, 22–27 (March 1959).

[8] R. H. Baker, "Symmetrical Transistor Logic," *Proc. WJCC, May, 1958*, **T-107**, 27–33 (March 1959).

[9] D. P. Masher, "The Design of Diode-Transistor NOR Circuits," *IRE Trans. Electron Computers*, **EC-9**, 1, 15–24 (March 1960).

[10] W. J. Poppelbaum, "Flow Gating," *Proc. WJCC, May, 1958*, **T-107**, 138–141 (March 1959).

[11] L. P. Retzinger, "High-speed Circuit Techniques Utilizing Minority Carrier Storage to Enhance Transient Response," *Proc. WJCC, May, 1958*, **T-107**, 149–155 (March 1959).

[12] G. Prom and R. Crosby, "Junction Transistor Switching Circuits for High-speed Digital Computer Applications," *IRE Trans. Electron. Computers*, **EC-5**, 4, 192–195 (December 1956).

[13] G. Booth and T. Bothwell, "Logic Circuits for a Transistor Digital Computer," *IRE Trans. Electron. Computers*, **EC-5**, 3, 132–138 (September 1956).

[14] K. H. Olsen, "Transistor Circuitry in the Lincoln TX-2," *Proc. WJCC*, 167–171 (February 1957).

[15] H. S. Yourke, "Millimicrosecond Transistor Current Switching Circuits," *IRE Trans. Circuit Theory*, **CT-4**, 3, 236–240 (September 1957).

[16] C. D. Florida, "Part I—Description and Analysis of a Nonsaturated Circuit," *IRE Trans. Circuit Theory*, **CT-4**, 3, 241-249 (September 1957).

[17] A. Rapp and J. Robinson, "Rapid-transfer Principles for Transistor Switching Circuits," *IRE Trans. Circuit Theory*, **CT-8**, 4, 454–461 (December 1961).

[18] D. Jarvis, L. Morgan, and J. Weaver, "Transistor Current Switching and Routing Techniques," *IRE Trans. Electron. Computers*, **EC-9**, 3, 302–308 (September 1960).

[19] D. C. Davies, R. B. Seeds, and S. H. Shou, "An Integrated Charge-control J-K Flip-flop," *IEEE Trans. Electron Devices*, **ED-11**, 12, 556–562 (December 1964).

[20] J. L. Walsh, "IBM Current Mode Transistor Logical Circuits," *Proc. WJCC*, May, *1958*, 34–36 (March 1959).

EXERCISES

NOTE: Unless otherwise noted, transistor and power supply characteristics are to be taken as specified in the Exercises of Chapter 5.

6.1 The input trigger circuit to a flip-flop is shown in Figure P.6.1. Assume transistor parameters are those specified in the Exercises of Chapter 5. Find the following:

(a) maximum R,

(b) minimum C for turning the OFF transistor completely ON by means of the 4-V negative trigger pulse,

(c) minimum period between trigger pulses,

(d) maximum worst-case fan-out.

6.2 A simplified input trigger circuit using a transistor amplifier to a flip-flop is shown in Figure P.6.2. The trigger current of 1 mA, if passed by the input transistor, is sufficient to set the OFF flip-flop to ON in 0.2 μs. The period between trigger pulses is 0.4 μs. The input transistor has a value of $\tau_N = 0.04$ μs, and β is large (recombination is negligible at these frequencies). Find I (assume constant current) and R (assume $V_{be} \cong 1.0$ V) for proper operation of the trigger circuit.

6.3 The RS flip-flop of Figure P.6.3 is to be triggered by a negative-going timing pulse at the input, which instantaneously transfers the capacitor charge (if any) to the OFF transistor. It is necessary to transfer enough charge to keep the previously OFF transistor within the saturation region while the previously ON transistor turns OFF. Use standard transistor parameters (see the Exercises of Chapter 5).

(a) Find R_1, R_2, and R_3 for a triggering time of 0.5 μs.

(b) Find the required value of C.

6.4 The capacitor input in Problem 6.3 (Figure P.6.3) is to be driven by a diode gate as in Figure P.6.4a. The waveform of the timing pulse train is shown in Figure P.6.4b.

(a) What is the required value of R in Figure P.6.4, if the input capacitor to the flip-flop is conservatively selected to be 500 pf?

(b) What is the fan-out of the flip-flop of Figure P.6.3 if it drives diode gates of this type?

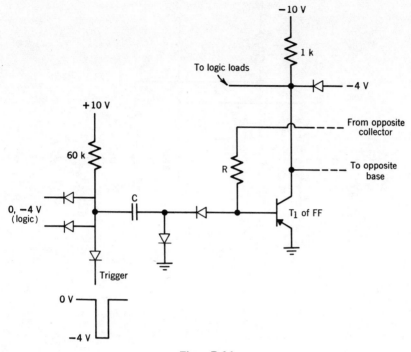

Figure P.6.1

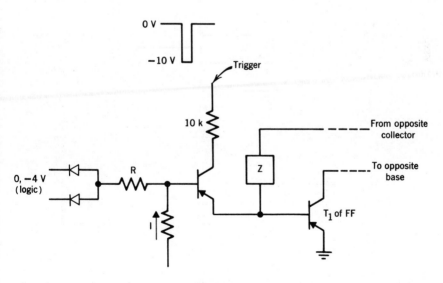

Figure P.6.2

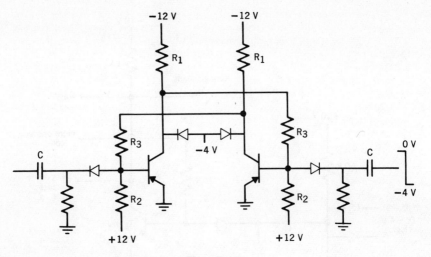

Figure P.6.3

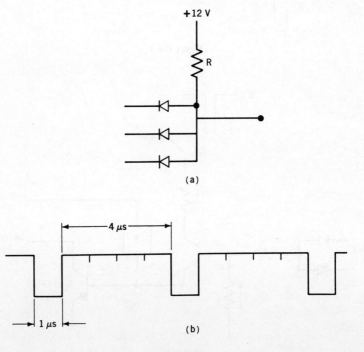

Figure P.6.4

6.5 The flip-flop of Figure 6.33 is to be triggered by a positive-going signal (clock) into the previously ON transistor, via the input capacitor C_T. The signal should be sufficient to turn the ON transistor OFF almost instantaneously. The flip-flop drives n gates similar to the input gate shown. Neglect OFF transistor leakage currents.

(a) Find R_1, R_2, R, C, and C_T for a triggering time of 0.2 μs. Use standard transistor and diode parameters.

(b) Find R_3 and flip-flop fan-out if the period between triggering pulses is 2 μs.

6.6 The RS flip-flop of Figure P.6.5 is to be triggered by turning ON the previously OFF transistor with a negative clock pulse. Assume that the clock pulse is short compared to transistor time constant. If the period between the trigger pulses is 2 μs and all voltage, capacitor, and resistor tolerances are ± 5 percent, find values for R, R_2, R_3, C, and C_T. In addition to those usually specified, the parameters are

$V_{be}(\text{ON}) = 0.4$ V (max),
r_b (base resistance) $= 100\ \Omega$,
$V_{ce}(\text{sat}) = 0.1$ V (max)
$I_{c0} = 20\ \mu$A (max),
V_D (diode drop) $= 0.8$ V (max).

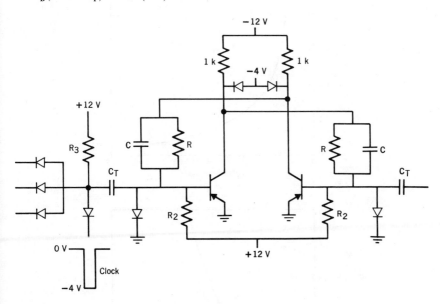

Figure P.6.5

6.7 A direct-coupled transistor flip-flop is illustrated in Figure P.6.6a.

(a) With no load, estimate V_{be} and V_{ce} of T_1, using standard transistor parameters.
The flip-flop is to be loaded at the collector of T_2 by direct coupling to the bases of similar transistors.

(b) If the characteristics of Figure P.6.6b and c describe the transistors, what is the maximum fan-out of the flip-flop?

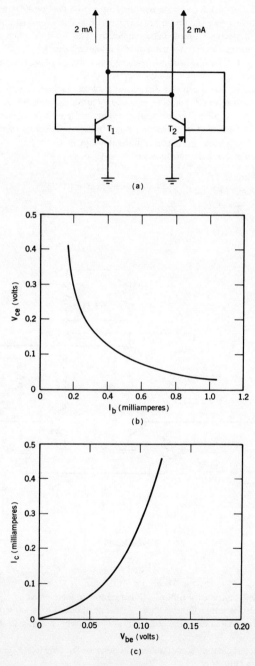

Figure P.6.6

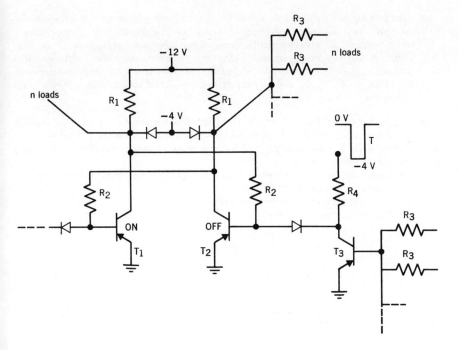

Figure P.6.7

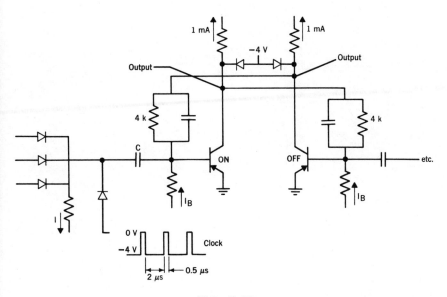

Figure P.6.8

6.8 The *RTL* flip-flop of Figure P.6.7 is to be triggered by turning the OFF transistor ON by means of a pulse that delivers enough charge to exceed that stored in the base of the previously ON transistor. The pulse is gated by means of the input NOR circuit shown. If the NOR transistor (T_3) is ON, the trigger pulse is shunted to ground via the collector of the saturated transistor (T_3). If the NOR circuit is OFF, the pulse is delivered to the base of the OFF transistor (T_2). Find R_1, R_2, R_3, R_4, and fan-out n of the flip-flop if the period between trigger pulses is very long.

6.9 The *RS* flip-flop of Figure P.6.8 is to be triggered by a positive-going clock pulse of 4 V amplitude, which will turn the previously ON transistor OFF. Assuming that the crossover capacitor is sufficiently large, and that the input trigger turns the previously ON transistor OFF almost instantaneously;
  (a) What is the maximum switching time of the flip-flop if standard transistors are used (assume I_B is negligibly small)?
  (b) What is the required minimum value of C?
  (c) What is the required value of I, and the fan-out?

Chapter 7

Magnetic Decision and Memory Elements

Any device that exhibits power gain is a potential candidate for mechanizing the decision operation in decision and memory elements. Semiconductors are certainly not the only materials possessing this property. In this chapter a second class of materials that can be made to exhibit power gain—namely, the ferromagnetic materials—is investigated.

Because of cost and speed disadvantages relative to semiconductor mechanizations, magnetic logic has found only limited use. Nevertheless, work continues in this area, motivated by the possibly attractive reliability of all-magnetic logic, and by the growing need for memory systems containing embedded logic. This chapter serves to introduce the basic design problems and set the stage for magnetic-core memories.

The Series Magnetic Amplifier

Perhaps the simplest magnetic circuit, and one that possesses the most similarity to other forms of amplifiers, is the conventional series magnetic amplifier. A basic form of this circuit is illustrated in Figure 7.1. Magnetic amplifiers of this form have been widely used in low-frequency power applications, at 60 and 400 c/s. In special versions they have been used in computer circuitry in the megacycle range.

Operation of the circuit is as follows. The power source e_1 is an ac carrier signal commonly of sine-wave form. The series diode permits power to be delivered to the load R_1 via the series winding N_1 only on positive half cycles of the carrier signal (full-wave versions of this circuit are common at lower frequencies). Depending on the magnetization state of the core at the beginning of each positive or power half cycle, more or less power is delivered to the load. The magnetic state of the core at the beginning of each power half cycle may be determined both by its state at the end of the preceding power half cycle, and by the signal e_2 applied to the input winding N_2 during the intervening negative half cycle. In general, the effect of both power and input signals on the state of magnetization of the core may extend over several cycles of the carrier frequency. Thus, in the most common forms of such a magnetic amplifier, the value of the power delivered to the load is allowed to build up in a transient manner over several cycles, in response to a change in dc or low-frequency input signal.

The state of magnetization of the core can, however, be completely established during a single half cycle by means of the signal on the input winding. This mode of operation was first described by Ramey [1], who pointed out its application to computer circuits.

Consider first the ideal square hysteresis loop as the characteristic of the core in Figure 7.1. Let the carrier power source e_1 be a square wave of amplitude E, as shown in Figure 7.2. The basic principles of circuit operation can best be seen by making two initial simplifying assumptions. First, let us assume that the diode in series with the output winding N_1 is a perfect diode and is always reverse-biased during negative half cycles of e_1. Reverse bias can always be assured by making e_1 unsymmetrical with respect to ground, so that on negative half cycles it is as large (in magnitude) as desired.

Second, let us assume that the input winding is always open-circuited during positive half cycles of e_1. This could also be accomplished with a series diode and a highly negative value of e_2 during positive half cycles of e_1.

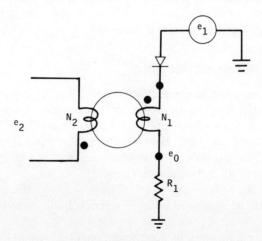

Figure 7.1

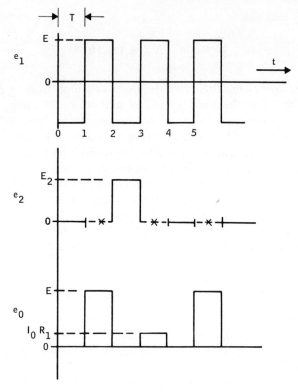

Figure 7.2

The half cycles during which the input is open-circuited are indicated by an x in the waveform of e_2 in Figure 7.2.

The main sequence of events can now be described as follows (see Figure 7.2). Assume the core is at $+\phi_s$ at t_0. During the initial negative half cycle from t_0 to t_1, $e_2 = 0$. No flux change occurs in the core because no voltage appears across it, and $\phi = \int e \, dt = 0$. Therefore at time t_1 the core is still at $+\phi_s$. When the positive voltage E is applied at t_1, no voltage can appear across the core, because it is already in saturation. Thus the entire voltage E appears across the load resistor, as shown by the e_0 waveform in Figure 7.2. During the period from t_1 to t_2, the core is at the point $\int e \, dt = \phi_s$, and $i = E/R_1$. At t_2, the carrier signal is removed, and the core returns to remanence at $+\phi_s$. During the negative half cycle from t_2 to t_3, a voltage E_2 appears across the input winding N_2 in such a polarity as to drive the core in the negative flux direction. The core is therefore driven down the hysteresis loop, and a current of magnitude I_0 flows in the input winding. If E_2 and N_2

are properly selected,

$$\int_{t_2}^{t_3} E_2 \, dt \geq N_2(2\phi_s) \times 10^{-8}, \tag{7.1}$$

and the core completely traverses the hysteresis loop from $+\phi_s$ to $-\phi_s$. At t_3, therefore, after the input is removed and before the carrier is applied, the core is at remanence at $-\phi_s$. Note that no current flows in the output winding between t_2 and t_3 because the series diode is reverse-biased. The output voltage is consequently zero. At t_3 the positive half cycle of the carrier is again applied to the core. This time, however, a voltage can appear across the winding because the core is not at positive saturation. The carrier voltage E, therefore, begins to drive the core in the positive flux direction, and a current I_0 flows in the load winding. The output voltage during this time is thus $I_0 R_1$. If E and N_1 are properly selected, the positive half cycle is just sufficient to drive to core from $-\phi_s$ to $+\phi_s$, and no more, during the period from t_3 to t_4; that is,

$$\int_{t_3}^{t_4} (E - I_0 R_1) \, dt = N_1(2\phi_s) \times 10^{-8}. \tag{7.2}$$

At t_4, therefore, after the voltage E is removed, the core is at the point $\int e \, dt = +\phi_s, i = 0$, the same point at which the sequence began. Any other desired sequence may follow.

There are several practical difficulties with the simple circuit of Figure 7.1. The first is the voltage $I_0 R_1$ appearing on the output during a half cycle when the core is switching. Ideally the output should be zero. In addition, the quantity $I_0 R_1$ is load sensitive, R_1 representing the particular loading configuration at a given time. To rectify this problem it is common practice to insert a clamping circuit in the load, as illustrated in Figure 7.3, with $I_1 > I_0$. If the core is switching, and passing a current I_0, the diode remains forward-biased by a current $I_d = I_1 - I_0$, and the output remains clamped to approximately ground by the small forward diode drop. If the core is saturated, the current rapidly builds up to a value greater than I_1, the clamp diode becomes reverse-biased, and the full output voltage appears at the load.

A second difficulty arises in connection with practical fan-out. Consider the arrangement of Figure 7.4 with the output of one core driving the input windings of n other cores. With a signal E, the n loads are being driven in the negative flux direction starting from an initial state at $+\phi_s$. However, we cannot expect all cores to be identical in total flux swing. Therefore, one core will reach $-\phi_s$ before the others do. When such a core reaches $-\phi_s$, it saturates and, ideally, no further voltage can appear across it. It will

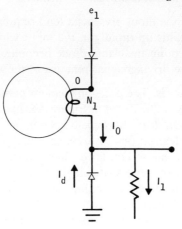

Figure 7.3

therefore short the driving signal to ground through a forward-biased diode. However, this has the effect of removing the driving signal from the other cores, which therefore may not reach $-\phi_s$. During a subsequent power half cycle, these cores will reach $+\phi_s$ too soon, violating the required relationship of the form of (7.2). To avoid this difficulty, it is the usual practice to insert a constant-current load and clamp diode D_4, in series with the input winding, as in Figure 7.5. If I_g is the current required to drive the input winding during switching, the clamp current is selected so that $I_2 > I_g$. During

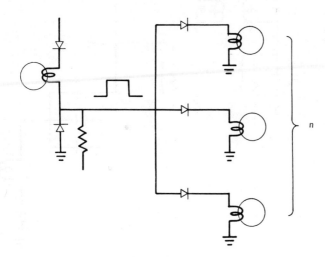

Figure 7.4

switching, therefore, the input presents a load requiring a drive I_g. If a core saturates, current builds up rapidly in the input winding until the value I_2 is reached, at which point the clamp diode becomes reverse-biased, and the load presented to the driving signal remains at the value I_2.

Basic network analysis. The decision element network resulting from the above considerations is illustrated in Figure 7.5. In analogy to the approach taken toward transistor decision element systems, let us analyze this circuitry using a number of idealizations. A finite core OFF impedance is adequately represented by an ideal square-loop characteristic, taking into account losses represented by the core switching constant S_w. Finite ON impedance must be attributed to the core winding inductance L_w. The assumed *B-H* loop is illustrated in Figure 7.6.

The requirement on a core in the high-impedance state is that the carrier signal drive the core from $-\phi_r$ to $+\phi_s$ during a power half cycle if the core was initially at $-\phi_r$. Defining the squareness ratio as $S_r \equiv \phi_r/\phi_s$ and neglecting diode forward drops, we obtain

$$ET \leq N_1\phi_s(1 + S_r) \times 10^{-8}. \tag{7.3}$$

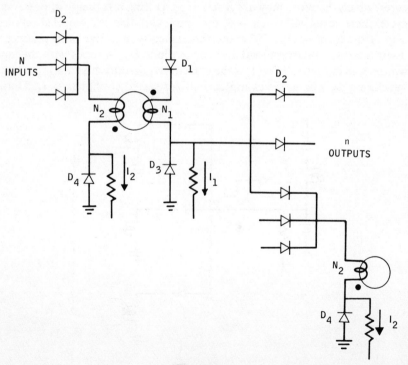

Figure 7.5

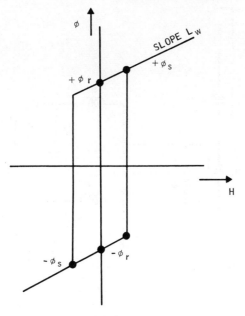

Figure 7.6

With respect to a core initially at $+\phi_r$ at the beginning of a power half cycle, a considerably different situation exists. Because of L_w, current in the winding N_1 cannot build up instantaneously. Because the clamp diode keeps the output at around 0 V until the diode becomes reverse-biased, the voltage across L_{w1} is nearly a constant E, and winding current increases approximately as

$$i_w \cong \frac{E}{L_{w1}}t = \frac{E}{N_1{}^2 L_w}t.$$ (7.4)

At what current level does the clamp diode become reverse-biased? Note that any tendency for the output voltage to rise will cause load diodes D_2 to become forward-biased. Consequently, some current will flow into the n loads during this time, but without causing switching to begin. Thus the load diodes may be regarded as forward-biased, and the equivalent load circuit during this time may be regarded as that of Figure 7.7. The saturated (or nonswitching) winding impedance, $N_2{}^2 L_w \equiv L_{w2}$, of the load cores is included in the circuit. Consider, now, the effect of i_w on this circuit. First, part of i_w goes toward reducing the forward diode current in D_3. Second, $1/n$ of the remainder goes through each L_{w2}, producing a voltage across D_4. However, D_4 must be reverse-biased before any significant voltage can

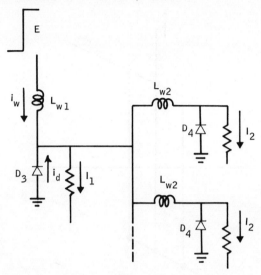

Figure 7.7

appear across it. Therefore, we may treat the circuit as if two time periods are involved. During the first, i_w reduces i_d to zero, requiring a time $T_1 \cong L_{w1}I_1/E$. Following T_1, with D_3 reverse-biased, i_w/n flows through each load, producing a voltage

$$e_2 = L_{w2}\frac{di}{dt} = \frac{L_{w2}}{L_{w1}}\frac{E}{n} = \left(\frac{N_2}{N_1}\right)^2\frac{E}{n}, \tag{7.5}$$

in which it is assumed that the output voltage is still small, and (7.4) continues to hold. The voltage e_2 is small for two reasons. One reason is the factor n; the other is the fact that $N_2 < N_1$, and winding inductance is proportional to N^2. Switching of the driven cores does not begin until current in each winding has reached a value $I_{02} \equiv I_0/N_2$, corresponding to the knee of the hysteresis loop. Therefore the second time delay is that required for i_w to build up from I_1 to $I_1 + nI_{02}$. Again using (7.4), we see that this is approximately $T_2 = nL_{w1}I_{02}/E$.

Finally, when the driven cores begin to switch, the input impedance rises sharply, and L_{w2} may be neglected. However, a finite time T_3 is still required for current to build up to the full switching value I_g, because L_{w1} is effectively in series with n parallel load circuits. Because the initial value of i_w at the beginning of this final transient is nI_{02}, the ac equivalent circuit is that of L_{w1} in series with R_{c2}/n. The equivalent core switching resistance, $R_{c2} \equiv N_2{}^2R_c$, is that value seen looking into the terminals of N_2. The

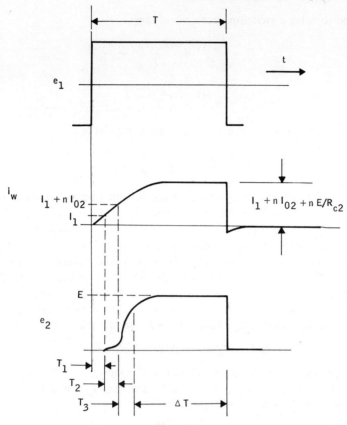

Figure 7.8

one-turn equivalent core resistance R_c is given by

$$R_c = \frac{0.4\pi\phi_s(1 + S_r)}{S_w p \times 10^8},$$ (7.6)

as derived in Chapter 3. The complete sequence of events is depicted schematically in Figure 7.8. It can be shown that the period T_3 is approximated by $T_3 = \tau \equiv nL_{w1}/R_{c2}$. Thus the voltage-time integral available for switching is equal to $E(\Delta T)$, with

$$\Delta T \equiv T - (T_1 + T_2 + \tau)$$

$$\equiv T - T_d,$$ (7.7)

in which T_d is defined as a pulse "delay" time. Using the derived values of T_1 and T_2, together with τ, and the limiting value of $I_1 \geq I_a$, the current required to drive a switching core, we obtain

$$T_d = \frac{L_{w1}I_1}{E} + \frac{nL_{w1}I_{02}}{E} + \frac{nL_{w1}}{R_{c2}}$$

$$= \frac{L_{w1}}{E}(I_a + nI_g). \tag{7.8}$$

In the above expression, the unit input drive current is

$$I_g = I_{02} + \frac{E}{R_{c2}}. \tag{7.9}$$

Equation (7.8) effectively means that the output power half-cycle pulse is delayed as if the carrier pulse were applied to the circuit of Figure 7.7 with $L_{w2} = 0$, $I_2 = I_g$, and $I_1 = I_a$. The waveforms of i_w, i_d, and the power half-cycle output for this circuit are illustrated in Figure 7.9.

We are now in a position to write the ON, OFF, and LOAD conditions for the network of Figure 7.5, and the core characteristic of Figure 7.6. The OFF condition has already been stated in (7.3). The ON condition may be taken to require that a core be driven from $+\phi_r$ to $-\phi_s$ by an input, produced by the output of a saturated core, of amplitude E and duration $T - T_d$, or,

$$E(T - T_d) \geq N_2\phi_s(1 + S_r) \times 10^{-8}. \tag{7.10}$$

The LOAD condition follows from (7.8). For simplicity, let us begin by neglecting I_a, which, because $N_1 > N_2$, must be less than I_g. Then $T_d \cong nL_{w1}I_g/E$. As a first estimate, take I_g as

$$I_g = \frac{H_a p}{0.4\pi N_2} = \frac{H_c p}{0.4\pi N_2}\left(1 + \frac{\tau_m}{T}\right). \tag{7.11}$$

This is clearly an optimistic estimate, because the time available for switching the input to a core is actually $T - T_d$, not T. However, the point of optimum operation is not greatly affected by refining this estimate. Combining (7.3), (7.10), and (7.11) at their limits and simplifying, we obtain

$$n \leq \frac{\phi_s'}{L_w I_0} \frac{1}{1 + \tau_m/T}\left[\frac{N_2}{N_1} - \left(\frac{N_2}{N_1}\right)^2\right], \tag{7.12}$$

in which

$$\phi_s' \equiv \phi_s(1 + S_r) \times 10^{-8},$$

$$I_0 \equiv \frac{H_c p}{0.4\pi}.$$

Fan-out can be optimized by setting the derivative of (7.12) equal to zero,

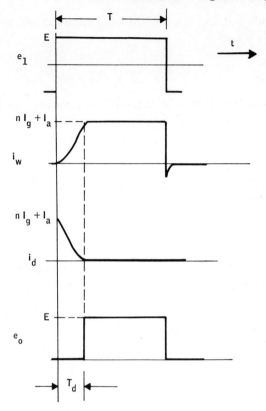

Figure 7.9

yielding $N_1 = 2N_2$. Note that $N_1 = 2N_2$ has the effect of matching impedances for maximum power transfer. Furthermore, from (7.3) and (7.10), $T_d \leq T(1 - N_2/N_1)$. Thus the limiting value of delay time is, for an optimum turns ratio, $T/2$.

The value of L_w can be found from the core characteristic of Figure 7.6 by an argument identical to that in Chapter 3. It is

$$L_{w1} = N_1^2 \left[\frac{0.4\pi\phi_s(1 - S_r) \times 10^{-8}}{H_c p} \right] \equiv N_1^2 L_w. \qquad (7.13)$$

Using this value in (7.12), together with the optimum turns ratio, the limiting fan-out becomes a function only of core parameters and switching time. Thus

$$n \leq \frac{1}{4} \frac{1 + S_r}{1 - S_r} \frac{1}{1 + \tau_m/T}. \qquad (7.14)$$

From this result, limiting system frequency response can be readily seen. At low frequencies, as $T \to \infty$,

$$n \le \frac{1}{4}\frac{1 + S_r}{1 - S_r} \equiv \frac{\gamma}{4}, \tag{7.15}$$

in which the quantity γ may be regarded as a measure of the OFF-to-ON impedance of the core. At high frequencies,

$$\frac{n}{\gamma} \le \frac{T}{4\tau_m}. \tag{7.16}$$

Defining the limiting gain bandwidth of the system at high frequencies as GB $\equiv n/T$, we have

$$\text{GB} = \frac{\gamma}{4\tau_m} \tag{7.17}$$

Note that (7.14) through (7.17) are functions of core parameters only. The system frequency response is plotted from (7.14) in Figure 7.10.

A more accurate appraisal of the optimum turns ratio can be obtained by noting that the time available for switching the driven cores is only $T - T_d$. Therefore (7.11) should be written

$$I_g = \frac{H_c p}{0.4\pi N_2}\left(1 + \frac{\tau_m}{T - T_d}\right). \tag{7.18}$$

Using this value of I_g for the LOAD condition, (7.10) for the ON condition, and (7.3) as before for the OFF condition, we obtain the limiting circuit fan-out,

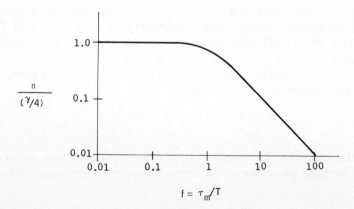

$$f = \tau_m/T$$

Figure 7.10

with L_w from (7.13), as

$$n \leq \gamma \left[\frac{(N_2/N_1) - (N_2/N_1)^2}{1 + (\tau_m/T)/(N_2/N_1)} \right]. \tag{7.19}$$

This function is plotted for various values of τ_m/T in Figure 7.11. Note that as $T \to \infty$, $\tau_m/T \to 0$, and the maximum value of the function is $\frac{1}{4}$, in accordance with (7.14). Note also that the value of N_2/N_1 at which n is maximum does not change greatly over a fairly wide range of τ_m/T.

Tolerance effects. Although tolerances on dc and carrier power supplies, resistor values, and nonideal diode characteristics all have a bearing on circuit performance, there is one tolerance that is of overriding importance.

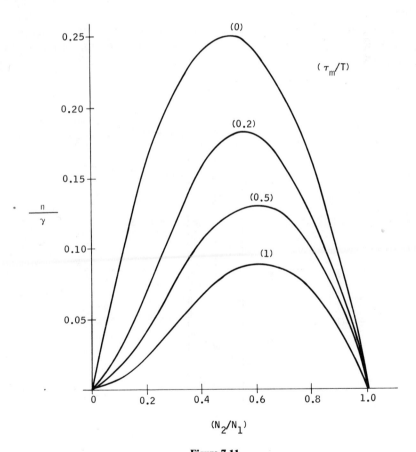

Figure 7.11

It is the range of minimum to maximum flux values from core to core. To understand the effect of this parameter, note that it is important for an OFF core that is switching from $-\phi_r$ to $+\phi_s$ during an output half cycle to produce no output signal. Therefore E, T, and N_1 must be selected so that even the minimum-flux core does not saturate. Let us denote the range of saturation flux values by $\phi_s(1 \pm \Delta_f)$. Then it is necessary that

$$\frac{ET}{N_1 \times 10^{-8}} \le \phi_s(1 + S_r)(1 - \Delta_f) = (\phi_s + \phi_r)(1 - \Delta_f). \qquad (7.20)$$

However, some cores in a system will possess the maximum flux range. When the voltage-time integral of (7.20) is applied to such a core (starting from $-\phi_r$), at the end of time T the core is left at a position below its value of $+\phi_s$ (or $+\phi_r$), as can be seen by the dotted path in Figure 7.12. If, on the subsequent input half cycle an input occurs and drives the core back down the loop to $-\phi_r$, no damage is done. If, however, no input occurs on the subsequent input half cycle, the core remains where it was left at the end of the output half cycle. Because this maximum core is not saturated, a voltage-time

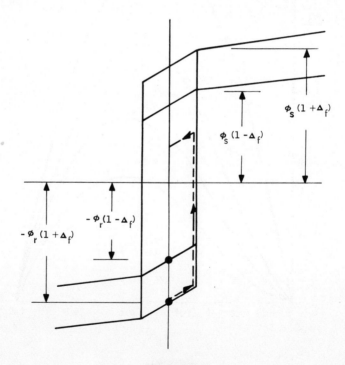

Figure 7.12

integral corresponding to the quantity

$$\Delta\phi = (\phi_s + \phi_r)2\Delta_f \qquad (7.21)$$

must be supplied before the core even begins the sequence of events described previously under the heading of delay time T_d. The effective delay time may therefore be severely increased. Two alternatives are available at this point: (a) accept the flux tolerance effect upon T_d and consequently on fan-out, or (b) provide a separate driving source that will ensure that the core is driven to its value of $+\phi_s$, and accept the inevitable effects of such a driving source upon fan-out. Although the choice between these alternatives is not immediately apparent, it has been found in practice that the second alternative furnishes a measure of noise immunity that is useful. The solution, therefore, is to provide a biasing voltage to the input winding via the clamp diode D_4, as illustrated in Figure 7.13. This voltage is commonly referred to as a *preset* voltage E_p. The circuit now operates as follows. With an input present, the voltage $E - E_p$ is applied to the input winding, and it must be sufficient to drive the maximum-flux core from $+\phi_r$ to $-\phi_s$. Thus

$$(E - E_p)(T - T_d) \geq N_2\phi_s(1 + S_r)(1 + \Delta_f) \times 10^{-8}. \qquad (7.22)$$

If an input is not present, the voltage E_p is applied to the winding in the

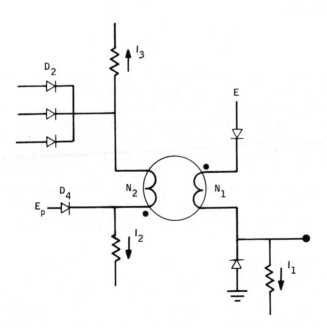

Figure 7.13

reverse direction and must be sufficient to drive the maximum-flux core by an amount corresponding to the maximum possible flux deficiency. This voltage is applied for the full duration of the period T. Thus from (7.21),

$$E_p T \geq N_2 \phi_s [2\Delta_f (1 + S_r)] \times 10^{-8} \tag{7.23}$$

Finally, the power half cycle must not drive the minimum core beyond saturation, in accordance with (7.20).

The quantity I_3 in Figure 7.13 requires some consideration. In the absence of an input pulse, the preset voltage E_p must furnish current to the core for changing flux by the amount (7.21). The time required to drive a core from negative remanence to positive saturation is given by $T = S_w/(H_a - H_c)$. The time required to drive a core only a fraction of this distance is consequently, to a first approximation, $T \cong (\Delta\phi/\phi)[S_w/(H_a - H_c)]$. Therefore

$$H_a \cong H_c \left(1 + \frac{\Delta\phi}{\phi} \frac{\tau_m}{T} \right) \cong H_c \left(1 + 2\Delta_f \frac{\tau_m}{T} \right). \tag{7.24}$$

However, in the absence of any other path the current corresponding to this field is blocked by the input diodes D_2. Therefore the path I_3 must be provided, of a magnitude $I_3 \geq H_a p/0.4\pi N_2$. This increases the load presented to an input signal, which must now supply both I_3 and I_2. The total gating current becomes

$$I_g \cong \frac{2I_0}{N_2} \left[1 + \frac{\tau_m}{2T}(1 + 2\Delta_f) \right], \tag{7.25}$$

which has the effect of approximately doubling the drive required at low frequencies, although making the response flat out to $\tau_m/T \cong 2/(1 + 2\Delta_f)$.

The design constraints are now as follows: (7.20) for output switching of an OFF core; (7.22) or (7.23) for the input switching of a core to the OFF or ON condition, respectively; and (7.8) for the LOAD condition, with the value of I_g from (7.25). Without belaboring the details, some idea of the effect of the preset bias can be obtained by solving the above equations, dropping second-order terms involving the incrementals. The result indicates that $N_1 = 2N_2$ remains nearly optimum, that low-frequency gain is reduced by the factor 2 appearing in (7.25), and that the frequency response is correspondingly extended.

Memory elements. In designing transistor memory elements we saw that a memory element flip-flop could be constructed by combining a pair of decision elements with appropriate time delays. In the case of carrier-operated magnetic decision elements a unit time delay is inherent in the operation of the basic element. Therefore no further consideration need be given to delays, and two magnetic decision elements, operating from appropriate carriers, can be directly combined to form the equivalent of the Eccles–Jordan flip-flop.

It is worth noting that although a unit delay is inherent in a single magnetic decision element, input and output times are incompatible. Therefore it is not possible to construct a one-element flip-flop without going through an additional carrier phase, or, alternatively, supplying passive signal storage.

A simplified two-element flip-flop based upon the foregoing type of decision element is illustrated schematically in Figure 7.14. C_1 and C_2 are two carrier power sources 180° out of phase, as indicated in the timing diagram of Figure 7.15. Operation is as follows. Suppose core 1 is ON (saturated), producing an output at 0_1 during positive half cycles of C_1. This output, applied to the input of core 2 turns the latter OFF. During the positive half cycle of C_2, therefore, core 2 traverses its hysteresis loop but does not produce an output at 0_2. With no input to core 1, it remains saturated, and therefore repeats the sequence. This pulse pattern, shown for the first two cycles of Figure 7.15, constitutes one state of the flip-flop; interchanging the roles of cores 1 and 2 constitutes the opposite state.

States may be interchanged by inserting a pulse input at the appropriate time. For example, if the input S occurs at the time indicated in Figure 7.15, core 1 is reset (turned OFF), and the subsequent pulse pattern is reversed, as shown. The system is thus equivalent to a simple RS flip-flop.

Parallel Magnetic Amplifiers

The series magnetic amplifiers analyzed in the foregoing section are characterized by several noticeable features. Among them are the

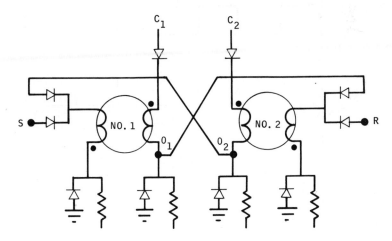

Figure 7.14

disadvantages of requiring a well-regulated carrier voltage source, the relatively large number of diodes used per stage (three plus coupling diodes; see Figure 7.13), and the relative expense of the high-quality core commonly used. Offsetting advantages are the relatively high gain and concomitant circuit flexibility. Where these advantages are not of great importance to the system, alternative approaches may be more attractive. One such application is in simple shift registers, for which frequent need arises. A shift register may be described as a one-input, one-output decision (and memory) element system. Hence circuit gain and flexibility are not at a premium, whereas low cost per bit, high reliability, and simple carrier circuitry are of importance.

Because of these considerations, magnetic amplifiers that are powered by constant-current, rather than constant-voltage, sources have found some

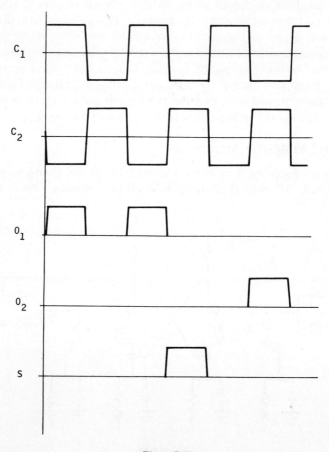

Figure 7.15

application. Because it is convenient to apply the constant-current drive by means of a winding in parallel with the output, such amplifiers have received the name *parallel magnetic amplifiers*. A simplified form of such an amplifier, illustrated in Figure 7.16, operates as follows. Time is divided into two periods. During the output period a constant-current pulse I_d of fixed duration flows in the drive winding N_d. In accordance with previous hysteresis loop conventions, if the core is initially at $+\phi_r$, the current pulse drives the core up the hysteresis loop to approximately $+\phi_s$, and only a small voltage-time integral corresponding to $\phi_s - \phi_r$ appears across the load. If the core is initially at $-\phi_r$, the core switches, and a large voltage-time integral corresponding to $\phi_s + \phi_r$ appears across the load.

During the second time period the current I_d is turned OFF, and N_d is open-circuited. During this time a signal may or may not appear across the input winding N_2. If a signal appears, the core is driven back down the hysteresis loop from $+\phi_r$ to $-\phi_s$. If no signal appears, the core is left at $+\phi_r$. For the winding polarities shown in Figure 7.16, the circuit is a NOT decision amplifier. Simple delay decision amplifiers can, of course, be produced by reversing winding polarity.

The effective driving signal is $N_d I_d$. Because the drive winding is the last to be designed in any system, no generality is lost in the following discussion if we let $N_d = N_1$.

One of the attractions of magnetic circuits is their potential for high reliability. Some motivation exists, therefore, to design all-magnetic circuits, avoiding the use of all other types of component with the possible exception of passive resistances. Because of the lack of either near-zero or near-infinite ON or OFF device impedance, however, such circuits possess many practical as well as analytical difficulties. These difficulties induce us to begin the investigation of parallel magnetic amplifiers by adding suitable nonlinear components, usually diodes, which correct major shortcomings. Let us

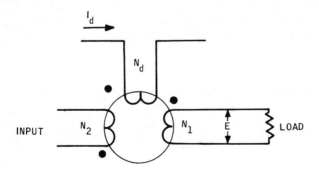

Figure 7.16

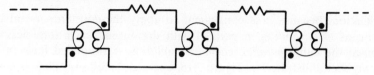

Figure 7.17

consider briefly some of the difficulties encountered without such nonlinear components.

All-magnetic logic. A one-input, one-output shift-register version of resistor magnetic logic is shown in Figure 7.17. Drive windings are not shown, it being implied that alternate cores are pulsed by the current I_d during alternate time periods. Such a system is called a *two-core-per-bit shift register*. This follows from the fact that to shift a bit of information from a core operated by one of the two clock cycles to another core operated by the same clock cycle requires an additional intermediate core and clock cycle.

To see some of the difficulties associated with this simple circuit, consider the events that occur when a core being driven by I_d is at its two possible initial states $-\phi_r$ and $+\phi_r$. In order to evaluate operation it is necessary to examine conditions up to two cores away from the driven core, in both directions. For this purpose, consider Figure 7.18, with C_0 initially at $-\phi_r$, and C_{-2}, C_0, and C_2 being driven by I_d. C_{-1} and C_{+1} are receiving no drive signal. The drive signal on C_0 causes it to switch toward $+\phi_s$, producing a voltage across N_1 in the polarity shown. A current flows in the direction of the solid arrow. This current (and the voltage appearing across N_2 of C_1) must be sufficient to switch C_1 from $+\phi_r$ to $-\phi_s$ before C_0 saturates. With C_1 being switched by the output from C_0, the voltages appearing across its input and output terminals must be of the polarities indicated, and the voltage across N_1 of C_1 will tend to drive a current in the direction of the solid arrow in the loop between C_1 and C_2. If C_2, which is being driven by I_d, is also switching, the voltage appearing across its input winding N_2 is in the polarity shown and will tend to drive a current through the output

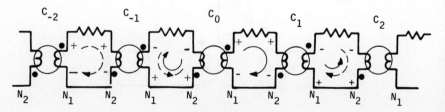

Figure 7.18

winding N_1 of C_1 in the direction of the dotted arrow. However, this current is in a direction to aid the desired switching of C_1, so for worst-case conditions C_2 should be considered initially saturated. Although C_2 will produce a small voltage across its input N_2 even when initially saturated, it is conservative design practice to ignore this slightly beneficial effect. In fact, it is most conservative to regard the saturated core C_2 as a short, thereby placing the greatest possible load on C_1.

Working now to the left from C_0, we see that its switching will produce the voltage indicated across its input winding N_2, and current will flow in the direction of the solid arrow in the loop between C_{-1} and C_0. Core C_{-1} is not being driven by I_d, but C_{-2} is. If C_{-2} is switching, voltages as shown will appear in the loop between C_{-2} and C_{-1}, and C_{-1} will be switching in a manner similar to C_1. The indicated voltage will therefore appear across the output winding of C_{-1}, tending to drive current in the direction of the dotted arrow in the loop between C_{-1} and C_0. However, this opposes the solid-arrow current and therefore reduces the load on C_0. Note that this reduction in load does not directly affect the switching of C_1, the driven core. It does, however, reflect into the drive current required for switching of C_0 and C_1. For maximum drive, therefore, let C_{-2} be initially saturated and producing a near-zero output. Therefore it may also be regarded conservatively as a short circuit.

Based on these considerations, the worst-case circuit reduces to that shown in Figure 7.19. Reflecting all impedances into the windings of C_0, and replacing cores (temporarily) by their equivalent current generators and loss resistors, we obtain the equivalent circuit of Figure 7.20. Note that everything to the left of C_0 represents an impedance in parallel with the load (C_1) and therefore enters only indirectly into the switching of C_1. Reflect this impedance into N_1 of C_0, introduce the drive current I_d, and include the effect of leakage inductance in the equivalent circuit of C_2. The resulting equivalent circuit is shown in Figure 7.21. I_d and all impedances in the circuit are fixed except Z. The lowest value of Z is given by the circuit to the left of C_0 in Figure 7.20. If C_{-2} is switching, however, a somewhat higher

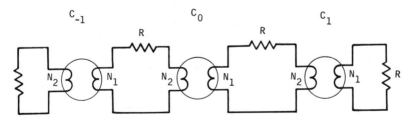

Figure 7.19

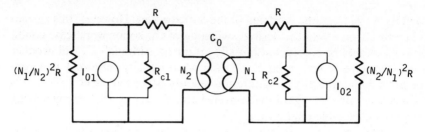

Figure 7.20

value of Z results. For the moment consider Z fixed, so that I_d flowing into the total circuit of Figure 7.21 produces some voltage E across N_1. With this simplification, it is more convenient to think in terms of E applied to the circuit, and to consider what flux changes occur in C_0 and C_1.

With E across C_0, the flux change in C_0 versus drive-pulse duration T is proportional to ET/N_1. For a step voltage E applied to the load of Figure 7.21 we have seen that the response can be divided into two periods. During T_1, current builds up from 0 to I_{02}, and $e_2 \cong 0$. Thereafter, voltage builds up as $1 - e^{-t/\tau}$. For reasonably large driving signals, the voltage-time integral across C_1 is therefore approximately

$$\int_0^T e_2 \, dt \cong E_2(T - T_d) \cong E_2[T - (T_1 + T_2)], \qquad (7.26)$$

in which

$E_2 =$ voltage across C_1 after it has been switching for some time,

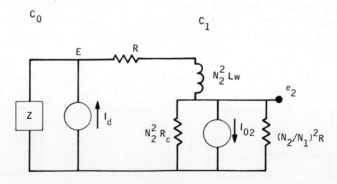

Figure 7.21

$$\cong E\left(1 - \frac{I_{02}R}{E}\right) \frac{1}{1 + R/R_{p1}}$$

$$T_1 \cong \frac{L_{w2}I_{02}}{E},$$

$$T_2 \cong \tau = \frac{L_{w2}}{R + R_{p1}},$$

$$R_{p1} = \text{parallel resistance of } R_{c2} \text{ and } \left(\frac{N_2}{N_1}\right)^2 R.$$

Denoting flux changes (times 10^{-8}) by ϕ_i, with i corresponding to the core numbering of Figure 7.19, the maximum possible flux change in C_0 is $\phi_0 = ET/N_1$. Substituting for the quantities in (7.26) and assuming a fixed operating level E, we see that the amount of flux switched in C_1 at time T is

$$\phi_1 = \frac{N_1}{N_2}\left(1 - \frac{I_0 R}{N_2 E}\right) \frac{1}{1 + R/R_{p1}}$$

$$\times \left[\phi_0 - \left(\frac{N_2 L_w I_0}{N_1} + \frac{N_2{}^2 L_w E}{N_1 R + N_1 R_{p1}}\right)\right]$$

$$\equiv G_f(\phi_0 - \phi_{Lf}), \tag{7.27}$$

in which all quantities are core or design parameters. In its simplified form (7.27) means that ϕ_1, the flux switched in C_1, is related to that switched in C_0 by a characteristic forward gain G_f and a forward flux loss ϕ_{Lf}. The latter is equivalent to the delay time, because (7.27) is a valid estimate only for $T > T_d$.

The importance of the form of (7.27) is that it is precisely the characteristic needed for reliable information transfer. It is sketched in Figure 7.22 and can be described as follows. If ϕ_0 is small, the amount of flux transferred is $\phi_1 < \phi_0$, and a 0 is propagated. If ϕ_1 is large, $\phi_1 > \phi_0$, and a 1 is propagated. The limiting design requirements are now clear. The minimum transfer flux associated with a 1 must lie above the point at which (7.27) crosses the unity-gain line. Similarly, the maximum transfer flux associated with a 0 must lie below the same point.

The transfer of a 0 can be investigated by considering the flux switched in core C_{-1} during the above process of transferring a 1 from C_0 to C_1. This "back path" flux will be subsequently transferred to C_0 from C_{-1} during the next pulse time. If the amount of flux in C_{-1} is less than the crossover point in Figure 7.22, a lesser amount will be switched in C_0 during the next pulse, and a 0 will therefore propagate. In this case the equivalent

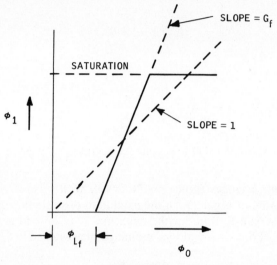

Figure 7.22

circuit under consideration is shown in Figure 7.23, with a back voltage of $E(N_2/N_1)$ appearing across the output winding of C_{-1}. Because C_{-2} should not be producing any output during this time (for 0 transfer), it is simply represented by the load R reflected into the output of C_{-1}. Repeating the same analysis as above, we see that the flux transferred back to C_{-1}

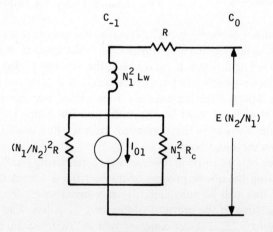

Figure 7.23

from C_0 is

$$\phi_{-1} = \frac{N_2}{N_1}\left(1 - \frac{I_0 R}{N_1 E}\right)\frac{1}{1 + R/R_{p2}}$$

$$\times \left[\phi_0 - \left(L_w I_0 + \frac{N_1 L_w E}{R + R_{p2}}\right)\right]$$

$$\equiv G_r(\phi_0 - \phi_{Lr}), \tag{7.28}$$

in which R_{p2} is the parallel resistance of $N_1{}^2 R_c$ and $(N_1/N_2)^2 R$ in Figure 7.23. This is an encouraging result, because it indicates that the flux transferred in the reverse direction can be made smaller than that transferred in the forward direction, a necessary condition for proper operation. Clearly, a desirable situation is one in which $\phi_1 \gg \phi_{-1}$, or

$$\frac{G_f}{G_r} \gg \frac{1 - \phi_{Lr}/\phi_0}{1 - \phi_{Lf}/\phi_0}. \tag{7.29}$$

These quantities can be evaluated as follows. Taking the ratio of (7.27) to (7.28), we obtain

$$\frac{G_f}{G_r} = \left(\frac{N_1}{N_2}\right)^2 f(p), \tag{7.30}$$

in which $f(p) < 1$ is a function of the parameters E, T, R, core material properties I_0, R_c, L_w, and turns N_1 and N_2. Its value is less than one, but by choosing the turns ratio $N_1/N_2 > 1$ the ratio of forward to reverse gain can be made sufficiently large. Similarly, an examination of the ratio ϕ_{Lr}/ϕ_{Lf} indicates that it is nearly equal to (N_1/N_2), and that the right-hand side of inequality (7.29) will be less than 1 for $N_1 > N_2$.

The flux transfer situation corresponding to the foregoing analysis is depicted in Figure 7.24. The realizability of (7.29) shows that indeed an operating point can be found (dotted line) such that forward transfer is amplified and reverse transfer attenuated. Thus, in principle, no more non-linearity than that introduced by the magnetic core itself is required, and simple all-magnetic logic is feasible.

The analysis of complete networks of this type is difficult, however, because of the extensive coupling permitted. More important, considerably improved performance can be obtained—at the expense of additional components—by eliminating undesirable coupling.

Nonlinear coupling. Unwanted coupling between cores other than that involved in information transfer can be avoided by the arrangement of Figure 7.25. Assume C_0 contains a 1 (is switching from $-\phi_r$ to $+\phi_s$ under

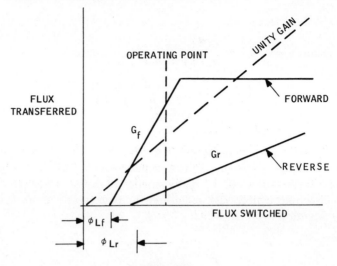

Figure 7.24

the influence of I_d in its drive winding). The drive winding on the other phase is an open circuit. Then a positive voltage appears across N_1 of C_0, driving current in the forward direction through diode D_3 and reverse-biasing D_4. Core C_1 is therefore caused to switch from $+\phi_r$ to $-\phi_s$. During switching of C_1, the voltage produced at its output winding N_1 is blocked by reverse-biased diode D_5, regardless of whether C_2 is switching or not. Switching of C_0 also produces a voltage at its input winding N_2. The resulting current is in the forward direction through D_2. Therefore only the small forward drop across D_2 can appear across the output winding N_1 of C_{-1}. This voltage is generally insufficient to produce any significant change in the state of C_{-1}.

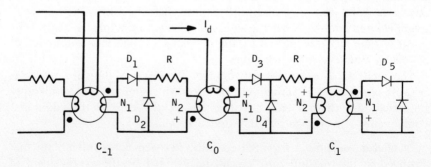

Figure 7.25

In addition if C_{-2} (and therefore also C_{-1}) is switching, diode D_1 is reverse-biased.

The net effect of the diodes, therefore, is to decouple the cores beyond those involved in the information transfer. Begin the analysis by considering the transfer of a 1 after switching has been initiated in both cores, during which time the effects of the winding inductance L_w can be neglected. Consider the magnetizing inductance L_m sufficiently large so that its effect may be treated as second order; that is, the actual voltages appearing across both C_0 and C_1 will possess a "droop" determined by the appropriate time constant, in the same manner as standard pulse transformers. Let us regard this effect as a correction to an otherwise square pulse response. Under the above conditions the equivalent circuit during switching reduces to that of Figure 7.26a, in which diode drops are neglected. When core parameters are inserted and all impedances reflected into the output winding of C_0, the result is Figure 7.26b.

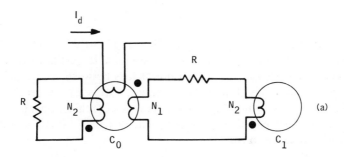

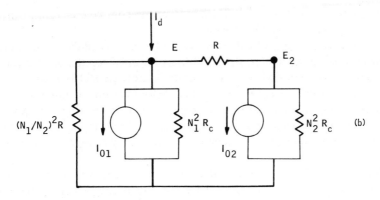

Figure 7.26

From Figure 7.26*b* the relationship between the voltage appearing across C_0 and that appearing across C_1 is

$$E_2 = \frac{E - I_{02}R}{1 + R/N_2{}^2 R_c}.$$

(7.31)

If the switching of the two cores proceeds at an identical rate, they must reach saturation at the same time T, and the voltage must also be related by

$$\frac{E_2}{N_2} = \frac{E}{N_1} = \frac{\phi}{T}.$$

(7.32)

Using this in (7.31) and substituting core parameters for I_{02} and R_c, we obtain

$$N_2{}^2 - N_1 N_2 + R\frac{I_0 T}{\phi}\left(1 + \frac{\tau_m}{T}\right) = 0.$$

(7.33)

Equation (7.33) has real solutions only for values of R such that

$$R \leq \frac{1}{4}\frac{N_1{}^2 \phi}{I_0 T(1 + \tau_m/T)} = \frac{1}{4}\frac{E}{I_{a1}} \equiv R_m.$$

(7.34)

In the limit, with equality prevailing in (7.34), $N_2 = N_1/2$. This result is similar to that for series magnetic amplifiers. It specifies that the optimum impedance match for minimum power loss between source and load is the value that makes both impedances equal; that is, EI_{a1} is precisely the power consumed in switching C_0. The identical power is also consumed in switching C_1. If the resistor value specified by equality in (7.34) is selected, the power consumed in the resistor is

$$P_R = \frac{(E/2)^2}{R} = EI_{a1} = P_{core}.$$

(7.35)

The minimum total power is consumed in the complete circuit when R is selected to have its maximum value.

The foregoing result was derived on the basis of a number of simplifying assumptions. In particular, the conditions required for the proper transfer of a 0 have been ignored. This necessitates a more careful analysis.

When the drive current is first applied, before currents in either core have an opportunity to build up to the value necessary for switching, the equivalent circuit may be drawn as in Figure 7.27. Because the cores have not yet begun to switch, internal losses and transformer action may be regarded as inoperative.

Simple considerations show that this circuit cannot possibly operate correctly with a step function of drive current. If I_d is a step, at the initial instant it must divide inversely with respect to the winding inductances

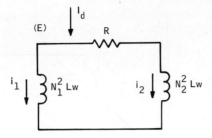

Figure 7.27

$N_1{}^2 L_w$ and $N_2{}^2 L_w$. However, with $N_1 > N_2$, the latter is smaller, and therefore more current will flow in N_2, causing core C_1 to begin switching before C_0. This can be verified by estimating the time required for the two cores to begin switching. If I_d rises rapidly (large di_d/dt), the voltage E is large, and the delay time can be approximated linearly from the rate of rise in the two inductors. This is equivalent to neglecting R during the initial instants. Using this approximation, we can readily derive the ratio of delay times as

$$\frac{T_{d0}}{T_{d1}} \cong \frac{N_1}{N_2}, \tag{7.36}$$

and, with $N_1 > N_2$, core C_1 begins switching first. However, we have already seen that there must also be flux gain from C_0 to C_1. Therefore the transfer functions of the two cores must appear as in Figure 7.28. Clearly, under these conditions, all configurations will eventually be driven to 1 (saturation).

The only stable pair of characteristics are those of Figure 7.29, in which small flux transfers are attenuated, and large ones are amplified. This same

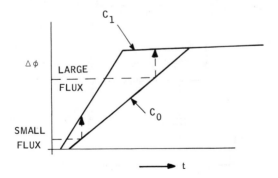

Figure 7.28

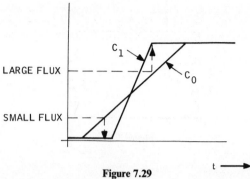

Figure 7.29

requirement is imposed on the all-magnetic networks of the preceding section; it is further compounded by the reverse-transfer problem. Thus some method must be provided to delay the beginning of switching in C_1.

An additional external inductance is sometimes inserted in series with C_1, as in Figure 7.30a to change the rate at which current builds up in C_1. This approach has been used by Crane and Van DeReet [9] in all-magnetic logic circuits. A small square-loop core is deliberately inserted into the transfer loop to subtract a fixed amount of flux from that transferred. The extra core is reset between transfer periods in order to be in the proper state

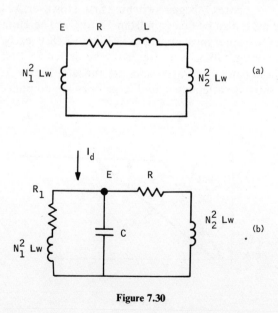

Figure 7.30

during transfer. Another technique is to parallel the output of C_0 with a capacitance, as in Figure 7.30b. The effect can be understood qualitatively by considering the capacitor large, so that voltage builds up very slowly. If a resistor R_1 (perhaps the winding resistance alone) is in series with N_1, it is only necessary that $I_{01}R_1 < E < I_{02}R$ for switching to begin in C_0 before it does in C_1.

Insertion of a flux-loss mechanism is in conflict with the choice of R, the coupling resistor. With R near its design limit, transfer gain is low, and no further attenuation in the form of flux loss can be tolerated. With R near zero, maximum tolerances to noise are achieved when flux gain is equal to 2. In this region of operation attenuation is deliberately introduced to suppress possible spurious signals, coupling between cores is strong, and possible fan-out is low. These basic concepts are illustrated in Figure 7.31.

A point of particular interest arises as $R \to 0$. Note that the operating characteristics of Figure 7.31 demand that C_1 complete switching before C_0. When this happens, the load presented by C_1 becomes effectively zero (saturation). If R is very small, this will have the effect of shorting out C_0, preventing it from completing the switching process. This effect, known as *tail-out*, means in practice that the drive to C_0 is reduced when the driven core saturates, and that its switching time is thereby prolonged. The additional time required for this reduced-speed switching must be allowed for in the design.

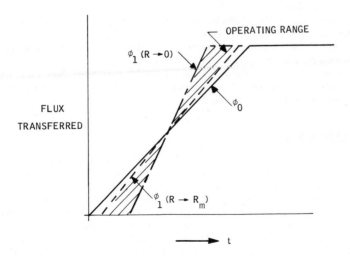

Figure 7.31

The problem is further compounded by flux tolerances. Thus the minimum-flux core C_0 must provide sufficient drive to complete the switching of a maximum-flux core C_1 within a time $T - T_d$. However a maximum-flux core C_0 will be left with a relatively large amount of flux to complete switching during tail-out if it is driving a minimum-flux core C_1 that saturates early.

REFERENCES

[1] R. A. Ramey, "The Single-core Magnetic Amplifier as a Computer Element," *AIEE Trans.*, I, **71**, 422–446 [1952 (January 1953 section)].

[2] R. A. Ramey, "On the Mechanics of Magnetic Amplifier Operation," *AIEE Trans.*, II, **70**, 2124–2128 (1951).

[3] D. G. Scorgie, "Fast Response with Magnetic Amplifiers," *AIEE Trans.*, I, **72**, 1214–1223 (1954).

[4] R. D. Torrey and T. H. Bonn, "A 2.5 Megacycle Ferractor Accumulator," *Proc. EJCC 1956*, 50–53 (1957).

[5] C. B. House, "Full-wave Reversible-polarity Half-cycle-response Magnetic Amplifiers," *AIEE Trans., Commun. and Electron.*, 20, 541–552 (September 1955).

[6] T. H. Bonn, "Magnetic Computer has High Speed," *Electronics*, **30**, 8, 156–160 (August 1957).

[7] H. D. Crane and E. K. Van De Reet, "Design of an All-Magnetic Computing System: Part I—Circuit Design," *IRE Trans. Electron. Computers*, **EC-10**, 2, 207–220 (June 1961).

[8] V. L. Newhouse and N. S. Prywes, "High-speed Shift Registers Using One Core per Bit," *IRE Trans. Electron. Computers*, **EC-5**, 3, 114–120 (September 1956).

[9] E. Block and R. C. Paulsen, "Magnetic Core Logic in a High-speed Card-to-tape Converter," *IRE Trans. Electron. Computers*, **EC-8**, 2, 169–181 (June 1959).

[10] A. Wang and W. D. Woo, "Static Magnetic Storage Delay Line," *J. Appl. Phys.*, **21**, 49–54 (1950).

[11] R. Kodis, S. Ruhman, and W. D. Woo, "Magnetic Shift Register Using One-core-per Bit," *1953 IRE Con. Rec.*, 7, 38–42.

[12] K. Broadbent, "A Thin Magnetic Film Shift Register," *IRE Trans. Electron. Computers*, **EC-9**, 3, 321–323 (September 1960).

[13] E. Goto, "The Parametron, a Digital Computing Element Which Utilizes Parametric Oscillation," *Proc. IRE*, **47**, 8, 1304–1316 (August 1959).

[14] A. J. Meyerhoff, *Digital Applications of Magnetic Devices*, Wiley, New York, 1960.

[15] R. Betts and G. Bishop, "Ferrite Toroid Core Circuit Analysis," *IRE Trans. Electron. Computers*, **EC-10**, 1, 51–56 (March 1961).

EXERCISES

7.1 A 20-V pulse is applied to the circuit of Figure P.7.1. Core parameters are $H_c = 1.0$ Oe, core loss resistance $R_c = 0.1$ Ω/turn^2, magnetic path length $(p/0.4\pi) = 1.0$ cm. Neglecting magnetizing inductance L_m, what is the approximate value of output voltage e_o?

7.2 The magnetizing inductance of a transformer, Figure P.7.2, is $L_m = 10 \times 10^{-6}$ H. The source impedance of the drive source is $R = 0.1$ Ω. For a 10-V, 5-μs pulse, what is the percentage "droop" of the output waveform (see Figure P.7.2b)?

7.3 The same core as in Problem P.7.2 is used in the blocking oscillator of Figure P.7.3 to produce a 10-Volt, 5-μs pulse. The core has a saturation flux density, $\phi_s = 50$ M. What is the minimum number of turns N to be used?

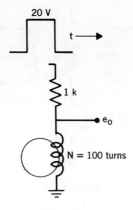

Figure P.7.1

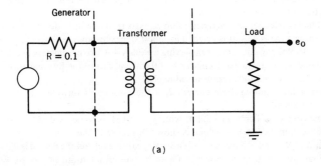

(a)

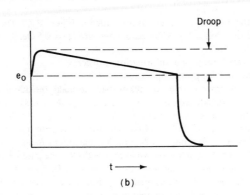

(b)

Figure P.7.2

−10 V

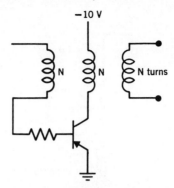

Figure P.7.3

7.4 The series magnetic amplifier of Figure P.7.4a has the following properties: $H_c = 0.25$ Oe, $S_w = 1.0$ Oe-μs, $p/0.4\pi = 1.0$ cm. If the core is initially at $-B_r$, the output waveform is as shown in b. If the core is initially at $+B_r$, the waveform is as in c.
   (a) Find the squareness ratio $S_r = B_r/B_s$.
   (b) If the magnetic amplifier is used to drive other circuits of the type shown in d, what is the limiting fan-out for a delay time $T_d \leq 1$ μs?

7.5 A simplified single loop of a magnetic shift register is shown in Figure P.7.5. The required pulse transfer time is 4 μs. It is decided for good signal-to-noise ratio to require a nominal output of 10 Volts from a core transferring a binary 1. Core parameters are $H_c = 0.5$ Oe, $S_w = 1.0$ Oe-μs, magnetic path length $p = 1.0$ cm, and saturation flux density $\phi_s = 100$ M.
   (a) Find N_1, N_2, and the maximum value of R.
   (b) What is the required drive current I_d if the number of turns on the drive winding $N_d = 10$ turns?

7.6 The same core as in Problem 7.5, but with $\phi_s = 10$ M, is to be used in a series magnetic amplifier, a simplified form of which is shown in Figure P.7.6a.
   (a) Find N_1, N_2, and R if the carrier is a 10-V (peak) signal and the period is 4 μs.
   (b) The output waveform is shown in Figure P.7.6b for a fan-out of $n = 10$. What is the saturation permeability μ_s of the core? The magnetic cross section of the core is 0.003 cm^2.

7.7 A form of parallel magnetic amplifier is illustrated in Figure P.7.7. The switching time for transferring information is 2 μs. Core parameters are $H_c = 0.5$ Oe, $S_w = 1.0$ Oe-μs, $2\phi_s = 20$ Mx, $p/0.4\pi = 2$ cm.
   (a) Find minimum I_d for proper information transfer.
   (b) Find maximum R for proper operation.

7.8 Two blocking oscillators are placed back-to-back to make the dc-to-ac converter shown in Figure P.7.8, with a half-cycle duration of 5 μs, and peak-to-peak output of 20 V. Transformer magnetic material parameters are $H_c = 0.5$ Oe, $S_w = 0.5$ Oe-μs, $p/0.4\pi = 2$ cm, $2\phi_s = 100$ Mx, and $L_m = 10^{-6}$ H. Maximum transistor collector current is limited to 100 mA, and $\beta = 50$. Find minimum N_1 and N_2, maximum R, and minimum R_L.

7.9 Part of a magnetic shift register is shown in Figure P.7.9. Core parameters are $H_c = 1.0$ Oe, $p/0.4\pi = 1.0$ cm, R_c (one-turn core loss resistance) $= 0.1$ Ω, $2\phi_s = 100$ Mx. Drive current is 300 mA. What is the time for transferring data from core 1 to core 2. Draw the equivalent circuit of the transfer loop during switching, and label component values.

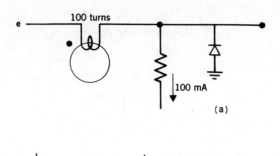

(a)

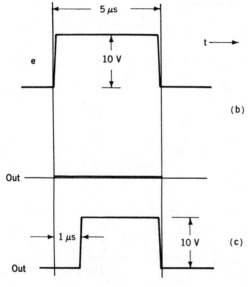

5 μs

t ──►

e

10 V

(b)

Out

1 μs

10 V

(c)

Out

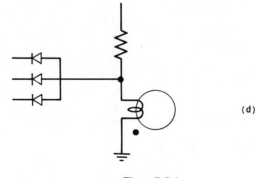

(d)

Figure P.7.4

373

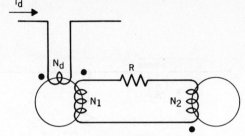

Figure P.7.5

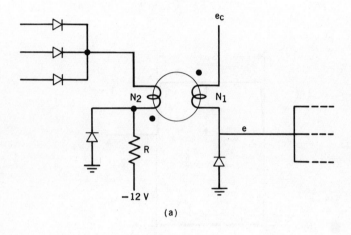

(a)

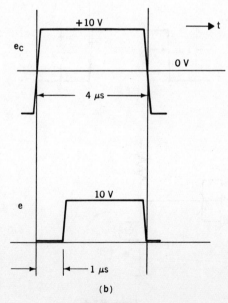

(b)

Figure P.7.6

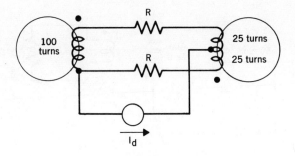

Figure P.7.7

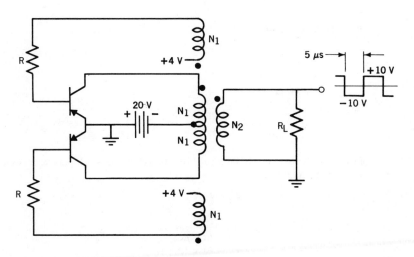

Figure P.7.8

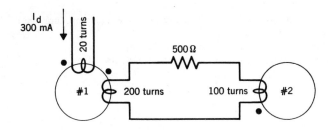

Figure P.7.9

375

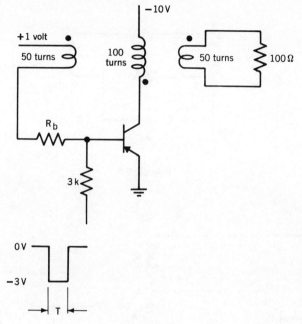

Figure P.7.10

7.10 The blocking oscillator circuit of Figure P.7.10 is triggered by a -3-V pulse as indicated. Core and circuit parameters are

$$H_c = 1 \text{ Oe,}$$
$$\frac{p}{0.4\pi} = 1 \text{ cm,}$$
$$S_w = 1 \text{ Oe-}\mu\text{s,}$$
$$L_w = 10 \times 10^{-6} \text{ H,}$$
$$R_b = 4 \text{ k}\Omega,$$
$$2\phi_s = 50 \text{ Mx,}$$
$$\beta = 50,$$
$$\tau_a = 1 \text{ }\mu\text{s.}$$

With the load resistor open-circulated, what is the minimum duration T of the input pulse for reliable triggering?

(b) What is the amplitude and duration of the output pulse?

(c) What is the maximum value of R_b for driving the 100-Ω load?

Chapter 8

Magnetic-Core Storage

In preceding chapters the design of single memory elements has been considered in some detail. These circuits, making up the bulk of the arithmetic and control unit of the digital computer, are characterized by high logic gain (fan-out) and flexibility of logic function (fan-in). They are useful wherever a variety of functions must be carried out by a limited number of logic elements. These elements suffer, however, from one serious disadvantage. They are too expensive to be used in those portions of a computer where requirements exist for the storage of large volumes of data. Such requirements are found in the storage unit of the computer in which it may be necessary to store from a few thousand to hundreds of millions of bits of data.

A complete storage unit may be regarded as a single logic element. All signals in and out of the storage unit are at logic levels compatible with the memory and decision elements of the arithmetic and control unit. The address of the location from which previously stored information is to be extracted (or into which new information is to be inserted) is presented on the address input line. In general, the address may be presented in either serial or parallel depending upon storage unit and arithmetic-unit mode of operation. Information to be stored is presented on the information input line. Information extracted from the storage unit appears on the information output line or lines, again in serial or parallel. Information appears at this output at some time delay following the read-write control signal. This time delay is a characteristic of the storage medium itself and is referred to as the *access time* of the storage unit. In this chapter we consider the design of storage units whose access times are on the order of microseconds. In the next chapter we shall study the characteristics of storage media whose access times are on the order of milliseconds.

A breakdown of functions within the storage unit by circuit type may be made as in Figure 8.1. The storage elements are the medium or components that perform the actual storage function. Selection circuits perform the task of making a connection between read-write circuitry and the particular storage element to be operated on. The selection function must, of course, be controlled by address information. Many memory units contain their own logic-level address registers, as indicated in Figure 8.1. Such registers are commonly made up of elements identical to those found in the arithmetic and control unit. We shall therefore relegate this portion of the storage unit to the realm of the logic designer and shall assume that all signals entering the selection circuitry are already at standard logic level. The read-write circuits must provide a means of matching between the standard logic signal level and the signal level associated with the particular storage element employed.

One method of classifying storage units is based upon the mode of operation of the selection system. Consider the problem of connecting the read-write circuits to any one or more of an array of bits stored somewhere in space. At least two methods exist for making the desired connection. One involves the provision of a separate circuit path between each bit and a common switching point. Thus the selection unit design is reduced to that of making one of a multiplicity of contacts according to the information contained in the address. Alternatively, the bits in storage may be caused to pass one by one in sequence past the read-write station. In this case the

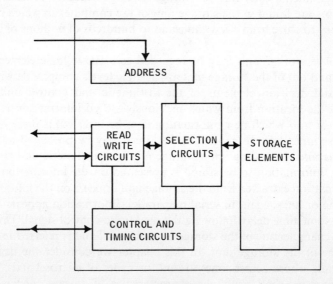

Figure 8.1

selection unit design is reduced to that of determining the time at which the desired information is in contact with the read-write station. These two types of selection have been termed S selection (for space) and T selection (for time), respectively. In this chapter we shall be dealing with the design of a purely S-selection storage unit. In the next chapter the characteristics of storage units combining S and T selection will be considered.

Another term commonly employed to define storage unit operation is *random access*. This is usually taken to mean that the access time to any information in the storage elements is independent of the address of that information. It has been applied, unfortunately, to systems in which this is not strictly true. The term is meaningful, therefore, only if accompanied by the corresponding access time associated with the storage unit as a whole. It does not follow that all random-access storage units have shorter access times than other types, nor that access time is always uniform. In this chapter we shall consider the design of a true random-access storage system in which the access time to any information is indeed uniform.

The unique characteristics of any storage unit are overwhelmingly determined by the characteristics of the storage elements. Each storage element may be regarded as a tiny circuit having logic properties. In particular, each storage element provides an output that is uniquely determined by its input at a previous time. It is therefore classifiable as a memory element according to the definition given in the first chapter.

Because of the large number of elements, each one of which must be individually accessible, the design of storage units is not so much a problem of designing well-behaved storage elements as it is a problem of designing appropriately large switches. Because some form of mechanization of the selection logic function cannot be avoided, the design problem becomes one of finding unconventional and economical techniques for decision element mechanization based upon taking advantage of the special requirements of storage systems and the special properties of storage elements. To this end, one commonly finds extensive use of (a) logic trees by means of which large fan-in or fan-out is achieved by repeated application of elements individually exhibiting limited fan-in or fan-out, and (b) logic decisions by the storage elements themselves.

The first technique involves straightforward decoding of complex logic to a level compatible with the storage element. Thus, in a coincident-current memory, address information is commonly decoded at the usual logic levels so that only one line in the x direction and one in the y direction carries a signal at a single time. The second technique rests on the fact that a storage element must be capable of some simple decision; that is, as a basic memory-type element, it possesses a nonlinear discrimination to signal levels. To the extent that this discrimination can be extended to more than two signal levels, the storage element itself can perform some simple logic. For example, the magnetic core in a coincident-current memory can be regarded as a linear input AND decision element of precisely the type considered in previous chapters.

Analysis of Single Core Switching

The physical property that makes magnetic-core storage systems possible is that of hysteresis. The characteristics of these materials and of certain basic circuits into which they may be incorporated have been studied in previous chapters. For purposes of magnetic-core storage it is only necessary to establish a convention for representing the binary bits 0 and 1 by appropriate conditions of a piece of magnetic material, to provide a means for applying the necessary power to drive the piece of material to one condition or another (write), and to provide a means for sensing a change in flux of the material (read).

Conceptually, therefore, the simplest possible magnetic-core storage element might be represented by the circuit of Figure 8.2. In this circuit the decision element, whose details need not concern us, is assumed to pass positive pulses from the generator whenever the address line is active, and it is assumed to pass negative pulses from the generator whenever both the address and information lines are active. Assume that the core is originally at $-B_r$, and that positive pulses drive it toward $+B_s$. During t_1 the core is therefore switched from $-B_r$ to $+B_s$, and an output is produced. This output may be amplified, clipped at logic level, and gated into an appropriate register. It furnishes the answer to the inquiry implied by the activity on the address line during t_1, i.e., that the core was previously storing a 1 rather than a binary 0. If it is now desired to rewrite a 1 into the core (indicated by the pulse on the information line during t_2), it is necessary to return it to the $-B_r$ state. This is accomplished when the negative pulse from the generator is gated into the core circuit. The fact that core switching during this time also produces an output is of no interest to the remainder of the logic circuitry; that is, no information is contained in this pulse because it is completely determined by the previous pulse sequence.

During t_5, after some inactivity, the events of t_1 are repeated as the previously written 1 is again read from the core. During t_6, the absence of a pulse on the information line indicates the desire to write a 0. Because the decision element does not pass a negative pulse during this period, the core receives no input and therefore simply remains at $+B_r$. Later, during t_9, the core is again driven to $+B_s$ by a positive pulse. Change in flux from $+B_r$ to $+B_s$ is very small, however, so that no appreciable output occurs at this time, and a 0 is considered to be read from the core.

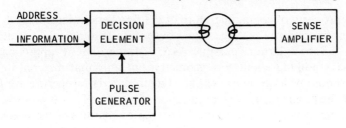

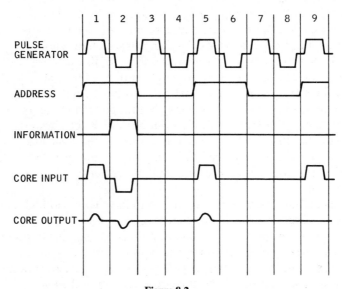

Figure 8.2

Although possible in principle, the above scheme suffers from the obvious defect of requiring a complex decision element and a sense amplifier for each core or bit of information. Circuitry of this type is on the same order of complexity as the one-bit memory elements already studied, so there is little to recommend such an arrangement. Leaving aside this overriding disadvantage, however, we may study the behavior of the core for the moment from the point of view of the light it can shed on the basic principles of operation of more practical and more complex storage element arrays.

Consider first the question of what characteristics are desirable with respect to core geometry. Because the amount of unwanted flux change obtained upon reading a 0 is dependent on the difference $B_s - B_r$, we wish the squareness ratio $S_r = B_s/B_r$ to be as near unity as possible. A toroidal, or at least gapless geometry is therefore called for, because this configuration has the property of containing more of the flux (less leakage flux) than any other configuration. For large signal output on reading a 1, it is desirable that $B_s + B_r$ be as large as possible. As $S_r \to 1$, this quantity approaches $2B_s$, which says nothing about core geometry, but only that material flux properties should be as good as possible. Other material properties are H_c and S_w. For the moment, H_c is of importance only as it affects power loss in the core during switching. Core power loss represents one of the limits upon switching rate, because a fixed quantity of energy is converted to heat in each switching cycle. If the heat cannot be dissipated rapidly enough, core temperature rises. As the temperature approaches the Curie point, magnetic properties change approximately as indicated in Figure 8.3. Because heat dissipation rates depend on mechanical construction, cooling provisions, and so forth, factors that are beyond the scope of this text, we shall assume that adequate caution has been exercised to prevent heating, and that core specifications are given for the worst-case environmental conditions.

The other material property that may be interpreted in terms of energy loss is the switching constant S_w. Aside from heating effects, however, we shall see shortly that the core time constant τ_m imposes the basic limitation upon memory cycle time. This limitation only arises, however, when the constraints of a coincident-current (or multiple-current) selection system are included.

For maximum 1 output, it is desirable to have core width W as large as possible in order to provide large cross-sectional area. This dimension is, however, limited by practical considerations in core manufacture and assembly of large arrays. In order to reduce required drive current it is desirable to have outer radius r_o as small as possible, because current is proportional to the mean path length around the circumference of the core.

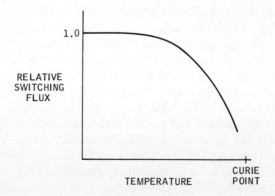

Figure 8.3

For the moment, variations in H_c from core to core are of no importance, because drive pulses are either full amplitude or not present at all. In arrays, where drive pulses other than zero or full amplitude may occur, the low H_c and high eddy-current losses of metal tape cores virtually eliminate them from consideration. The high volume resistivity of ferrite materials means that their apparent coercive force is more nearly a constant independent of switching time. Ferrite materials possess another important advantage; namely, they are readily manufactured to close tolerances by mass production methods, resulting in low unit cost. Ferrite materials are therefore used almost exclusively in the magnetic-core storage field. The hysteresis loop for a typical memory core is presented in Figure 8.4.

A gross estimate of the signal-to-noise ratio may be made from this characteristic by noting that the flux change when reading a 0 is $B_s - B_r$, whereas the change when reading a 1 is $B_s + B_r$. The ratio of flux change associated with a 1 to that associated with a 0 is thus

$$\frac{S}{N} = \frac{1 + S_r}{1 - S_r}. \tag{8.1}$$

Because S_r is commonly above 0.9 the ratio is usually above 20:1. It will be

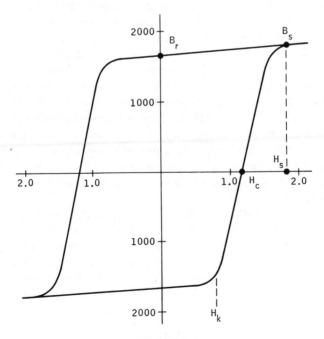

Figure 8.4

seen in the following discussion of output waveforms that further advantage may be taken of the fact that the peak output signals for a 0 and a 1 do not occur at the same time with respect to the driving pulse wavefront. In general, if the signal output of a core is sampled or strobed some appropriate time after the rise of the driving pulse, discrimination ratios on the order of 50:1 may be achieved.

Consider now the conditions that are imposed upon the driving pulse. The current drive pulse will be characterized by its amplitude I_d, duration T, and rise time T_r. I_d must be sufficient to drive the material in the region of the outermost radius to saturation within the desired switching time. Substituting in the switching time equation, we obtain

$$I_d \geq \frac{5H_c r_o}{N}\left(1 + \frac{\tau_m}{T}\right).$$

(8.2)

Equation 8.2 holds only over a range of values of T for which the switching-time equation is valid. Below a certain value of T, domain inertia effects become predominant, preventing switching from occurring within extremely short periods of time. In particular, it has been shown that the state of a core may be sensed by the application of extremely short pulse durations without causing any reversal of magnetization to occur. The minimum value of the rise time T_r is limited only by the impedance of the memory element array and the power available from the pulse-generating source.

Specifications on the details of core output under carefully defined drive conditions are of great value to the circuit designer. Manufacturers commonly supply recommended operating conditions for each type of core intended for memory storage application. These include (a) drive pulse amplitude (I_d), width (T), and rise time (T_r); and (b) output signal turnover time (T_o), peaking time (T_p), turnover amplitude (E_o max), maximum peak noise signal (E_n), and signal-to-noise ratio at peaking time. These quantities are defined by reference to Figure 8.5, showing the output of a typical ferrite memory core. Curves of signal-to-noise or squareness ratio, and of 1-signal output or B_s versus driving field strength or ampere turns, are also commonly supplied (see Figure 8.6).

Such information does not, however, yield much insight into the effects to be anticipated from variations of available parameters. We would like, for example, to be able to predict the effect of changes in drive current or core geometry upon switching time and output voltage. For this purpose a simple first-order approach that is consistent with previous models of core behavior can be employed. The switching time of a thin band of material at radius r, driven by current I_d, may be written

$$T(r, I_d) = \frac{\tau_m}{NI_d/5H_c r - 1}.$$

(8.3)

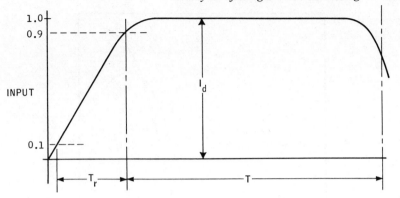

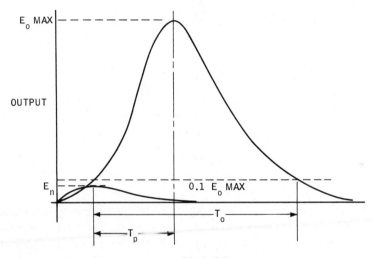

Figure 8.5

Now consider the toroid made up of a series of small incremental bands of width dr, subjected to a field proportional to I_d/r. In order to obtain the output voltage waveform, assume that $d\phi/dt$ is constant. For each incremental band, $d\phi/dt$ is therefore given approximately by

$$\frac{d\phi}{dt}(r, I_d) \cong \frac{\Delta\phi}{T} = \frac{(B_s + B_r)W}{\tau_m}\left(\frac{NI_d}{5H_c r} - 1\right) dr. \qquad (8.4)$$

The value of $d\phi/dt$ is therefore greatest at the inner radius, where switching

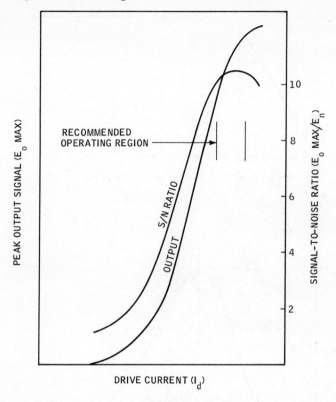

Figure 8.6

is completed most rapidly, and least at the outer radius. The output voltage may be regarded as the sum of the contributions from each of the incremental bands, as shown in Figure 8.7, where t_1 corresponds to the switching time at the inner radius and t_2 corresponds to the switching time at the outer radius. The output voltage during the period $0 < t < t_1$ may therefore be written as

$$e \cong N \times 10^{-8} \int_{r_i}^{r_o} \frac{d\phi}{dt}(r, I_d). \tag{8.5}$$

During the period $t_1 < t < t_2$, the output voltage may be written as

$$e \cong N \times 10^{-8} \int_{r_s}^{r_o} \frac{d\phi}{dt}(r, I_d), \tag{8.6}$$

where r_s represents the radius at which switching has already been completed.

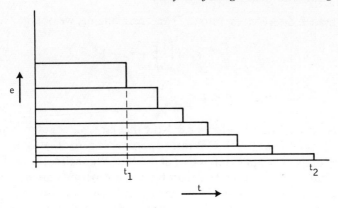

Figure 8.7

It is given by

$$r_s(t) = \frac{NI_d}{5H_c(1 + \tau_m/t)}. \tag{8.7}$$

Substituting (8.4) into (8.5) and performing the indicated operations, we obtain, for the period $0 < t < t_1$,

$$e = \frac{N(B_s + B_r)W(r_o - r_i)}{\tau_m \times 10^8}\left[\frac{NI_d}{5H_c(r_o - r_i)}\ln\left(\frac{r_o}{r_i}\right) - 1\right]. \tag{8.8}$$

For the interval $t_1 < t < t_2$

$$e = \frac{N(B_s + B_r)Wr_o}{\tau_m \times 10^8}\left[\frac{NI_d}{5H_cr_o}\ln\left(\frac{1 + \tau_m/t}{NI_d/5H_cr_o}\right) + \frac{NI_d}{5H_cr_o(1 + \tau_m/t)} - 1\right]. \tag{8.9}$$

These equations may be placed in a more useful form by defining the quantities

$$r_d \equiv \frac{NI_d}{5H_cr_o} \quad \text{(overdrive ratio)},$$

$$r_r \equiv \frac{r_i}{r_o} \quad \text{(ID–OD ratio)},$$

$$E_i \equiv \frac{N(B_s + B_r)Wr_o}{\tau_m \times 10^8}. \tag{8.10}$$

Note that r_d is the ratio of the drive current applied to that current required to produce a field of value H_c at the outer radius r_o. It will also be called the *current drive ratio*. The value of r_d must be greater than 1, whereas r_r must be less than 1. Substituting into (8.7) and rearranging, we obtain

$$\frac{e}{E_i} = r_d \ln\left(\frac{1}{r_r}\right) + r_r - 1,$$

$$0 < t < \frac{\tau_m}{r_d/r_r - 1}. \tag{8.11}$$

The function is plotted in Figure 8.8. Note that large values of output voltage may be obtained as the ID–OD ratio approaches zero; that is, as the inner diameter of the core approaches that of the wire. We shall see in the following section that ID–OD ratios less than 0.5 are not permissible in coincident-select memory arrays. Output voltage also increases with increasing current drive ratio. This ratio must likewise be restricted within narrow limits for practical arrays. Equation 8.9, giving the output voltage as a function of time following saturation of the core in the region of the inner radius, can be treated in a similar manner and written as

$$\frac{e}{E_i} = r_d \ln\left(\frac{1 + \tau_m/t}{r_d}\right) + \frac{r_d}{1 + \tau_m/t} - 1, \tag{8.12}$$

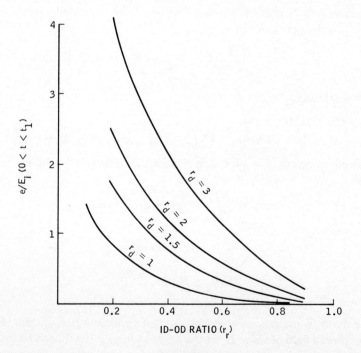

Figure 8.8

for

$$\frac{1}{r_d/r_r - 1} < \frac{t}{\tau_m} < \frac{1}{r_d - 1}.$$

This function, for one value of ID–OD ratio corresponding to that of a practical core, is presented in Figure 8.9. Notice the extremely slow decay of the output voltage as the current drive ratio approaches unity.

We may now compare our predictions with the output observed on switching a real core. This is shown in Figure 8.10, where predicted values are calculated from manufacturer's data given for a typical ferrite core. The

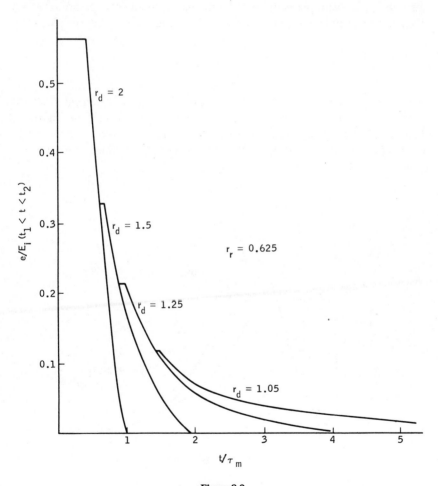

Figure 8.9

recommended current drive ratio for this core is 1.35. We see from Figure 8.10 that the model roughly describes the output of the core, approximating both peak amplitude and switching time. The observed output on the rising edge of the pulse is, however, not accurately predicted. This is partially a result of the finite rise time on the drive current to the observed core. The model may be modified to take account of this effect, as shown by the dotted curve of Figure 8.10.

Note the reasonably good agreement between observed and predicted behavior, despite the highly simplified model of constant rate of change of flux during switching under constant-current drive conditions. Considerably better agreement can be obtained by making $d\phi/dt$ a function of ϕ. Betts and Bishop [1], for example, obtained excellent agreement between predicted and measured waveforms using a core switching function of the form

$$\frac{d\phi}{dt} = f(\phi)(H_a - H_c),$$

with $f(\phi)$ given by a gaussian distribution function normalized such that $e^{-(\phi)^2}$ is a maximum as ϕ goes through zero. As mentioned in Chapter 3, Gyorgy [2] has closely approximated rotational switching behavior by a model that is effectively equivalent to the above switching function with $f(\phi) = [1 - (\phi/\phi_s)^2]$.

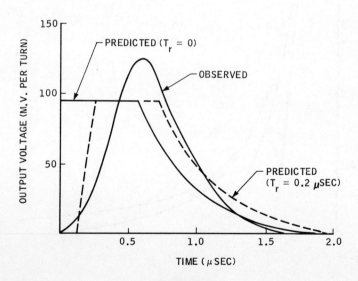

Figure 8.10

Coincident-Current Memories (3D)

In the preceding section we considered the behavior of a single storage element, leaving in abeyance the question of the economics of the complex drive circuitry. If the core storage element is to be an economically feasible component of storage units, the number of drive circuits must be minimized, and a few circuits must suffice to control a large number of elements. An early scheme that followed naturally from diode matrices is illustrated in

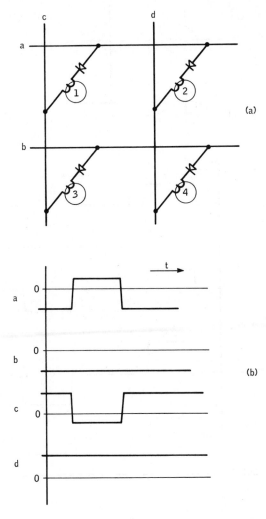

Figure 8.11

Figure 8.11*a*. In this arrangement a combination of signals on two lines, shown in Figure 8.11*b*, causes forward bias to appear across only one diode out of the entire population. In this way a single core is selected. The requirement placed upon the signal levels is that the most positive value of the signal on line *a*, for example, must be less than the steady-state value on all but one vertical line (e.g., line *c*), and that the most negative value of the signal on line *c* must be greater than the steady-state value on all but one horizontal

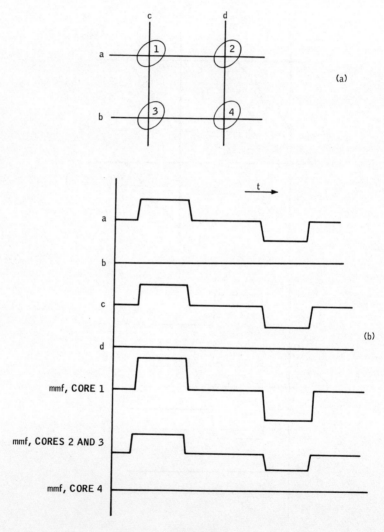

(a)

(b)

Figure 8.12

line. In this manner core 1 is selected. Assuming a square array, we see that the number of lines (and therefore drive circuits) required for a memory capacity of Q bits is $D = 2\sqrt{Q}$. Because current can flow in only one direction in the schematic shown, a duplicate drive matrix must be provided in order to reverse core magnetization when writing information into the cores. Note that this scheme has the advantage of imposing very few restrictions upon the magnetic properties of the core other than a detectibly high remanent flux level. It was therefore of particular interest at a time when cores having suitably square hysteresis characteristics were not readily available. It possesses a serious economic disadvantage, however, in that a diode whose cost is many times that of the core is required for each storage element. It was therefore natural to search for some means of eliminating the diode.

The outgrowth of this search was the coincident-current selection principle illustrated in Figure 8.12. Current pulses, each of an amplitude insufficient to switch a core, are delivered on one horizontal and one vertical line. The sum of these currents, to which only one core is subjected, is sufficient to drive that core past the knee of the hysteresis loop and therefore cause switching. This core is termed the *selected core*. Other cores, which receive one or the other of the drive current pulses, but not both, are described as *half-selected cores*. Cores that are subjected to neither drive current pulse are termed *unselected cores*. These terms may be visualized with reference to Figure 8.12, in which core 1 is selected, cores 2 and 3 are half-selected, and core 4 is unselected.

It is apparent that the coincident-current principle permits highly simpli-fied construction of large storage units. Particularly, note that magnetization in either direction can be accomplished by means of the same set of drive windings. The presence of half-amplitude pulses on some lines, however, imposes much more severe restrictions on the hysteresis loop and geometry of the core, as well as on the drive current itself.

In order to analyze these restrictions let us represent the hysteresis-loop measurements taken on an actual core by the schematic parallelogram of Figure 8.13a, where primes indicate measure values; that is, the core par-ameters are obtained from drive current measurements by

$$H'_c \equiv \frac{I_c}{5(r_i + r_o)/2}$$

$$H'_s \equiv \frac{I_s}{5(r_i + r_o)/2} \cdot$$

$$H'_k \equiv \frac{I_k}{5(r_i + r_o)/2}, \tag{8.13}$$

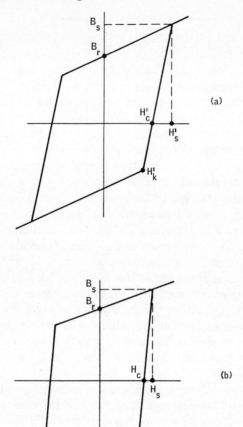

Figure 8.13

in which

$I_c \equiv$ current required to produce $B = 0$,

$I_s \equiv$ current required to saturate the core (low frequency),

$I_k \equiv$ current at which irreversible switching is initiated.

The above are measured quantities. It is desirable to remove the parameters that are due to the accidents of geometry, and describe core properties in terms of intrinsic thin-band material properties. This can be done by defining

$$H_c \equiv H_c',$$

$$H_s \equiv \frac{I_s}{5r_0} = H_s' \frac{r_i + r_o}{2r_o},$$

$$H_k \equiv \frac{I_k}{5r_i} = H_k' \frac{r_i + r_o}{2r_i}. \tag{8.14}$$

The resulting equivalent characteristic of the intrinsic material is therefore as illustrated in Figure 8.13b. All intrinsic quantities may be related to H_c by letting

$$H_s \cong H_c + \frac{B_s}{\mu_i} = H_c(1 + \Delta_B),$$

$$H_k \cong H_c - \frac{B_s}{\mu_i} = H_c(1 - \Delta_B), \tag{8.15}$$

in which

$$\mu_i \equiv \frac{B_s}{H_s - H_c},$$

$$\Delta_B \equiv \frac{B_s}{\mu_i H_c}.$$

The required drive currents for a particular core are directly related to these intrinsic quantities, and to the particular geometry. Only two levels of drive current are of interest. One is that sufficiently small to prevent irreversible switching of the core. In terms of the intrinsic parameters and geometry, it is

$$I_k = 5H_k r_i = 5H_c r_i (1 - \Delta_B). \tag{8.16}$$

The other level of drive current is that sufficiently large to cause complete reversal of the flux in the core. To produce complete switching at the outer radius within a time T, the required field is $H_a = H_s + S_w/T$. Using (8.15), we see that this current is

$$I_s = 5H_s r_o \left(1 + \frac{\tau_m}{T}\right) = 5H_c r_o (1 + \Delta_B)\left(1 + \frac{\tau_m}{T}\right). \tag{8.17}$$

The intrinsic coercive force of the material, as well as the precise dimensional values, may vary from core to core. Let

$$\Delta_H \equiv \text{tolerance on } H_c,$$
$$\Delta_r \equiv \text{tolerance on } r_i \text{ and } r_o.$$

Neglecting second-order effects, these terms enter in an additive manner into the parentheses of (8.16) and (8.17). Finally, the quantity of greatest general interest is the ratio of I_s to I_k. Redefining the current drive ratio r_d in terms of the above parameters, and including all tolerances, we obtain

$$
\begin{aligned}
r_d &\equiv \frac{I_s}{I_k} = \frac{(1 + \tau_m/T)}{r_r}\left(\frac{1 + \Delta_H}{1 - \Delta_H}\frac{1 + \Delta_r}{1 - \Delta_r}\frac{1 + \Delta_B}{1 - \Delta_B}\right) \\
&\equiv \frac{(1 + \tau_m/T)}{r_r}(1 + 2\Delta),
\end{aligned}
\tag{8.18}
$$

in which total tolerance effects Δ are defined by (8.18). For small tolerances, second-order terms may be neglected, and $\Delta \cong \Delta_H + \Delta_r + \Delta_B$.

We are now in a position to derive some useful relationships among the tolerances in terms of r_d. To do this, consider the complete cycle of events in a coincident core memory.

Read. During read, three cases have been identified; see Figure 8.14. Let

$$x \equiv \text{tolerance on } x\text{-axis drive current } I_x,$$
$$y \equiv \text{tolerance on } y\text{-axis drive current } I_y.$$

Then under worst-case conditions it is required that

Selected core

$$I_x(1 - x) + I_y(1 - y) \geq I_s. \tag{8.19}$$

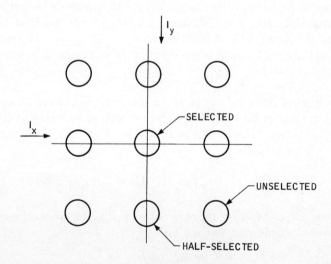

Figure 8.14

Half-selected cores

$$I_x(1 + x) \le I_k,$$

$$I_y(1 + y) \le I_k. \tag{8.20}$$

Unselected cores
No constraints.

Values of I_s and I_k are given by (8.16) and (8.17), with possible additional terms due to Δ_H and Δ_r.

Write. During write the same core selected during read is again driven by currents of magnitude I_x and I_y, but this time in the negative direction. If a 1 is to be written, no inhibiting current is present on a z line common to all the cores on the plane, and the selected core is to be returned to the $-B_r$ state. Half-selected and unselected cores are to be unaffected. Therefore, the same conditions as (8.19) and (8.20) apply.

To write a 0, a current I_z is used to cancel part of the sum $I_x + I_y$ at the selected core, leaving it in the $+B_r$ state (see Figure 8.15). Let

$$z = \text{tolerance on } z\text{-drive current } I_z.$$

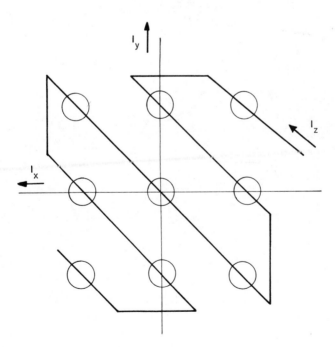

Figure 8.15

Waveforms of the complete coincident-current read-write cycle are given in Figure 8.16. Under worst-case conditions the constraints upon the drive currents are, for the previously identified cores;

Selected

$$I_x(1 + x) + I_y(1 + y) - I_z(1 - z) \le I_k, \tag{8.21}$$

Unselected

$$I_z(1 + z) \le I_k, \tag{8.22}$$

Half-selected
No additional constraints.

Note that for the last case, half-selected cores may receive as much drive as $I_x(1 + x) - I_z(1 - z)$, but this is covered by constraints (8.20) and (8.21).

Taking the limits in (8.20), it is clear that x- and y-axis drive currents are limited only by an equal I_k and by their respective tolerances. In the usual case, the x and y drive current generators are similar circuits, subject to the same tolerance conditions. Therefore let $I_x = I_y$, $x = y$, and

$$I_x = I_y = \frac{I_k}{1 + x}. \tag{8.23}$$

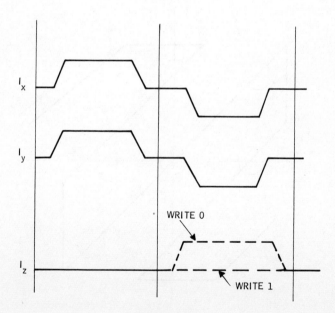

Figure 8.16

Using (8.19), (8.21), (8.22), and (8.23), one obtains the relationship

$$r_d \frac{1 + x}{1 - x} - \frac{1 - z}{1 + z} \leq 1. \tag{8.24}$$

This relationship may be taken to define the basic design trade-off between memory switching time and core parameters (implicit in r_d), and current driver requirements. To explore this relationship, simplify by letting z-driver tolerances be the same as others; that is, $z = x = y$. Then combining (8.18) and (8.24) and dropping second-order terms, we can obtain the constraint on the core geometry that can be selected as

$$\frac{r_i}{r_o} \equiv r_r \geq \tfrac{1}{2}\left(1 + \frac{\tau_m}{T}\right)[(1 + x)/(1 - x)](1 + 2\Delta)/(1 - x). \tag{8.25}$$

Generally, however, mechanical strength considerations limit the value of ID–OD ratio that can be achieved. Therefore the limiting tolerances imposed upon the current drivers constitute the design constraints. Maintaining equal tolerances as before, we see that the solution for (8.24) is

$$\frac{\Delta I}{I} \equiv x \leq \left(1 + \frac{1}{r_d}\right)\left[\left(1 + \frac{2/r_d - 1}{(1/r_d + 1)^2}\right)^{1/2} - 1\right]. \tag{8.26}$$

Under the range of allowable values of r_d, $1 < r_d < 2$, (8.26) reduces to a much simpler approximate expression, namely,

$$x \lesssim \frac{1}{2}\frac{2 - r_d}{1 + r_d}. \tag{8.27}$$

This function is plotted in Figure 8.17. It may also be regarded as imposing a constraint on r_d, namely,

$$r_d < 2\frac{1 - x}{1 + 2x} < 2. \tag{8.28}$$

Practical values for $1/r_r$ are, however, more nearly 1.6. It is apparent from Figure 8.17, therefore, that coincident-current core memories must inevitably operate within very narrow driver tolerance limits.

Another way of looking at the design constraints is to note that (8.18) and (8.24) effectively impose a limit on memory cycle time, namely,

$$T \geq \frac{\tau_m}{2r_r(1 - x)/[(1 + x)/(1 - x)](1 + 2\Delta) - 1}. \tag{8.29}$$

Even if we force all tolerances to approach zero by careful circuit design and manufacturing control, the speed of a coincident-current type of memory

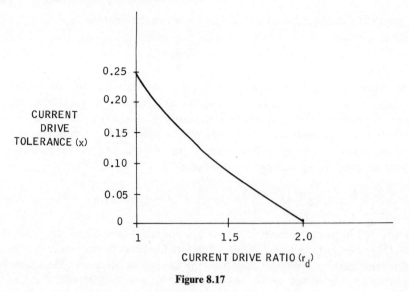

Figure 8.17

selection system therefore imposes the inherent limitation

$$T \geq \frac{\tau_m}{2(r_i/r_o) - 1}. \tag{8.30}$$

Although (8.29) and (8.30) are slightly pessimistic as a result of extremely simplified assumptions about the nature of the switching process, they correctly predict the general character of the relationships among the design variables. It will be found, in fact, that many practical core memories operate very close to the limits imposed and are possible only because of close manufacturing control.

The coincident-current type of selection system also causes at least one other constraint to be imposed upon the design of the storage unit. This constraint has to do with the signal-to-noise ratio during the read portion of the cycle. It arises because of the possibly additive effects of small differences between the signals produced by half-selected cores in different states. To understand these effects, we must examine in detail the outputs from all cores in the array under worst-case conditions for the two possible states of the selected core.

First, we can (roughly) predict by means of (8.11), or measure, the minimum output from a selected core storing a 1. Let

$V_s \equiv$ minimum signal output of a selected core storing a 1.

We have noted previously that the maximum output of a core storing a 0 is generally reached well before that of a core storing a 1. The output does not die to precisely zero, however, by the time the peak 1 output signal arrives. It is convenient to measure the largest value of signal from a 0-storing core under expected drive conditions, and to employ this figure in determining array size limits. Therefore let

V_z = maximum signal output of a selected core storing a 0, at the time of peak signal output of a core storing a 1.

If the selected core alone is considered, clearly the minimum signal-to-noise ratio is V_s/V_z. However, the selected core is not the only one producing signals during read. In general, a large number of cores strung on the same sense winding are also receiving half-select drive at this time, and are consequently producing some small noise signals. Again, analytical treatment is difficult, and the best source of data for the magnitude of half-select signals is direct measurement under appropriate drive conditions.

The source of the signals can be seen by reference to Figure 8.18*a* for the case of a stored 1, and to Figure 8.18*b* for a stored 0. Immediately after storing a 1, the state of the core is at $-B_r$ (see Figure 8.18*a*). If this core now receives a half-select read signal, it will traverse a path (1) first to the right under the drive $I_d/2$, then back to $H = 0$ as the drive is removed. The resulting change in flux is described as that due to an "undisturbed, half-selected 1." The term *undisturbed* refers to the fact that it has received no half-select signals between the time of storing a 1 and the time of interest. This flux change is symbolized as $\phi_{hs}(u1)$. Under the influence of subsequent half-select drive signals, the core will traverse a series of paths progressively closer to $\phi = 0$, until coming to rest at a final position $-B_1$, greatly exaggerated in Figure 8.18*a*. All further half-select disturbances produce only reversible flux changes about this point, a fortunate circumstance for memory purposes.

The initial half-select flux change of an undisturbed core tends to be much larger than any subsequent changes. This is simply due to the fact that all easily reversed domains will switch with the first application of reverse field, but once having switched will remain until the state of the core is changed. In order to avoid this large initial change, it is common practice to provide some means of half-selecting or disturbing all cores before they can be read the first time. The easiest way to ensure that this is done is to disturb the core immediately after a write operation. Such a process is termed *post-write disturb*. It can be accomplished, for example, by allowing a period following write for deliberately introducing a disturb signal on all cores in the positive-field direction, as illustrated in Figure 8.19.

If it is assumed that some arrangement is made for post-write disturb, then the change in flux of the half-select core storing a 1 can be symbolized

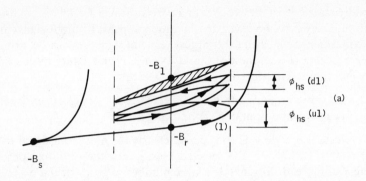

(a)

ϕ_{hs} (dl)

ϕ_{hs} (ul)

(1)

$-B_1$

$-B_r$

$-B_s$

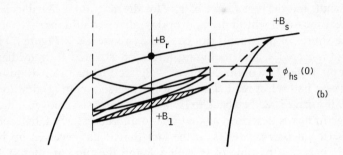

(b)

$+B_s$

$+B_r$

ϕ_{hs} (0)

$+B_1$

Figure 8.18

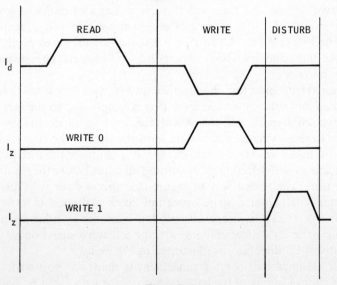

READ   WRITE   DISTURB

I_d

WRITE 0

I_z

WRITE 1

I_z

Figure 8.19

by $\phi_{hs}(d1)$. The corresponding output voltage, as a noise signal during read time, is approximately proportional to the flux change and may be written as

$V_{hs}(d1)$ = noise output of half-selected disturbed core storing a 1, measured at peak 1 signal output time.

All stored 0's are already disturbed, by the nature of the write process. This is illustrated in Figure 8.18*b*; that is, prior to write, a read half-cycle has driven the core to $+B_s$ and let it return to $+B_r$. During the write half-cycle for a 0, it is driven in the reverse direction and allowed to return to $H = 0$. This point is denoted by B_1 in Figure 8.18*b*. On subsequent half-select disturbances, the 0-storing core is driven slightly toward $+B_s$ and allowed to return. The resulting flux change, denoted by $\phi_{hs}(0)$, is nearly reversible. A few half-select cycles will return the core to a point $+B_1$ from which further disturbances are exactly reversible, corresponding to the point $-B_1$ for the 1-storing core. The resulting output voltage may be defined as

$V_{hs}(0) \equiv$ noise output of a half-selected core storing a 0, measured at peak 1 signal output time.

The fact that limits the size of achievable coincident-current arrays is that the above noise signals are not exactly equal; that is

$$\phi_{hs}(0) \neq \phi_{hs}(d1),$$

and

$$V_{hs}(0) \neq V_{hs}(d1).$$

Qualitatively, the reason for this difference can be seen by noting that a disturb signal in the read direction drives a 1-storing core away from maximum flux magnitude whereas the same signal drives a 0-storing core toward maximum flux magnitude. Because the resistance to increasing stored energy is larger in the latter direction, slightly less flux change occurs. Another way to say this is to note that the minor hysteresis loop, although reversible, is not quite symmetrical about $H = 0$. Greatly exaggerated, this effect is illustrated in Figure 8.20.

The difference between the output signals produced by half-selected cores at maximum 1 signal output time is called the *delta voltage*. It is

$$V_\delta \equiv V_{hs}(d1) - V_{hs}(0).$$

With the delta voltage of a particular core type known, the limits upon coincident-current array size are readily determined.

Let the sense winding be arranged to sense half the cores in one polarity and the other half in the opposite polarity. In a square array of dimension n, for every core on a given drive line there must be another sensed in the

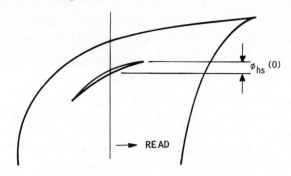

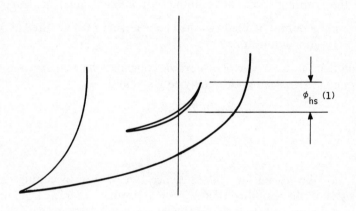

Figure 8.20

opposite direction. The selected core and its mate on one drive line therefore produce a signal $V_s - V_{hs}$. The remaining $n - 2$ cores on the same line produce signals that in the worst possible case cancel within V_δ. Similar considerations apply to the other drive line. The output for a stored 1 is, therefore,

$$V_1 = V_s(d1) - 2V_{hs} \pm (n - 2)V_\delta. \tag{8.31}$$

The output for a stored 0 is

$$V_0 = V_s(0) - 2V_{hs} \pm (n - 2)V_\delta. \tag{8.32}$$

For worst-case conditions, the alternative signs may be taken to be negative

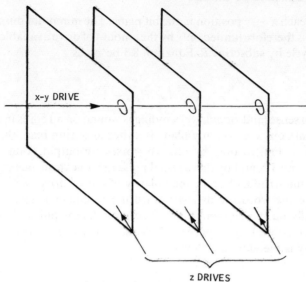

x-y DRIVE

z DRIVES

Figure 8.21

in (8.31) and positive in (8.32). This is because the configuration of information (stored 1's and 0's) in the $2(n-2)$ half-select cores may change between reading a stored 1 at one time and a stored 0 at another (in particular, the configuration may precisely reverse).

Combining (8.31) and (8.32) with a signal-to-noise ratio defined by $S_n \equiv V_1/V_0$, we find that the limiting size of array becomes

$$n \le \frac{V_s(d1) - S_n V_s(0) + 2V_{hs}(S_n - 1)}{V_\delta(S_n + 1)} + 2. \tag{8.33}$$

Typically, V_δ is an order of magnitude smaller than V_{hs}. Using typical values $V_\delta = V_{hs}/10$, $V_s(0) \cong V_{hs}$, $V_s(d1) = 20 V_{hs}$, and $S_n = 2$, we see that n must be less than 33, or an array size of, say, 1,024 (32 × 32).

Linear-Select Memories (2D)

Some applications place a premium upon memory speed and reliability and can tolerate smaller total memory capacity and/or increased cost per bit of storage. In these applications the linear-select type of memory drive organization is a strong contender.

In the linear-select memory (also called *word-select*), only one word receives a drive signal during the read cycle. All other cores are undisturbed. The arrangement, shown schematically in Figure 8.21, requires one drive circuit for each $x - y$ position in the bit plane. The minimum duration of the read cycle is therefore limited only by the amount of drive available. Denoting the read cycle by subscript R, Equation 8.3 becomes

$$T_R = \frac{\tau_m}{I_R/5H_c r_o - 1}. \tag{8.34}$$

A common sense (and/or z-drive) winding is wound on all cores in a bit plane. Because only one core on any plane is switched during read, the signal-to-noise ratio is limited only by relative worst-case output voltages of cores storing 0's and 1's, and by drive signal noise due to extraneous coupling.

During the write half cycle, the polarity of the xy driver is reversed, and all cores in the word are driven toward the 1 state. To store information differentially on the different planes, the z winding is pulsed in a direction either to aid (store 1) or oppose (store 0) the xy drive. The net drive to cores in the array is therefore as follows:

	I_{xy}	I_z	Net I	
Unselected	0	-1	$-I_z$	Disturbed
	0	$+1$	$+I_z$	
Selected	$+1$	-1	$I_{xy} - I_z$	Store 0
	$+1$	$+1$	$I_{xy} + I_z$	Store 1

In order to analyze this case, the same basic constraints as in coincident-current switching may be applied. Thus, the net drive must be less than some "knee" value I_k in order to prevent switching, and the net drive must be greater than some switching value I_s to ensure switching within a given time T. Including driver tolerances as before, with the tolerance on I_{xy} denoted by x, we obtain the constraints

Selected core,

$$\text{Write } 1: I_{xy}(1 - x) + I_z(1 - z) \geq I_s,$$

$$\text{Write } 0: I_{xy}(1 + x) - I_z(1 - z) \leq I_k. \tag{8.35}$$

Unselected cores,

$$I_z(1 + z) \leq I_k. \tag{8.36}$$

Combining these constraints, we obtain the limiting values of the xy and z drive currents as

$$I_{xy} = \tfrac{1}{2}(I_s + I_k),$$

$$I_z = \frac{I_s(1 + x) - I_k(1 - x)}{2(1 - z)} = \frac{I_k}{1 + z} \tag{8.37}$$

The values of I_s and I_k in terms of core parameters are given by (8.16) and (8.17). The constraint imposed by (8.37) on the drive current ratio is

$$r_d < \frac{3 - [z + x(1 + z)]/3}{(1 + x)(1 + z)}, \tag{8.38}$$

in which r_d is given by core parameters and switching time, as in (8.18). To compare this with previous results for the coincident-current scheme, let $x = z$. Then the positive solution of (8.38) is

$$x = 2\left(\frac{1}{r_d + 1}\right)^{\frac{1}{2}} - 1. \tag{8.39}$$

Equation 8.39 is plotted in Figure 8.22. Note the substantially higher tolerance limits permitted at a given value of drive ratio, as compared to Figure 8.17. Alternatively, observe that at a given value of tolerance limit, the drive ratio can be much larger, leading to substantially shorter switching times compared to the coincident-current technique. This, together with the fact that in linear-select memories the read-cycle time is limited only by available drive, by (8.34), means that much higher-speed storage systems can be achieved.

Equation 8.39 also implies a constraint on the allowable core geometry, in a manner similar to the result (8.25). Using (8.18) in (8.39) and solving for the ID–OD ratio yields a limiting value of $r_r = r_i/r_o$ given by $\frac{1}{3}$ (rather than $\frac{1}{2}$) times the variable quantity in (8.25). Relationships between switching time, geometry, and tolerances similar to (8.29) can also readily be obtained.

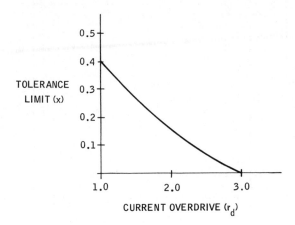

Figure 8.22

In the limit, as tolerances go to zero, the memory switching time is

$$T \geq \frac{\tau_m}{3(r_i/r_o) - 1}. \tag{8.40}$$

Compare this result with (8.30). Note that this limiting switching time applies only to the write half cycle, the read-cycle time being limited only by available drive. The total memory cycle time of a linear-select memory can therefore be made substantially less than that of its coincident-current counterpart. The price paid, of course, is that of the relatively large number of drive circuits required, one for each word in the memory.

Memory Array Organizations

The basic geometrical parameters of an array are ultimately established by the system specifications of the memory unit. The particular items of interest to the circuit designer at this stage of the design include (a) total number of words to be stored in the memory unit, (b) number of bits per word, and (c) number of bits in parallel to be written into or read from the memory. It is the job of the circuit designer to manipulate these parameters in such a way that a reliable balance is achieved between the number of drive circuits required and the complexity of each drive circuit.

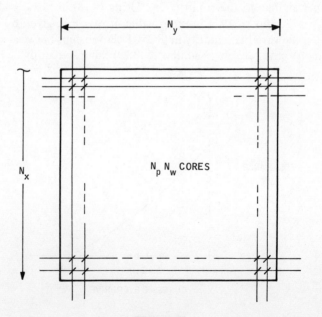

Figure 8.23

Consider the simplest case in which the number of bits to be handled in parallel is equal to the number of bits per word. Term this quantity N_p, and term the total number of words in memory as N_w. The most straightforward method of constructing an array might appear to be a rectangular arrangement containing a total of $N_p N_w$ cores, as in Figure 8.23. The number of drive lines entering the array along the x axis is N_x, and the number of lines entering along the y axis is N_y. The total number of drive circuits to operate this array is $N_x + N_y$, where $N_x N_y = N_p N_w$. Under the most fortunate circumstances $N_p N_w$ may be a perfect square, in which case the minimum number of drive lines in each direction is $N_x = N_y = \sqrt{N_p N_w}$. Notice that in this arrangement any core may be individually selected, and that reading or writing may be accomplished by means of pulses of the proper polarities on the appropriate pair of drive lines.

A substantial reduction in the total number of drive circuits required may be made, however, by noting that N_p x lines and N_p y lines must be

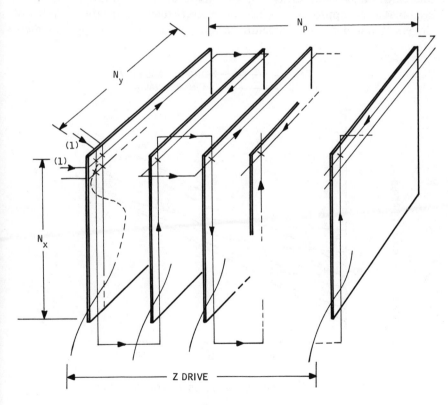

Figure 8.24

simultaneously active in order to read N_p bits in parallel from the memory. It is possible, of course, to arrange the matrix in such a way that all N_p cores required in any given reference to the memory happen to be strung upon a single x line and a single y line. Then only one x line and one y line need be active in any given memory cycle. The result is the classical three-dimensional form of the memory core storage array (see Figure 8.24). In this arrangement the matrix is broken up into a series of N_p arrays, each containing N_w cores. The x and y drive lines are strung continuously through one plane after another.

The total number of drive circuits required for this arrangement now becomes $N_x + N_y + N_z$, where $N_z = N_p$, and $N_x N_y = N_w$. Systems specifications will usually permit the selection of N_w as a perfect square. In this case the minimum number of total drive circuits required for the storage array becomes $2\sqrt{N_w} + N_p$. In Figure 8.25 a comparison is made between the total number of drive circuits required by the two-dimensional and three-dimensional arrays for various values of N_p and N_w. It can be seen that as N_w approaches values much larger than N_p, the total number of drivers required for the three-dimensional array becomes approximately

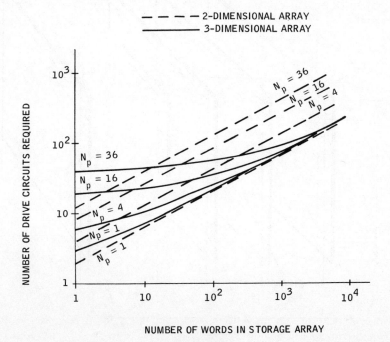

NUMBER OF WORDS IN STORAGE ARRAY

Figure 8.25

$2\sqrt{N_w}$ while the number of drivers required for the two dimensional array becomes $\sqrt{N_p}$ times greater. There is thus usually a decided advantage in favor of the three-dimensional array except in the trivial case when $N_p = 1$.

Systems specifications almost invariably call out the number of words in a memory unit to be some power of the radix of the computer; that is, $N_w = r^{n_w}$. This is done in order to simplify the addressing of words in memory and to make maximum use of the available address bits of the computer. It might appear at first glance that if n_w is even, the selection of N_x and N_y would be almost automatic. If we let $N_x = r^{n_x}$ and $N_y = r^{n_y}$, then $n_x + n_y = n_w$. Provided n_w is even, a solution resulting in minimum components is $n_x = n_y = n_w/2$. Although this is certainly a possible solution, and one that results in the minimum number of drive circuits, it may not be optimum from the standpoint of other memory circuit considerations. For example, the rise-time requirements may be considerably different for the x versus the y drive pulses. Because the rise time that a circuit of given power output is capable of producing is related to the load impedance, and the impedance presented by a line of the array is related to the number of cores strung on it, it is often desirable to sacrifice a few drive circuits in the interest of circuit standardization; that is, it may be possible to use the same drive circuits throughout the memory if it is possible to arrange their load impedances such that memory lines which require fast rise times have low impedances and vice versa. The number of cores on an x drive line, X_c, and corresponding quantities for y and z drive lines are:

$$X_c = N_y N_p,$$
$$Y_c = N_x N_p,$$
$$Z_c = N_w. \tag{8.41}$$

In terms of the number of cores (and therefore impedance) per memory driveline we may wish to impose additional restrictions upon the values of X_c and Z_c, such as

$$X_c = aZ_c = bY_c, \tag{8.42}$$

in which a and b define the relative impedances of the drive lines. These constraints make possible an explicit solution for N_x and N_y; that is,

$$N_x = \frac{1}{b}N_y = a\frac{N_w}{N_p}. \tag{8.43}$$

An organization intermediate between the linear-select two-dimensional (2D) and coincident-current three-dimensional (3D) has been developed to meet special system requirements. Known as a $2\frac{1}{2}$D memory organization, it is characterized by decoding efficiency intermediate between 2D and 3D memories, and by the elimination of the z or inhibit drive line. It can be derived from the arrangement of Figure 8.24 by keeping x drive lines intact (threading all planes), but breaking y drive lines at each plane and providing separate drive for each. Because all cores are now individually selectable, z drive may be eliminated and information inserted by making y current conditional during the write cycle. In a three-wire $2\frac{1}{2}$D system, the z line is retained for use as a sense line during the read cycle. In a two-wire system, the back voltage produced by a switching core is sensed on one of the drive lines. Gilligan [16] has summarized the arguments in favour of the $2\frac{1}{2}$ organization, pointing out in particular that the absence of an inhibit drive line threading a large number of cores means that recovery time following a write cycle is reduced, permitting more rapid and simpler sensing during a subsequent read cycle.

Drive-Circuit Considerations

The conditions imposed on the driving-current waveform in order to produce the desired output voltage, and to select or half-select cores within their tolerance conditions, have been explored above. Before investigating circuits for producing these currents, it is necessary to consider the impedance through which the current must flow. This is primarily a function of the type and number of memory cores strung on a single wire that threads the array. Other considerations involve the geometry of the drive wire in relation to other wires within the array.

Drive-line impedance. Consider the situation schematically pictured in Figure 8.26a. The impedance facing the generator consists of a large number of memory cores and some terminating or other distributed resistance, considered to be lumped in R_T. If this is an x or y drive line then, of all of the cores on the line, N_p cores may or may not switch, depending upon the information previously stored in them. None of the remaining $X_c - N_p$ or $Y_c - N_p$ cores on the line will switch. They will, however, be disturbed by the half-amplitude current flowing in the drive line. A model of these cores consisting of a pure linear inductance describes their behavior to a reasonably good approximation. As a first estimate, therefore, consider the drive line as represented by Figure 8.26b. The box represents the impedance of N_p selected cores, and L_c is the inductance of each of the half-selected cores. A good estimate of the value of L_c can be obtained by relating the measured peak noise voltage to the rate of change of drive current,

$$L_c = \frac{e}{d_i/dt} \cong \frac{V_{hs}(pk)T_r}{I_d}. \tag{8.44}$$

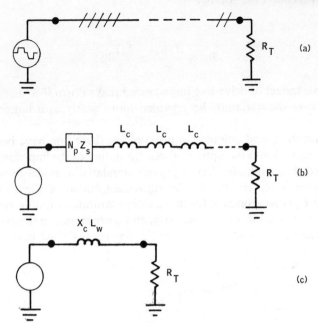

Figure 8.26

Note that $V_{hs}(pk) \neq V_{hs}$, because the former is measured at a time that does not correspond to maximum 1 output time. L_c is exactly equivalent to core winding inductance L_w, for which other estimating equations have been previously derived. Because relatively few cores on any given line are selected, their impedance will usually be relatively low compared to that of the rest of the line. The simplest possible equivalent circuit of the drive line therefore reduces to that of Figure 8.26c.

The drive-signal source is usually limited to some maximum operating voltage, commonly the peak inverse voltage of a transistor if a direct-drive system is used. This places a limit on the number of cores that can be driven from a single source; that is, maximum voltage occurs across the circuit of Figure 8.26c during the rise time T_r. The latter, in turn, must not exceed some minimum for proper core switching. Therefore (approximately),

$$X_c < \frac{E_{max}T_r}{I_d L_w}. \qquad (8.45)$$

Another way of looking at this constraint is to observe that in order to achieve a specified rise time, approximately given by the time constant of the circuit,

$$R_T > \frac{X_c L_w}{\tau} \cong \frac{X_c L_w}{T_r}. \qquad (8.46)$$

This simple model of drive-line impedance breaks down if X_c is very large, in which case the line must be regarded more nearly as a lumped-circuit delay line.

More accurate, and complex, models of the drive line have been investigated using various assumptions about the nature of the impedances being represented. For example, Weeks [3] has simulated a model in which the memory core is represented by the equivalent circuit of Figure 8.27a. In this circuit L_w corresponds to the normal core saturated winding inductance. R_w is taken to be a loss or heating term, and corresponds to the areas of the shaded regions enclosed by the minor loops of disturbed cores, pictured

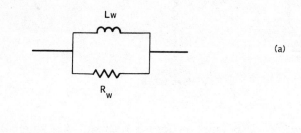

(a)

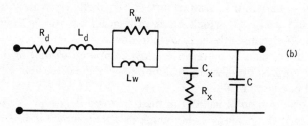

(b)

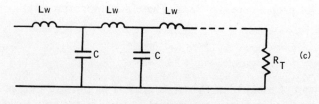

(c)

Figure 8.27

in Figure 8.18. When combined with the properties of the line itself, a section of the equivalent drive-line impedance appears in Figure 8.27b. R_d and L_d are drive-line resistance and self-inductance. C_x corresponds to capacitive coupling to other drive lines, of effective impedance R_x, and C represents capacitive coupling to a common ground. Weeks shows that multiple sections of this type can be represented by transmission-line equations, and obtains good agreement between predicted and measured waveforms using that approximation.

Generally, however, considerably less detailed models are sufficiently accurate for design purposes. For example, Allen, Bruce, and Councill [4] report that even in a rather large memory, all windings approximate lossless transmission lines, and can therefore be reduced to the equivalent of Figure 8.27c. In this case the characteristic impedance is simply

$$Z_0 \cong \left(\frac{L_w}{C}\right)^{1/2}. \tag{8.47}$$

Values of Z_0 on the order of $100\,\Omega$ are typical, corresponding to expected values of L_w in the range of 10^{-8} H, and of C in the range of 1 pF. To avoid undesirable reflections it is necessary to terminate the line in its characteristic impedance, thus establishing the value $R_T = Z_0$.

In large memories the propagation time of pulses may become important, requiring special timing considerations. In the lossless transmission line the delay time per section is on the order of $\sqrt{L_w C}$, and the total pulse delay down the line is therefore estimated on the conservative side by

$$T_d \cong X_c\sqrt{L_w C}. \tag{8.48}$$

Rise time will also be affected, although it is generally limited by that of the driving source. For a step input, the rise time of n sections of a loss-less transmission line is approximately $n^{1/3}$ times the rise time for a single section, which in turn is on the order of the delay time per section. In large arrays it may be necessary to divide the drive lines into segments, with separate drivers for each segment, to avoid excessive delay or rise-time degradation. This is most commonly done in the case of the Z axis (inhibit) line, because in (8.41), Z_c is frequently much larger than either X_c or Y_c.

Drive-circuit requirements. For the core storage array the most straightforward drive arrangement is simply to supply a separate amplifier for each drive line. These drive amplifiers can be addressed directly and the appropriate pulse pattern gated from master x, y, or z timing signals. The driver therefore consists of a decision element and an amplifier (one-input decision element). The decision element passes master timing pulses only when the address, in the case of x or y, or the information, in the case of z, is of appropriate configuration. The amplifier converts this signal from logic level to the current pulse whose characteristics have been previously considered. Before examining the contents of this black box in more detail it is necessary to establish the characteristics of the master timing pulses with greater precision.

It would appear that waveforms on the drive lines as shown in Figure 8.28a would satisfy the requirements of the core storage array. Several problems are associated with such an arrangement of drive pulses. First, the timing of the z pulse may be extremely critical. If it occurs too early, it may partially cancel some of the read-cycle drive leaving cores not wholly switched to $+B_r$. If it occurs too late, cores that should remain at $+B_r$ may be partially switched toward $-B_r$ by the two negative-going x and y pulses. A second problem is the large noise signal that will be induced upon the sense winding by the simultaneous rise of both x and y pulses during the read cycle. This noise can be reduced considerably by applying one of the pulses early, allowing the noise signal to die down, and then applying the other. The third difficulty is the short time between the readout signal from the cores, which may occur toward the end of the read cycle, and the beginning of the z pulse. Thus the readout signal must be amplified, stored, and employed to set gates in an appropriate manner; all this happens prior to the beginning of the z pulse. The result of these various considerations is the drive waveform timing illustrated in Figure 8.28b. This pattern may be regarded as the classical form of the magnetic-core storage-unit memory cycle, although a large number of variations have been practiced. In this sequence the x drive, which normally threads a larger number of cores than the y drive, is allowed to rise first at t_1. It may be permitted to have a rise time considerably longer than that required for optimum core switching, because full drive is not applied to any core during this period. Between t_1 and the rise of the y pulse at t_2, noise induced on the sense winding by the large number of half-disturbed cores threaded by the x drive is allowed to subside. The y pulse must rise within time T_r, as previously discussed, and its duration must be sufficient to permit complete switching of all cores. Both x and y pulses are terminated at t_3. During the write cycle, symmetrical x and y pulses occur between t_4 and t_7, and between t_6 and t_7, respectively. During this same period the z pulse, lasting from t_5 to t_8, must either allow, by its absence, or prevent, by its presence, the switching of the appropriate core bit. Although t_5 and t_8 are normally coincident with t_4 and t_7, respectively, this is not necessarily required. It is only necessary that t_5 occur prior to t_6, and t_8 subsequent to t_7, in order that the z pulse mask the full drive. Notice that x and y drives represent three-level signals: $+I_d/2, 0$, and $-I_d/2$. Because it is not usually convenient to provide for three-level gating, a push-pull arrangement is normally employed with alternate amplifier inputs gated by two-level logic signals. Master timing pulses are therefore illustrated in

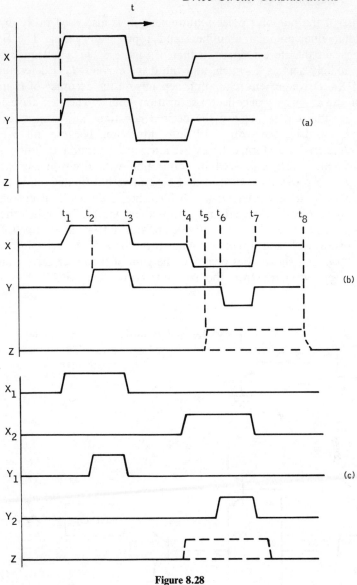

Figure 8.28

Figure 8.28c. One of the major variations upon this minimal scheme is that involving the post-write disturb. This is illustrated by the extension of the z drive pulse in Figure 8.28b, between t_7 and t_8.

The most straightforward drive arrangement is simply direct coupling via an amplifier. This type of drive is particularly appropriate for the z line because of the unipolar pulse requirement. It is also used in *xy* drive, by providing dual unipolar amplifiers and appropriate gating. The simplest drive circuit therefore reduces to that of Figure 8.29*a*. The transistor amplifier merely acts as a switch, and the drive current, I_d is determined by E and R_T. Drive-current regulation depends on the accuracy of E and R_T, and on the uniformity of collector saturation voltage. The base drive current must be sufficient to achieve the desired rise time, and consequently, in general, will drive the switch well into saturation. Because this may lead to intolerably long storage times, various antisaturation techniques such as those previously discussed in connection with decision elements are employed to provide heavy transient base drive but limited static drive. A method of particular interest, which furnishes additional current gain for matching to logic-level signals, is illustrated in Figure 8.29*b*. When the input signal goes positive, the upper drive transistor acts as an emitter follower, with current gain and direct coupling to the base of the drive-line switch. When the input signal goes negative, the gain of the lower drive transistor helps to remove charge from the base of the drive-line switch.

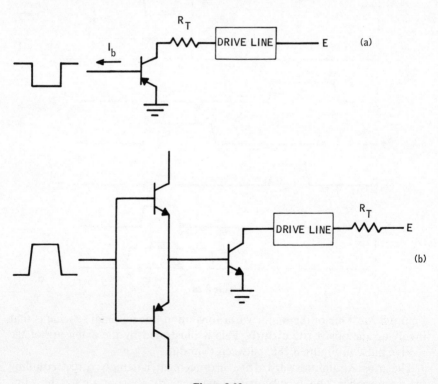

Figure 8.29

For generating bipolar xy pulses by means of direct drive, two switches are necessary. A highly simplified version of one type of circuit is illustrated in Figure 8.30. Input signal levels are OFF, greater than 0 V, and ON, less than $-E$. Operation of the circuit is as follows. If the input to T_1 is at 0 V, it is cut off; if the input is below $-E$, the base is clamped to approximately $-E$ by the base-emitter junction, and the base drive becomes E/R_1. Assuming this is sufficient to drive T_1 into saturation, it passes a current approximately E/R_T through the drive line in a positive sense from right to left. If the input to T_2 is at 0 V, it is also cut off. If its input falls below $-E$, assuming T_2 saturates, the base drive again becomes E/R_1, and the collector current becomes E/R_T. Supply voltages and resistors may, of course, be adjusted to compensate for saturation voltage drops across the switches.

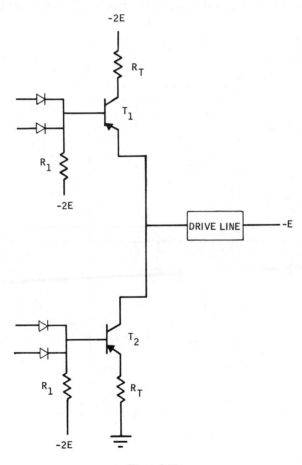

Figure 8.30

Because of the relatively large number of x and y drive lines, it is frequently advantageous to gate drive currents at the lines, rather than at the switches. A highly simplified schematic of one such arrangement is illustrated in Figure 8.31, in which drive lines are regarded as arrayed two-dimensionally. The coincidence of drive on the appropriate pair of coordinates selects the line to be driven.

Although the x and y drive circuits must produce signals of both polarities, the waveform has the advantage of being symmetrical about zero. Therefore it is sometimes convenient to employ a pulse transformer in push-pull arrangement for coupling to the drive line. Generally, the number of x and y drive circuits is much greater than that for z, and circuit simplicity is at some premium. Therefore it is also sometimes convenient to generate the required drive at a single source and route it by means of simple switches to the

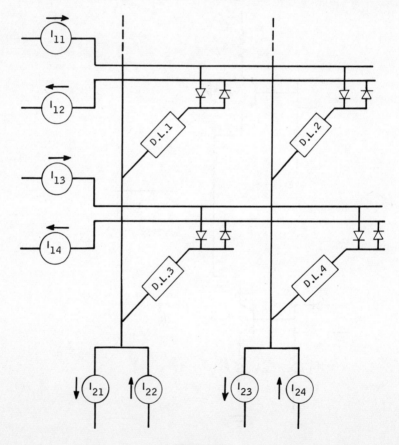

Figure 8.31

appropriate line. The switches must meet the following requirements:
(a) pass at least the nominal drive current I_d, (b) not produce an excessive
drop in the load circuit, and (c) close and open within times compatible with
memory cycle timing. These requirements can usually be met with the simple
arrangement shown schematically in Figure 8.32. If, for example, the drive
signal is supplied in common to all transformer center-taps, it will flow
through whichever leg is switched ON. The switch must be turned ON with a
signal sufficient to keep the transistor in saturation under worst-case conditions. The drop across the switch can be considered to be a small impedance
lumped with that of the load, and, in particular, variations in the drop
constitute even smaller variations in total load impedance. In some cases
the switch need meet only modest transient requirements, because, if timing
permits, it is possible to turn it ON before the arrival of the drive pulse, as
indicated in Figure 8.32.

The drive-line transformer must (a) support the necessary voltage-time
integral that appears across the drive line, (b) prevent excessive "droop"
in the current waveform on the drive line, and (c) provide the necessary rise
time. The transformer circuit is indicated in Figure 8.33*a*, in which the impedance of the switch has been neglected. The transformer equivalent circuit
is shown in Figure 8.33*b*, in which L_w and L_m are the winding (including
primary and secondary leakage) and magnetizing inductances of the transformer, respectively, and I_0 and R_c are equivalent transformer losses. A
unity turns ratio is assumed without loss of generality.

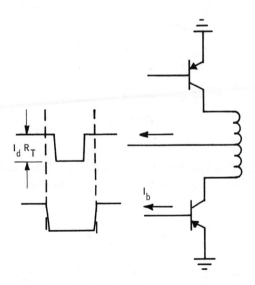

Figure 8.32

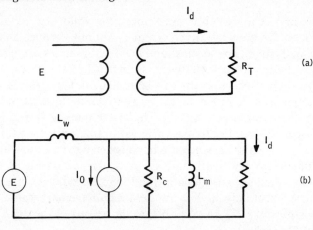

(a)

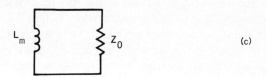

(b)

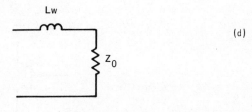

(c)

(d)

Figure 8.33

The supporting flux constraint imposes a lower limit on the number of turns on the transformer. Because the applied voltage must be approximately, $E = I_d Z_0$,

$$N > \frac{I_d Z_0 T}{\phi \times 10^{-8}}, \tag{8.49}$$

in which T is the drive-pulse duration and ϕ the transformer flux swing. It is generally also desirable to minimize losses by making $R_c \gg Z_0$, thereby requiring

$$N^2 \gg \frac{S_w p Z_0}{0.4\pi\phi \times 10^{-8}}.$$ (8.50)

With this condition met, the approximate ac equivalent circuit during the flat top of the pulse is shown in Figure 8.33c. The amount of droop is $\Delta E/E \cong T/\tau$, with τ being the time constant of the equivalent circuit. Using the approximate expression for L_m previously derived in terms of toroidal core material properties and geometry and letting $\mu_p \equiv (B_s + B_r)/H_c$, we obtain

$$\frac{\Delta E}{E} \cong \frac{5Z_0 T}{N^2 \mu_p W \times 10^{-8}},$$ (8.51)

in which W is the width of the transformer toroid.

During the rise time T_r, with $L_m \gg L_w$, the former may be neglected, and the equivalent circuit reduces to that of Figure 8.33d. Equating T_r to the circuit time constant and using the approximate expression for L_w in terms of the squareness ratio of the transformer material, we obtain

$$T_r \cong \tau \cong \frac{N^2 \mu_s W \times 10^{-8}}{5Z_0},$$ (8.52)

in which $\mu_s \equiv (B_s - B_r)/H_c$. With maximum values of both fractional droop (Δ) and rise time (T_r) specified, (8.51) and (8.52) impose a constraint upon the properties of the transformer core material, namely,

$$\frac{\mu_p}{\mu_s} > \frac{T}{T_r\Delta}.$$ (8.53)

Read-Circuit Considerations

When considering the memory unit read amplifiers, the circuit designer emerges from the somewhat self-contained region of magnetic storage and drive arrays to make contact once more with the outside world of the overall digital computer. Logic signal levels, timing, and matching to standard circuits must be considered. Although embodying no new principles in circuit techniques, design of the read amplifier encompasses an interesting combination of circuit problems that illustrate some of the concepts previously considered.

Because the sense winding is likely to thread alternate cores or alternate halves of the bit plane in opposite polarity directions, the ideal waveform to be amplified and discriminated may be any one of the four patterns shown in Figure 8.34. We may assume the logic to be such that if the output is a 1,

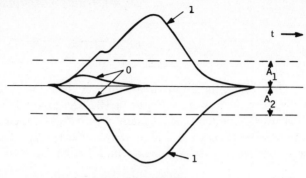

Figure 8.34

a pulse is to be generated or some decision element enabled. If the output is 0, no action is to occur. It would appear from the illustrated waveform that rectification of the signal, amplification, and clipping at some level between maximum 0 and minimum 1 output would satisfy the requirements.

Unfortunately the output waveform does not appear in such a pure environment as that of Figure 8.34. Despite every precaution to cancel drive-pulse noise or eliminate coupling among drive lines, some irreducible noise is induced on the sense line, its shape and magnitude being dependent upon the size, geometrical configuration, and winding arrangement of the storage array. Typical read-line signals may therefore appear as in Figure 8.35, in which the desired signal occurs in the presence of, but not at the same time as, noise that may be greater in amplitude. If two discriminators are located at levels $A1$ and $A2$ and their outputs are clipped at two levels, the resulting possible waveforms are as shown. It is apparent that noise may be eliminated and signal selected by making up the OR decision function $S(A_1 + A_2)$, where S is a strobe or sample pulse occurring at the peaking time of the output signal. The Z or information flip-flop, which must control the Z driver during the write cycle, can then be turned on in the presence of, or remain off in the absence of, the decision element output. The logic of the Z flip-flop must be such that it is cleared to the 0 state at or prior to the initiation of the read cycle.

Some of the operations to be performed by the read amplifier may now be diagrammed as in Figure 8.36, although these operations are by no means always performed in the order illustrated. An inverter is inserted in one leg of the dual circuit arrangement in order to permit the use of identical subsequent stages for operating on both positive and negative portions of the signal. Limiters are inserted to prevent saturation—and consequent slow recovery of the following circuits—by noise that may be larger in amplitude than the desired information. A discriminator distinguishes between 0 and 1

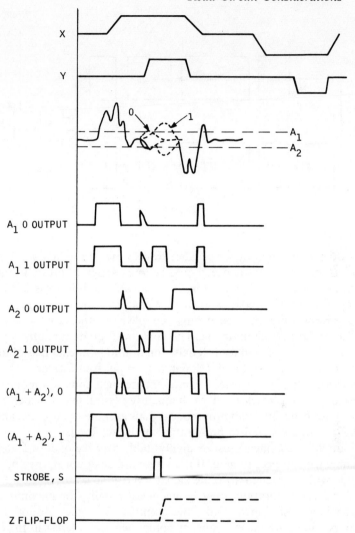

Figure 8.35

signal levels, and after amplification, clipping places the signal at logic level. Once the signal is at logic level, OR and AND functions may be performed by standard decision circuits.

This idealization does not, however, complete the enumeration of sense amplifier requirements. The amplifier input impedance must be sufficiently high to avoid loading memory cores on the sense line with what would

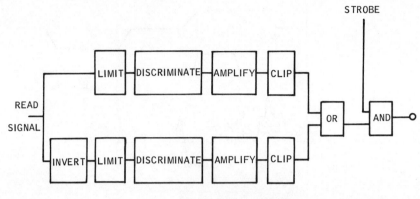

Figure 8.36

otherwise amount to a low-impedance transformer secondary winding on a switching core. The impedance should be low enough to prevent unwanted reflections due to line termination. A reasonable compromise is to match the amplifier input impedance to the line characteristic impedance Z_0.

The sense winding may be strongly coupled to either x or y drive lines, as well as to the z drive inhibit line. This coupling is predominantly capacitive and can result in large induced signals common to both channels of the dual amplifier of Figure 8.36. Careful design of an input transformer—which also serves the function of inverter—can attenuate such common-mode signals considerably. Nevertheless, it is frequently desirable to make specific provision for canceling common-mode signals at some point in the amplifier chain, if maximum signal-to-noise ratio is to be achieved.

The amplifier channels must be highly stable over the operating temperature range with respect to gain. If the stages are dc-coupled, they must also be stable with respect to absolute signal level; if ac-coupled, usually some means of level restoration must be included. Finally, for maximum noise rejection it is desirable to make the amplifier frequency response band-limited, passing only those frequency components of interest in the desired read signal, and rejecting other frequency components that may be lumped as noise.

Impedance matching to the line, signal inversion, and common-mode rejection are frequently accomplished by means of transformer coupling between the sense line and an emitter-biased feedback amplifier, as illustrated schematically in Figure 8.37. The impedance looking into the bases of the transistors is

$$R_{in} \equiv \frac{\partial e_b}{\partial i_b} \cong \frac{R_2}{1 - \alpha_N}. \qquad (8.54)$$

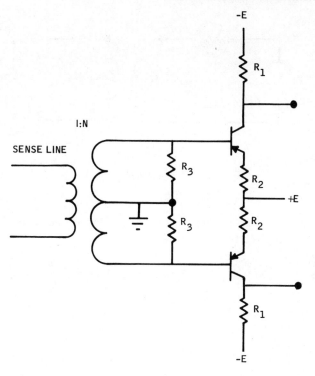

Figure 8.37

This is generally quite large, and the line impedance-matching requirement is therefore satisfied by R_3.

For a perfect transformer, common-mode noise rejection will be complete, because the signal voltage appearing on the secondary is the turns ratio times that across the primary. Thus, from Figure 8.38,

$$E_2 = \frac{N_2}{N_1}(-E_{n1} + E_s + E_{n2}) \cong NE_s, \qquad (8.55)$$

in which E_s and E_n are the signal and noise voltages, respectively, provided the noise voltages induced on the two legs of the sense winding are indeed nearly equal; that is, $E_{n1} \cong E_{n2}$. In a practical transformer some capacitive coupling will also exist, as indicated by the dotted component of Figure 8.38. Because this signal appears at the input to both channels of amplification in Figure 8.37, it can be reduced by similar transformer coupling in a later common-mode rejection stage. The standard differential amplifier, illustrated in schematic form in Figure 8.39, is also sometimes used for

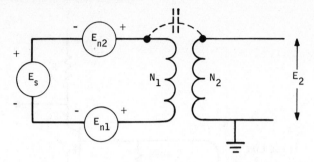

Figure 8.38

common-mode rejection. The ac signal appearing at one of the collectors of this amplifier is given by

$$e_{c1} = \alpha_1 \frac{R_{c1}}{R_E} \frac{1}{1 + \Delta_1 - 1/(1 + \Delta_2)} \left(e_{b1} - \frac{1}{1 + \Delta_2} e_{b2} \right), \qquad (8.56)$$

in which

$$\Delta_1 \equiv (1 - \alpha_1) \frac{R_{b1}}{R_E},$$

$$\Delta_2 \equiv (1 - \alpha_2) \frac{R_{b2}}{R_E},$$

and subscripts refer to the two sides of the amplifier. Provided $\Delta_1 = \Delta_2 = \Delta \ll 1$, and $R_{b1} = R_{b2} = R_b$, (8.56) reduces to the approximation

$$e_c \cong \frac{\alpha R_c}{2 R_b} [e_{b1} - (1 - \Delta) e_{b2}]. \qquad (8.57)$$

If $e_{b1} = e_{b2} = e_b$, the net signal is proportional to Δe_b. If $e_{b1} = -e_{b2} = e_b$, the net signal is proportional to $2e_b$. Therefore common-mode signals are attenuated relative to difference signals by the ratio $\Delta/2$.

Input transformer coupling provides one point at which some degree of bandpass action can be achieved. Additional shaping of the amplifier bandpass characteristic can be achieved in subsequent amplifier stages by appropriate bypass and peaking circuits.

Perhaps the major problem in sense amplifier design is that of discriminating against the large noise signals induced on the sense line by drive signals, while passing the relatively small signal of a switching core. Although these do not occur at the same time, some overlap is inevitable. As

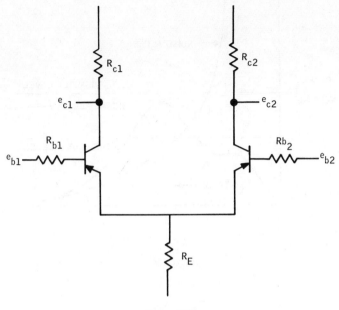

Figure 8.39

Tsui [13] has pointed out, signal-alone and noise-alone waveforms tend to appear as in Figure 8.40, implying two requirements on the amplifier. The first is that the amplifier must be capable of recovering from the noise signal in time to pass the core switching signal. The second is that the desired signal must be sensed at a particular time (strobe; see Figure 8.40) in order to maximize signal-to-noise ratio.

It is difficult to design an amplifier with a large enough dynamic range to accommodate the noise signals, and therefore a means for limiting the signal swing without sacrificing response time is usually provided. The ideal value at which to limit the signal is a voltage near the peak of the minimum 1 readout level. Because limiting is commonly accomplished by diodes, however, and because the memory signal tends to be an order of magnitude smaller than the normal forward drop across a diode, some amplification must be performed prior to limiting. On the other hand, to avoid relatively long transistor recovery times, it is desirable to keep the gain modest, so that noise signals still lie within the linear dynamic range of the stage. Because noise signals tend to be about an order of magnitude larger than core switching signals, these conflicting requirements are frequently met by a first-stage gain of approximately one order of magnitude. This can be readily achieved in a circuit of the type of Figure 8.37, in which the voltage gain is approximately

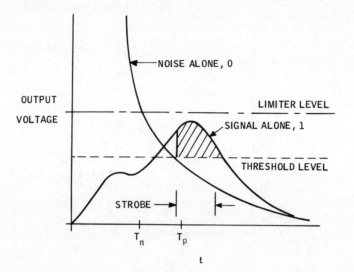

Figure 8.40

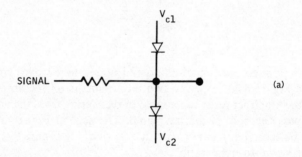

(a)

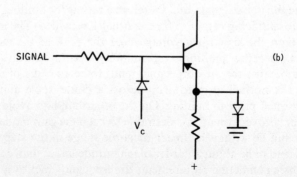

(b)

Figure 8.41

430

$G \cong R_1/R_2$. Both bipolar and single-sided limiting have been employed. In Figure 8.41a, the signal is limited between clamp voltages V_{c1} and V_{c2}. In some cases, specific voltages with $V_{c1} < V_{c2}$ have been used to allow a wide signal swing: in others, $V_{c1} = V_{c2}$, and the forward threshold voltages of the diodes themselves are used to provide the allowed signal swing. In Figure 8.41b, positive signals can do no more than cut off the amplifier stage, so limiting is provided only in the negative direction via the diode to V_c. If dc amplifiers are used, careful attention must be given to operating-point stability in order to ensure that the desired signals will indeed lie within the range of the limiter circuit.

Strobing is invariably accomplished by pulsing some point in the channel that is normally cut off or clamped to a fixed level. Two illustrative schemes are shown in Figure 8.42a. In one case E_2 is a fixed value, say 0 V. With E_1 normally negative, the transistor is cut off, and $I = 0$. When E_1 goes positive, if $E_s > E_2$, the transistor remains cut off, and I flows through the output diode where it may act as a trigger signal for a logic-level flip-flop. If $E_s < E_2$, the transistor conducts, and I flows through it without unblocking the output diode. A slightly different version of this circuit operation occurs when E_1 is a fixed positive voltage, and the strobe pulse appears on E_2. In this case, with E_2 negative, the transistor is cut off and the output diode is blocked. When E_2 goes positive, if $E_s > E_2$ the transistor remains off, and I flows into the trigger circuit. If $E_s < E_2$, the transistor conducts and the output diode remains blocked. Note that the strobe function is essentially that of AND logic, as indicated in Figure 8.36. This is particularly apparent if Figure 8.42a is redrawn as in Figure 8.42b, in which the base-emitter junction of the transistor becomes one input to a two-input positive diode AND gate.

In most cases it is difficult to design a high-gain dc amplifier with sufficient stability to ensure that signals will be correctly discriminated. Thus most amplifiers involve some kind of ac coupling. As pointed out by Goldstick and Klein [12], ac coupling introduces special problems that must be given careful attention. In particular, signal levels become sensitive to the pulse sequence, and it is usually necessary to provide some method of dc restoration if discrimination is to be reliable. The crux of the problem is illustrated in Figure 8.43. The sense signal, after limiting, may be regarded as of the ideal form of Figure 8.43a, in which a large noise signal limited to E_L precedes a smaller sense signal that may be of either polarity. The time T_n is defined in Figure 8.40 as that during which the noise induced by x-y drive-signal rise is greater than the limiter level. The time T_p is approximately the peaking time of the memory core. Because the low-frequency response of the amplifier must be sufficient to pass the readout signal, but does not extend to dc, the actual signal at a stage following limiting will appear as in Figure 8.43b and

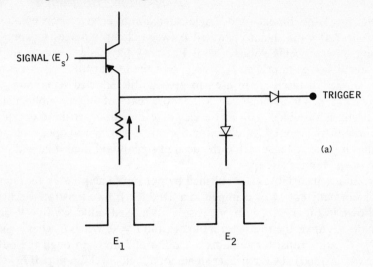

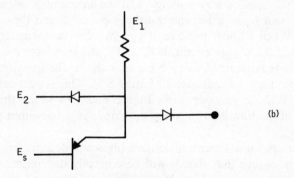

Figure 8.42

c for positive and negative sense signals, respectively. Because $T_p - T_n$ is generally short, and the time constant of the circuit is substantially longer than T_n, the approximate peak signal magnitudes for the two cases are

$$|E_+| \cong E_s - E_L \frac{T_n}{\tau},$$

$$|E_-| \cong E_s + E_L \frac{T_n}{\tau}. \tag{8.58}$$

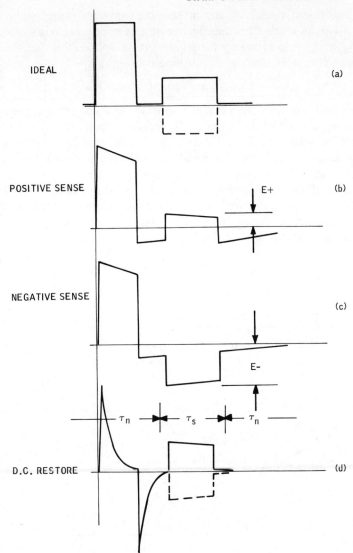

IDEAL   (a)

POSITIVE SENSE   E+   (b)

NEGATIVE SENSE   (c)

E−

τ_n τ_s τ_n

D.C. RESTORE   (d)

Figure 8.43

Methods for alleviating this problem are commonly termed *dc restoration*, implying a return toward the ideal signal by clamping to a reference voltage at some reference time. An equivalent description might be *frequency-response manipulation*, because most methods involve a deliberate change in circuit time constant.

The equivalent circuit of one ac-coupled amplifier stage is shown in Figure 8.44*a*, in which R_0 is the effective output impedance of the amplifier. With $R_0 \ll R_L$, it is necessary for good signal amplification that the read-cycle time T be related to the circuit time constant by

$$T \ll \tau_s \cong R_L C. \qquad (8.59)$$

However, in order to be sure that the output is at a known (and fixed) level prior to the arrival of the signal, it is necessary to fix the charge on the capacitor between T_n and T_p. To accomplish this, the circuit time constant must be less than $T_p - T_n$. The usual method for accomplishing this is to clamp the output to a known level, except during the signal time, by means of one or more diodes and pulse sources, as illustrated schematically in Figure 8.44*b*. Provided the diode forward resistance is small compared to R_0, the ac equivalent circuit, during the time the clamp signal is negative, is that of Figure 8.44*c*, and it is desired that

$$T_p - T_n \gg \tau_n \cong R_0 C. \qquad (8.60)$$

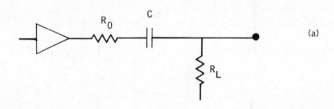

(a)

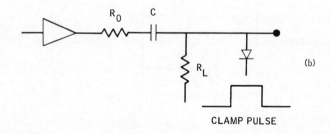

(b)

CLAMP PULSE

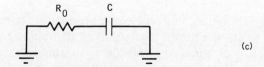

(c)

Figure 8.44

The requirement imposed upon the output impedance of the amplifier and the input impedance to the following stage is therefore, from (8.59) and (8.60),

$$\frac{R_0}{R_L} \ll \frac{T_p - T_n}{T}. \tag{8.61}$$

With this condition satisfied, the resulting waveform for the ideal input of Figure 8.43*a* is illustrated in Figure 8.43*d*, in which the time constant of the circuit is switched from τ_n to τ_s and back during appropriate portions of the memory cycle.

REFERENCES

[1] R. Betts and G. Bishop, "Ferrite Toroid Core Circuit Analysis," *IRE Trans. Electron. Computers*, **EC-10**, 1, 51–56 (March 1961).

[2] E. M. Gyorgy, "Rotational Model of Flux Reversal in Square Loop Ferrites," *J. Appl. Phys.*, **28**, 1011–1015 (September 1957).

[3] W. T. Weeks, "Computer Simulation of the Electrical Properties of Memory Arrays," *IEEE Trans. Electron. Computers*, **EC-12**, 5, 874–887 (December 1963).

[4] C. Allen, G. Bruce, and E. Councill, "A 2.18 Microsecond Megabit Core Storage Unit," *IRE Trans. Electron. Computers*, **EC-10**, 2, 233–237 (June 1961).

[5] A. J. Meyerhoff (ed.), *Digital Applications of Magnetic Devices*, Wiley, New York, 1960.

[6] A. Robinson, et al., "A Digital Store Using a Magnetic Core Matrix," *Proc. IEE*, **103**, B, suppl. 2, 295–301 (April 1956).

[7] R. Minnick and R. Ashenhurst, "Multiple Coincidence Magnetic Storage Systems," *J. Appl. Phys.*, **26**, 575–579 (May 1955).

[8] E. L. Younker, "A Transistor-driven Magnetic-core Memory," *IRE Trans. Electron. Computers*, **EC-6**, 1, 14–20 (March 1957).

[9] O. Landsverk, "A Fast Coincident Current Magnetic Core Memory," *IEEE Trans. Electron. Computers*, **EC-13**, 5, 580–585 (October 1964).

[10] B. D. Jimerson, "Designing a Small Core Memory for Economy and Reliability," *Solid State Design*, **5**, 4, 31–34 (April 1964).

[11] A. Melmed and R. Shevlin, "Diode-steered Magnetic Core Memory," *IRE Trans. Electron. Computers*, **EC-8**, 4, 474–478 (December 1959).

[12] G. Goldstick and E. Klein, "Design of Memory Sense Amplifiers," *IRE Trans. Electron. Computers*, **EC-11**, 2, 236–253 (April 1962).

[13] F. F. Tsui, "Improving the Performance of the Sense-amplifier Circuit through Pre-Amplification Strobing and Noise-matched Clipping," *IRE Trans. Electron. Computers*, **EC-11**, 5, 677–683 (October 1962).

[14] R. K. Richards, *Digital Computer Components and Circuits*, Van Nostrand, Princeton, N.J., 1957.

[15] J. Millman and H. Taub, *Pulse and Digital Circuits*, McGraw-Hill, New York, 1956.

[16] T. J. Gilligan, "2½D High Speed Memory Systems—Past, Present and Future," *IEEE Trans. Electron. Computers*, **EC-15**, 4, 475–485 (August 1966).

[17] J. R. Brown, Jr., "First- and Second-Order Ferrite Memory Core Characteristics and their Relationship to System Performance," *IEEE Trans. Electron Computers*, **EC-15**, 4, 485–508 (August 1966).

EXERCISES

8.1 The memory cores in a coincident-current memory have the following properties:

$$H_c = 1.0 \, \text{Oe},$$
$$S_w = 0.5 \times 10^{-6} \, \text{Oe-}\mu\text{s},$$
$$B_s = 2,000 \, \text{G},$$
$$B_r = 1,900 \, \text{G},$$
$$\text{OD} = 30 \, \text{mils},$$
$$\text{ID} = 20 \, \text{mils},$$

(a) What is the nominal drive current if driver tolerances are ± 3 percent?

(b) What is the memory cycle time if all drivers are producing nominal currents?

8.2 A linear-selection memory uses the same cores as in Problem 8.1. In this case, however, driver tolerances are ± 10 percent over the operating range.

(a) What are the nominal values of word-select (xy) and bit-plane (z) current drive pulses for a minimum write-cycle time?

(b) If the word-line driver is limited to 1 A, what is the worst-case minimum read-cycle time?

8.3 A linear-select memory is to be designed around cores possessing the following characteristics:

$$H_c = 1.0 \, \text{Oe},$$
$$S_w = 1.0 \, \text{Oe-}\mu\text{s},$$
$$5r_o = 0.5 \, \text{cm},$$
$$\text{OD/ID} = 1.6.$$

The desired total memory cycle time is $3 \, \mu\text{s}$, but the maximum drive current from any source is limited to 1 A.

(a) What is the minimum read period time?

(b) What must be the limiting tolerances on drive currents during write?

8.4 The following data results from measurements taken on a series of magnetic memory cores:

$$uV_1 = 65 \, \text{mV},$$
$$rV_{h1} = 4 \, \text{mV},$$
$$rV_1 = 60 \, \text{mV},$$
$$rV_{hz} = 3.5 \, \text{mV},$$
$$rV_z = 6 \, \text{mV}.$$

What is the maximum size of a (square) memory plane in a coincident-current memory for a worst-case signal-to-noise ratio of $2:1$?

8.5 The parameters of cores chosen for a coincident-current magnetic core memory are as follows:

$$H_c = 1.0 \, \text{Oe},$$
$$S_w = 0.2 \, \text{Oe-}\mu\text{s},$$
$$5r_o = 0.5 \, \text{cm},$$
$$\text{OD/ID} = 1.6.$$

(a) What is the minimum memory cycle time if perfect current drivers are used (zero tolerances)?

(b) What is the value of drive current corresponding to this case?

(c) What are the limiting tolerances on drive-current values (all equal) if the desired memory cycle time is $2 \, \mu\text{s}$ (read plus write)?

8.6 A 4,096-word coincident-current memory is planned using the following cores:

$$H_c = 1.0 \text{ Oe},$$
$$\phi_s = 2.00 \text{ M},$$
$$\phi_r = 1.95 \text{ M},$$
$$5r_o = 0.31 \text{ cm},$$
$$\text{ID/OD} = 0.60,$$
$$S_w = 0.25 \times 10^{-6} \text{ Oe-}\mu\text{s}.$$

(a) Find the worst-case minimum memory cycle time if all driver tolerances are ± 3 percent.

(b) What is the nominal drive current (x, y, or z)?

(c) Estimate the minimum output voltage on the sense line if the selected core stores a 1 and $V_\delta = 0.1$ mV.

(d) What is the approximate inductance of the z drive line?

8.7 In a word-select memory system the word drivers can produce at most 450 mA during read. Core parameters are $H_c = 1$ Oe, $5r_i = 0.1$ cm, $5r_o = 0.15$ cm, $S_w = 1.0$ Oe-μs.

(a) What are the word and bit-plane drive currents during the write cycle under zero-tolerance conditions?

(b) What is the memory cycle time (read plus write)?

8.8 A coincident-current core memory is to use the same cores as in Problem 8.7, but with improved $S_w = 0.5$ Oe-μs. Additional core parameters are $V_1 = 50$ mV, $V_z = V_{hs} = 5$ mV, $V_\delta = 0.2$ mV.

(a) What is the memory cycle time (read plus write)?

(b) What is the worst-case signal-to-noise ratio if the memory of 8,192 words is arranged in planes of 64×128 cores?

8.9 Design a drive circuit of the type illustrated in Figure 8.30 for the coincident-current memory of Problem 8.6. Required rise time for driving the specified core is a minimum of 0.2 μs. Take available power supplies to be ± 24 and ± 12 V, as needed. Transistors available for this purpose have minimum current gain $\beta \geq 30$ and switching constant $\tau_a \leq 0.4\,\mu$s.

8.10 Design a first-stage preamplifier of the type illustrated in Figure 8.37 of this chapter for the coincident-current memory of Problem 8.6. Take available power supplies and transistors to be the same as those in Problem 8.9.

Chapter 9

Magnetic-Surface Recording

Recording of information by making marks on a thin layer of material is as old as the invention of writing. Today the handwritten or typed records of the pre-computer era have their closest counterpart in the punched card and punched paper tape storage systems. These storage media are capable of very low costs per bit of information. They suffer from the disadvantage of requiring separate mechanical motions for writing, access, and reading, which for the information usually required, demands a total time on the order of seconds. They are also incapable of erasure, in the sense that the old media must be discarded and new records generated. At the other end of the memory unit scale is the magnetic-core storage of the previous chapter. Although providing access to information at microsecond rates, it is not economically justifiable when large quantities of data are involved. Magnetic-surface recording lies at approximately the geometric mean between these two extreme storage methods with respect to both access time and cost.

Although the magnetic surface has been organized into a variety of different geometrical configurations, the most widely used are the drum and tape configurations shown in Figure 9.1*a* and *b*. In the first system read and write heads are positioned permanently, close to the circumference of a continuously rotating drum whose surface is coated with magnetic material. Although access is inherently serial, the system may be organized into a parallel-serial or parallel storage by taking outputs from several heads at one time. The functions of recording and readback may be performed by a single head, or separate write and read heads may be positioned on the same track. In the former case maximum access time to any information is one revolution, and in the latter case, maximum access time is determined by the relative angular positions of the two heads. The physical size of the drum

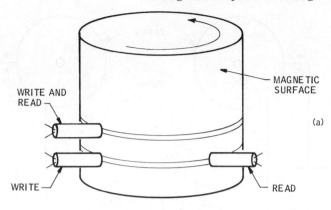

(a)

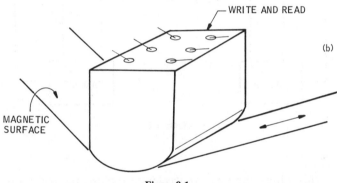

(b)

Figure 9.1

is normally limited by dimensional stability under the given environmental conditions. The head-to-drum spacing is determined by the drum and the structure holding the heads, which are normally in contact only at the two bearing points located at either end of the drum. Drum sizes have ranged from an inch to several feet in diameter and from an inch to several feet long.

In magnetic tape recording a long strip of coated tape (commonly 1,200 to 2,400 ft) is passed over a multichannel record-read head. Information is impressed on the magnetic material across the width of the tape (commonly on the order of 1 in.) as shown in Figure 9.1b. Tape is caused to move past the head in one direction or the other by one of two electromechanical pinch rollers that press the tape against continuously rotating capstans (see Figure 9.2). In order to decouple the small inertia of a short length of tape from the large inertia of a full reel, some slack-takeup device is commonly

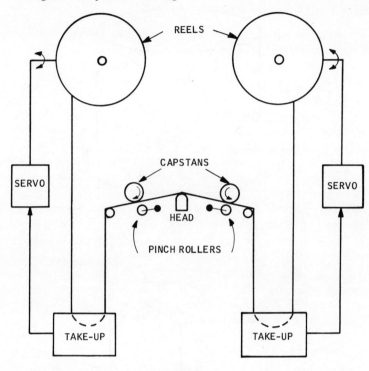

Figure 9.2

provided. This allows the tape to accelerate rapidly, the reel more slowly. A servo, driven by sensing the amount of tape in takeup, maintains this amount approximately constant and must therefore be capable of driving the reels at an angular velocity corresponding to a linear velocity greater than that of the tape over the head. The time to accelerate the tape to normal velocity or vice versa, is called the *start-stop time*, though these are not necessarily the same. In order to insure good tape-to-head contact, slight tension may be placed on both loops of tape in takeup. The surface recording medium, a magnetic iron oxide, is extremely hard and will normally cause wear on the head, the extent depending upon tape, speed, tension, head material, and environmental conditions.

The list of practical problems associated with both drums and tapes is almost inexhaustible. The brief discussion above is presented as an indication to the circuit designer that close collaboration with the mechanical design of a storage system is of overriding importance to successful operation of the unit.

Memory Unit Characteristics

The components making up a surface-recording memory unit are illustrated in Figure 9.3. Although shown separately in the illustration, the write and read functions may be performed by the same head and the two selection systems may be common. Incoming logic-level signals to be recorded must be amplified or matched to the head impedance. This is accomplished by an appropriate pulse amplifier. The selection unit, ideally, simply acts as a multiple-pole switch, connecting one of many heads to the amplifier. The write head must produce sufficient field strength in a portion of the recording medium to leave it in an unambiguous state corresponding to given information, irrespective of the previous state of the medium. The head must perform this operation without disturbing other information in surrounding areas of the medium. When the desired information passes under the read head, magnetic flux from the medium must link the windings. The time rate of change of this flux must produce a voltage above noise level and corresponding in polarity or wave shape unambiguously to the information recorded. After selection from one of many heads the readback voltage must be amplified and further processed (e.g., sampling, cancellation, and logic decision) to reproduce the original information at a level compatible with the remainder of the logic.

The most confining design constraints in this sequence of transformations are imposed by the head-to-medium and medium-to-head stages. Once these designs have been established to satisfy system requirements, the remainder of the circuitry in the storage loop may be determined with comparative facility. Therefore it behooves the circuit designer to understand

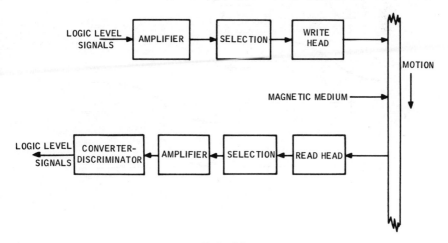

Figure 9.3

the basic relationships underlying the recording and reading processes and the limitations thereof. Our discussion of circuits begins with these processes.

The recording and readback sequence can be described by reference to the schematic illustrations of Figure 9.4a. The record head, with an air gap close to the magnetic medium, is driven by a magnetomotive force $0.4\pi N_w i_w$. This produces a field distribution H in the region of the gap, and more particularly within the medium itself. Because recording must erase previous information, the medium must be driven to saturation, corresponding

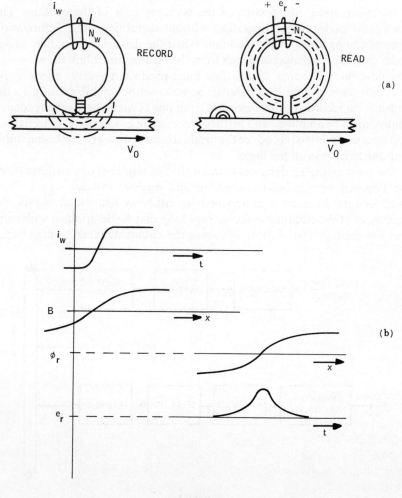

Figure 9.4

to a field strength H_s. When the medium passes from the influence of the record head, it is therefore left with a certain remanent magnetization B determined by the shape of its hysteresis loop and the demagnetizing effect of surrounding nonmagnetic paths. An illustrative curve of B as a function of x, the distance along the medium, is shown in Figure 9.4b. When this magnetization passes under the read head at a later time, the reluctance of the path through the high-permeability magnetic material of the head is less for some of the flux than the reluctance of the path outside of the head. Therefore, a portion of the remanent flux will link the read winding. This flux, $\phi_r(t)$, is a function of the position of the magnetized spot with respect to the center of the air gap. The induced voltage e_r in the read winding will be proportional to the time rate of change of the linked flux; that is, $e_r = N_r(d\phi_r/dt) \times 10^{-8}$. Because x and t are related by the scale factor V_o, the surface velocity of the medium, the open-circuit output voltage will appear as in Figure 9.4b. Notice that if the change of i_w had been from plus to minus, rather than from minus to plus, the polarity of e_r would have been reversed.

The Write Head

In order to derive some useful relationships by means of which we may study the effects of various parameters upon the complete record-reproduce process, it is feasible to divide the study into three distinctly separate inquiries. These are (a) the magnitude and direction of the field strength vector H in the space constituting the neighborhood of a write head carrying current i_w in its windings, (b) the magnitude and direction of magnetization M produced in a magnetic medium that traverses a portion of that space, and (c) the flux produced in a read head in the vicinity of a magnetic medium possessing magnetization M. In this section we shall deal with the first study, which involves determining field strengths in a magnetic circuit containing an air gap.

As we have seen in Chapter 3, the exact analysis of magnetic circuits with air gaps can become excessively complicated for engineering purposes. We must therefore select some appropriate idealizations that permit the understanding of significant relationships, while not too greatly reducing the accuracy of design predictions. The first of these idealizations is that the material of which the record head is constructed possesses very large permeability, approaching infinity. This is justified by the relatively high permeability of most head core materials. For such materials the range of permeabilities is typically $10^2 < \mu < 10^4$. The idealization that $\mu \rightarrow \infty$ allows us to draw field distribution plots with greater facility, and reduces the complexity of calculating air-gap magnetomotive force.

A second idealization is that the presence of the magnetic recording medium in the vicinity of the head has negligible effect on the field distribution. This is equivalent to saying its permeability is unity, equal to that of air. Such an idealization is justified by the considerations that (a) in usual practice (particularly noncontact recording), the magnetic medium occupies only a small percentage of the total volume of space in the vicinity of the head gap, (b) actual permeabilities are in the range of 2 to 3, and (c) it errs on the conservative side. This assumption has the merits of simplifying field plots and permitting the separate treatment of head and medium.

Certain other idealizations can be made, based upon usual head geometry. Referring to Figure 9.5a, we see that head construction is shown schematically to be an assembly of two cores each having a winding $N_w/2$ and front and back air gaps. Because care can be exercised to ensure a close fit at the back gap, its effects may usually be neglected in comparison to those of the front gap. Referring to the detail of Figure 9.5b, we may further assume that all other head dimensions are large in comparison to gap length in the direction of motion, $2g_x$, and head-to-surface spacing S. In particular, g_x and S are small compared to the gap height g_y and gap width g_z constituting the dimensions of the facing pole pieces. For simplicity, it is further assumed that there is no variation in field strength in the direction of gap width (values of z up to $\pm g_z$). The determination of field strength therefore reduces to a two-dimensional problem.

A final observation is that the head dimensions are generally small compared to the ratio of the velocity of propagation of electromagnetic radiation to the switching times of interest. The problem may therefore be treated as a static one in which the write current is constant; that is, $i_w = I_w$.

Although the above set of idealizations appears to restrict unduly the applicability of any conclusions to which we may come, it has come to be regarded as the classical approach to the study of recording, although there is by no means uniformity in the various attacks that have been made on the problem. It will be found, also, that correction terms may be generated for those cases in which one or more assumptions are violated.

An approximate determination of the function $H(x, y, I_w)$ for typical head geometries can be made by means of a field plot. Applying the principles outlined in a previous chapter, the following rules may be used as a guide in preparing such sketches: (a) lines of constant H (flux lines) must intersect the surface of the head at right angles; (b) for symmetrical placement of the write winding, the field plot will be symmetrical about a plane through the centerline of the gap ($x = 0$ plane); (c) for parallel facing pole pieces (planes $x = \pm g_x$) the field in the region between pole pieces will be almost perfectly rectangular; (d) for a symmetrical write winding, lines of constant magnetomotive force must approximately divide the space outside the gap into equal

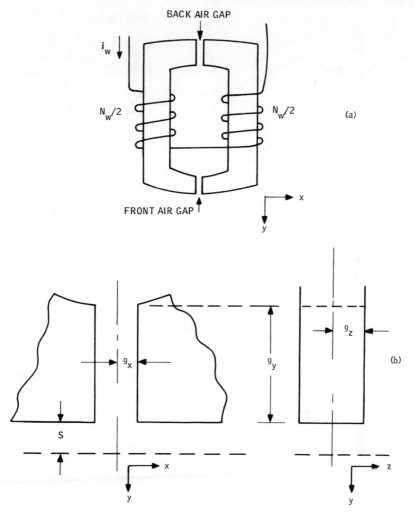

Figure 9.5

angular segments. The rules for right-angle intersections between lines of constant magnetomotive force and constant field strength must of course be observed, and it is visually convenient to plot on the basis of a distorted square.

The plot of Figure 9.6 represents an approximate realization of these rules for a typical head in which the gap makes a right angle with the part of the head facing the magnetic medium. In order to evaluate H at any point,

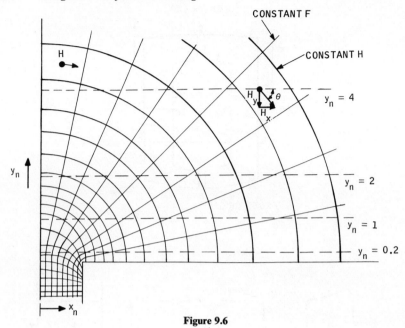

Figure 9.6

a path of integration may be taken along a line of constant H from one pole piece to the other. Thus

$$\oint H \, dp = H p_a + H_i p_i = 0.4\pi N_w I_w, \tag{9.1}$$

in which

    H = field strength along (uniform) path outside head,
    p_a = length of path in air,
    H_i = field strength within head (assuming a uniform cross section),
    p_i = path length in head.

To the extent that the permeability of the head is sufficiently greater than that of air, $\mu_i \gg \mu_0$, then $H_i = B/\mu_i \to 0$. Consequently (9.1) reduces to

$$H p_a \cong 0.4\pi N_w I_w. \tag{9.2}$$

In particular, the field strength in the approximately rectangular field distribution region well within the gap is

$$H_g = \frac{0.4\pi N_w I_w}{2 g_x}, \tag{9.3}$$

and for a path outside the gap,

$$\frac{H}{H_g} = \frac{2g_x}{p_a}. \tag{9.4}$$

It is therefore only necessary to measure p_a in any convenient units to obtain the field strength at a point on the plot. If x and y components of the field strength, H_x and H_y, are desired, the angle θ made by the line of constant H with the horizontal may also be measured. For the coordinates chosen in Figure 9.6, $H_x = H \cos \theta$ and $H_y = -H \sin \theta$. For values of $x < 0$, H_x will be identical, and the sign of H_y will be reversed. The above quantities are graphed in Figures 9.7 through 9.10, in which dimensions are normalized to the gap width according to

$$x_n \equiv \frac{x}{g_x},$$

$$y_n \equiv \frac{y}{g_x}. \tag{9.5}$$

Crossplots of field strength versus distance from the head are presented in Figures 9.11 through 9.13.

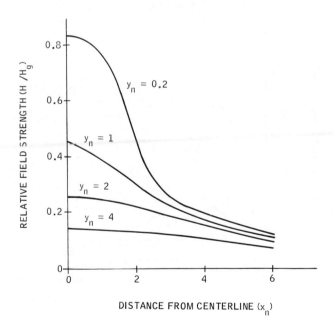

Figure 9.7

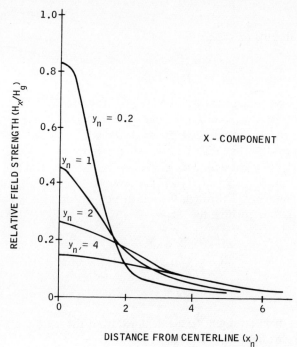

Figure 9.8

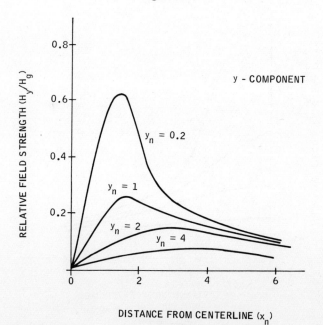

Figure 9.9

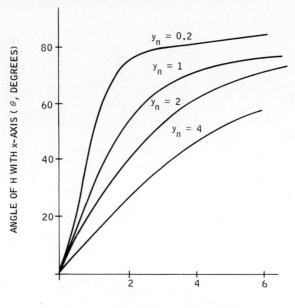

Figure 9.10

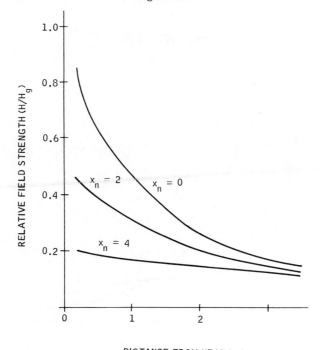

Figure 9.11

449

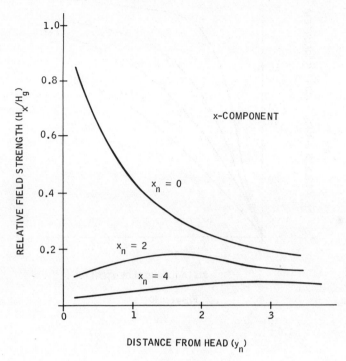

Figure 9.12

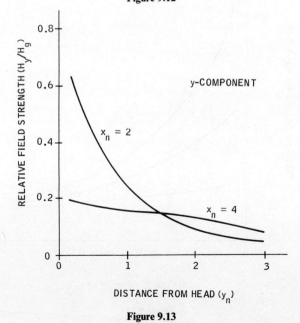

Figure 9.13

Some useful qualitative observations can be made on the basis of the various graphs. It is apparent that field strength is roughly inversely proportional to distance from the head. This holds more closely at the centerline ($x = 0$) and at distances greater than g_x from the head. Note also the sequence of field strengths to which a point traveling along a line parallel to the x axis is subjected. As the point approaches the gap, it will first encounter a strong y field. As it reaches and passes the gap, this will give way to a somewhat stronger x field. As it proceeds away from the gap, it will again come under the influence of a strong y field (of opposite polarity). This sequence holds true whether the path is near or far from the head, the only difference being in magnitude of field strength. Notice that for any given distance from the head, the peak x field is somewhat stronger than the peak y field. Another observable effect having to do with the amount of "spread" of a recording can be demonstrated by means of Figures 9.7 and 9.11. Consider a plane of finite thickness lying between y_1 and y_2, and suppose it is desired to produce some given field strength at the farther boundary (y_2) from the head. This might be the field strength required to saturate the recording medium, in order to eliminate all effects due to previously written information. If such a field strength is applied so as to just saturate the material on the far boundary along the centerline of the head, material on the near boundary some distance from the centerline will likewise be saturated. The distance away from the centerline will be greater for either (a) a thicker plane ($y_2 - y_1$ larger), or (b) greater distances from the head (y_1 larger).

Generally, the quantity of major interest is the x component of field strength. This is because most magnetic recording surfaces are fabricated with very small dimensions in the y direction (thin surface). Therefore any components of magnetization in the y direction are subjected to a large demagnetizing field and will be substantially attenuated relative to x components of magnetization, which are in a direction along the medium.

For digital recording, it is virtually mandatory to record to saturation of the medium throughout the medium. Thus the saturation field strength H_s must be produced at least directly in front of the gap at a distance $r = S + D$, with

$$S = \text{head-to-surface spacing,}$$

$$D = \text{recording medium thickness.}$$

The required ampere-turns for saturation recording are therefore

$$N_w I_w = \frac{H_s(S + D)}{0.4} \tag{9.6}$$

for $S + D$ sufficiently large that $p_a = \pi r$. The field strength as a function of x

and y, from (9.4), is

$$\frac{H(x_n, y_n)}{H_g} = \frac{2}{\pi} \frac{1}{\sqrt{x_n^2 + y_n^2}}. \tag{9.7}$$

This function may be regarded as the "far-field" approximation, applying only to distances relatively far from the gap. Although this is a good approximation for normalized distances $\gg 1$, it has the annoying property of going to infinity at $x = y = 0$. This can be corrected by taking the center of the radius vector, r, to $H(r)$ to be located a distance g_x below $y = 0$, and normalizing (9.7) to unity at $x = y = 0$. This is equivalent to redefining the coordinate center, or replacing y_n by $y_n + 1$. Thus

$$\frac{H(x_n, y_n)}{H_g} \cong \frac{1}{\sqrt{x_n^2 + (y_n + 1)^2}}. \tag{9.8}$$

This approximating function is plotted along the head centerline in Figure 9.14, where it can be compared with the result from the graphical method.

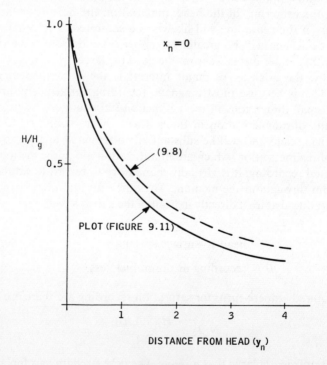

Figure 9.14

The important x component of field strength is given by $H_x = H \cos \theta$, with θ defined in Figure 9.6. Using the approximation (9.8), we obtain

$$\frac{H_x(x_n, y_n)}{H_g} \cong \frac{y_n + 1}{x_n^2 + (y_n + 1)^2}. \tag{9.9}$$

A more elegant approximation to the x component of field strength has been used by Stein [12]. It is

$$\frac{H_x(x_n, y_n)}{H_g} = \frac{1}{\pi} \left(\tan^{-1} \frac{1 + x_n}{y_n} + \tan^{-1} \frac{1 - x_n}{y_n} \right). \tag{9.10}$$

These approximating functions are plotted in Figure 9.15, along with the results of the graphical method. All field strengths are adjusted to be equal on the gap centerline. A crucial point to observe is that all of the functions of Figure 9.15 possess maximum slopes at approximately the same distance from the gap centerline. This will be of importance later in determining the point at which the peak readback signal occurs.

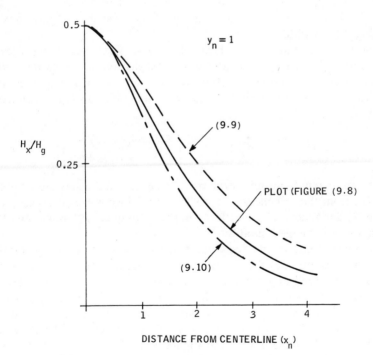

DISTANCE FROM CENTERLINE (x_n)

Figure 9.15

The head permeability has so far been regarded as infinite, and the gap field strength has been given simply by (9.3). A first-order correction can be made in this expression, with results indicating the desirability of other head geometry changes. With finite head permeability, the field strength in the gap becomes approximately

$$H_g \cong \frac{0.4\pi N_w I_w}{2g_x(1 + A_g g_i/\mu_i A_i g_x)} \tag{9.11}$$

in which

A_g = area of (assumed parallel) facing pole pieces,
A_i = cross-sectional area of (assumed uniform) magnetic head,
g_i = magnetic path length in the head.

Equation (9.11) neglects fringing effects by assuming that all flux in the head is also confined to the gap, and it may be corrected by an appropriate fringe factor. With other variables fixed by required recording density and available head core material, it is apparent that an increase in H_g may be obtained by reducing the effective gap area A_g. This can be accomplished by means of pole-piece geometries such as those illustrated in Figure 9.16a and b. Shew [14] has collected data indicating the value of such configurations for reducing pulse width.

Static fields surrounding the head have so far been considered. Clearly there must be some time variation in the field distribution, because a finite time is required for current to build up in the write winding. Including finite head permeability, the inductance of the write winding is given approximately by

$$L = N\frac{\Delta\phi}{\Delta i} \times 10^{-8} \cong N_w{}^2 \left[\frac{0.2\pi A_g \times 10^{-8}}{g_x(1 + A_g g_i/\mu_i A_i g_x)} \right]. \tag{9.12}$$

The rate of current build-up is primarily dependent on the maximum voltage available from the driving source. As frequency increases it is necessary to reduce N_w and increase I_w in order to maintain frequency response and stay within a given voltage limitation.

A number of unwanted effects occur in conjunction with the generation of a field distribution at the recording gap. These are termed *crosstalk* and may be categorized under two main headings: (a) direct coupling between heads by means of the field, and (b) effects on records in adjacent tracks. Two types of direct coupling are common. In drum memory systems recirculating delay lines can be formed by placing a read head following a write head on the same track, as in Figure 9.17a. If the length of the line is short, the field produced by the continuously recording write head may link the windings on the read head, appearing as noise on the readback signal. A field sketch for the recording head, in the absence of any surrounding structures, is shown in Figure 9.17b, in which the dotted flux line links the read winding. Actual field distributions will be greatly affected by the presence of the supporting structure, the presence of the read head itself, and

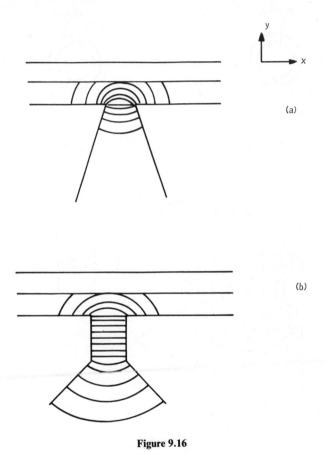

(a)

(b)

Figure 9.16

by its geometry. In practice, magnetic shielding (dotted box) can usually reduce this type of coupling to such a point that it is feasible to mount pairs of heads as close as mechanical considerations will permit. A more serious problem is encountered in parallel recording by a multiple-head assembly. It is commonly desirable to make the center-to-center track spacing D_t as small as possible, but the amount of flux thereby coupled to the adjacent head may be greatly increased. Illustrative flux lines and linkages for two adjacent heads are shown in Figure 9.18. Although coupling can be reduced substantially by means of a high-permeability shield placed between adjacent heads, simultaneous reading and writing on different tracks of a multiple-

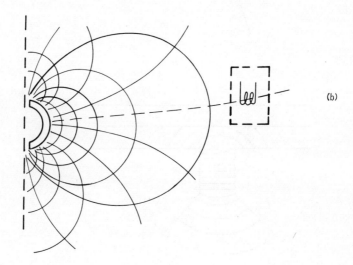

Figure 9.17

head assembly is seldom practiced because of the inherently close coupling involved.

The effect of record-head field on records in adjacent tracks is primarily due to the lack of squareness of the recording-medium hysteresis loop. Consider the field at one side of the record head, as shown in Figure 9.19*a*.

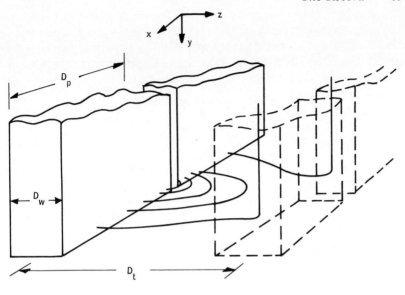

Figure 9.18

The lines are constant H corresponding, for example, to the field strength H_s required to saturate the medium. If the medium possessed a perfectly square loop, $H_c \cong H_s$, and the volume bounded by the surface generated by the $H = H_s$ lines would be driven to saturation. Regions outside the surface would be partially disturbed ($H < H_c$), but no switching and therefore no change of state would occur. Looking at the field plot along the x axis in Figure 9.19b, it would appear that it is only necessary to ensure that tracks are sufficiently separated to prevent intersection of saturation regions (indicated by dotted envelope). Therefore the minimum center-to-center spacing must at least be $D_t - D_w > 2(S + D)$. Unfortunately, this simple model is too idealized, because magnetic hysteresis loops are far from square, and practical spacings must be correspondingly greater.

The Record

In the preceding section it was noted that any element of magnetic medium passing in the vicinity of the record head is subjected to a continuous sequence of values of field strength. For any given write current, head geometry, head-to-surface spacing, and medium thickness the magnitude and direction of H as a function of t can be determined with reasonable accuracy. For uniform motion of the medium in the direction of the x axis this function is given reasonably well by plots such as those of Figure 9.8, by replacing x by

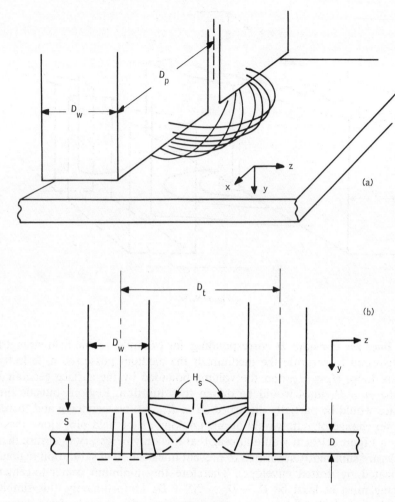

Figure 9.19

$V_0 t$. The question now is: As a result of these events, what will be the flux density within the medium at some later time when the record returns to the vicinity of a read head?

The flux-density distribution, or magnetization, M is determined by three factors: (a) the hysteresis loop of the medium, (b) the geometry of the record itself, and (c) the field strength distribution due to the write head. Regarding the first factor, a material possessing high stored energy properties is desirable because the effect of the demagnetizing field is thereby reduced. For any given geometry of magnet the demagnetizing field is a linear function of flux density. The amount of remanent magnetization can therefore only be increased by increasing either H_c or B_r of the material, as illustrated in Figure 9.20. An increase in B_r has the greatest effect upon weak demagnetizing fields, whereas an increase in H_c has more effect upon B in the presence of strong demagnetizing fields. Because most magnetic recording involves strong demagnetizing fields, considerable effort has been devoted to developing materials that possess extremely high coercive force and moderate saturation flux density. One of the most widely used recording mediums is

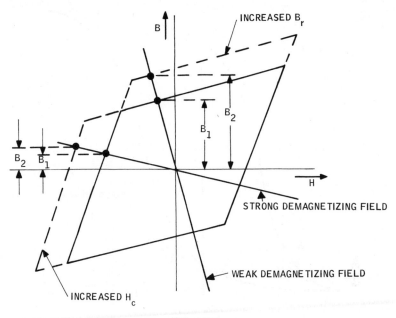

Figure 9.20

red iron oxide, Fe_2O_3, commonly prepared in powdered form mixed with a binder and coated onto the surface. A representative hysteresis loop of this material is shown in Figure 9.21. A variety of other hysteresis-loop shapes can be obtained by variations of the powder particle size (usually about 1 μm), the percentage of powder contained within the binder (usually between 25 and 50 percent by volume), or by changes in the magnetic characteristics of the ferromagnetic material itself (intrinsic coercive force usually between 100 and 250 Oe). Magnetic recording surfaces may also be prepared by controlled growing or rolling of various alloy steels, or by plating of the same alloy on a base material.

Demagnetizing fields. In general, the demagnetizing field may be evaluated by means of two approaches: (a) a mathematical solution to certain limiting geometries, and (b) a graphical or field plot technique. Both are based on the fact that the integral of field strength around any closed path not enclosing current is identically zero. Both methods depend on the derivation of two functions relating flux density and field strength within the magnetic medium. One function is determined from the geometry of the path of integration external to the medium. The other is the hysteresis loop itself. The simultaneous solution of these two functions, or the intersection of their plots,

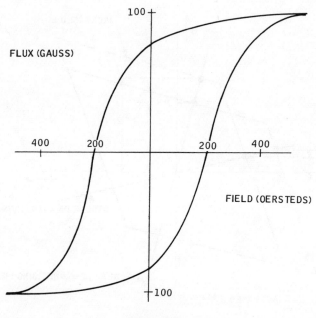

Figure 9.21

yields the values of field strength and flux density at which the medium may be expected to come to rest upon removal of the magnetizing field. Although the methods are in theory quite general, they are difficult to apply unless restricted (as in the following discussion) to paths within a magnetic medium having uniform cross section and high permeability.

Equation (9.1) defines the field strength integral both inside and outside the medium. The shape of the field (and flux) distribution outside the magnetic medium is independent of intensity and is only a function of geometry, or position in space. Therefore, denoting the flux density in air and magnetic material by B_a and B_i, respectively, we obtain $B_a = B_i f(p)$. Because in air

$H_a = B_a$, and assuming a uniform field within the magnetic material, (9.1) may be rewritten as

$$B_i = -\frac{p_i}{\int f(p)\,dp}\left(H_i - \frac{0.4\pi NI}{p_i}\right),\qquad(9.13)$$

and for $I = 0$, $B_i = -\mu_d H_i$, with μ_d defined as the demagnetizing factor.

The second of the simultaneous equations of interest is represented by the hysteresis loop itself, which may be described by the relationship $B_i = f_h(H_i)$. The simultaneous solution of these two equations, shown graphically in Figure 9.22, is the desired value of flux density and field strength within the magnetic material for any particular case. Because for any given material the hysteresis loop is fixed, the value of μ_d and, consequently, geometry alone determines this point. In most cases the graphical solution illustrated is useful and easily visualized. It has been shown, however, that the hysteresis loop may also be approximated over a limited region by the equation

$$B = B_r\frac{1 + H/H_c}{1 + (H/H_c)(B_r/B_s)}.\qquad(9.14)$$

Defining $\mu_i \equiv B_r/H_c$, $S_r \equiv B_r/B_s$, and setting (9.14) equal to $-\mu_d H$, we

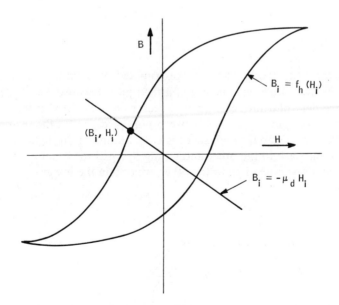

Figure 9.22

obtain

$$H = -\frac{B_s(\mu_i + \mu_d)}{2\mu_i\mu_d}\left\{1 - \left[1 - \frac{4S_r\mu_d\mu_i}{(\mu_d + \mu_i)^2}\right]^{1/2}\right\}. \qquad (9.15)$$

The case of greater interest is that for $\mu_d \rightarrow 0$ (strong demagnetizing fields). Using the approximation $\sqrt{1 - x} \cong 1 - x/2$, for $x \ll 1$, (9.15) can be reduced to

$$B \cong B_r\frac{\mu_d}{\mu_i}. \qquad (9.16)$$

The evaluation of the integral of which μ_d is a function is often difficult in particular cases, and resort may therefore be made to a graphical or field plot solution. Because $\mu = 1$ in air, the magnetomotive force across the path in air may be regarded as the summation of the product of a flux function and an incremental reluctance R_a along the path. Thus

$$\int H_a(p)\,dp = \int \phi_a\,dR_a. \qquad (9.17)$$

If the incremental reluctance encompasses all space surrounding the magnetic medium, ϕ_a is simply a constant (i.e., total flux) and is equal to B_iA_i, where A_i is the cross section of the magnetic material. We may therefore write

$$B_i = -\frac{p_i/A_i}{\int dR_a}H_i. \qquad (9.18)$$

The integral $\int dR_a$ can be closely approximated by a finite sum, which in turn can be evaluated with the aid of a field plot. Because $\Delta R_a \cong \Delta p/A(p)$, in a square field plot the reluctance of each square is unity. If the number of squares in the plot in parallel between pole pieces is S_p, the incremental reluctance of each step through space is $1/S_p$. Summing through space from one end of the path to the other and letting S_s equal the number of squares in series from one pole piece to the other, we obtain the integral

$$\int dR_a \cong \sum_{i=1}^{S_s} \Delta R_a = \sum_{i=1}^{S_s} \frac{1}{S_p} = \frac{S_s}{S_p}. \qquad (9.19)$$

The demagnetizing factor is therefore

$$\mu_d \cong \frac{p_i}{A_i}\frac{S_p}{S_s}. \qquad (9.20)$$

This function is particularly easy to evaluate for geometries that have uniform field plots in the z direction, that is, in which the dimension of the magnetic material in the z direction is large compared to either length or thickness. Because this is usually true of magnetic-surface recording, end effects may be neglected and a simple two-dimensional plot prepared, resembling Figure 9.6. It will be found that μ_d is primarily a function of the length-to-diameter (L/D) ratio of the recording.

The effect of the demagnetizing factor will be to reduce the effective remanent flux of each record as the pulse packing-density increases. Basic hysteresis loops, such as that of Figure 9.21, are normally taken on the medium as it will be employed in actual recording, that is, with an air-gap flux path. However, the curve corresponds to low packing density, with $L/D \to \infty$. An appropriate factor, such as that of (9.16) should therefore be applied to expected flux values if a packing density resulting in a low L/D ratio is to be used.

Although the demagnetizing field enters into the eventual recorded flux pattern, interacting with the recording-head field, it is assumed in the following section that this effect can be treated as a correction to the ideal recording at low packing densities.

The write field and net magnetization. Expressions for the spatial distribution of the write-head field have been previously developed. The question now is: What is the resultant magnetization when this field is impressed upon a material such as that of Figure 9.21? In order to handle the resulting nonlinear situation, various approximate approaches have been taken.

The most simplifying assumption to make is that the magnetization curve can be approximated by linear expressions over various regions of operation. In particular, the hysteresis loop is approximated as in Figure 9.23. From this figure, the assumptions amount to

$$\frac{M}{B_s} = \pm 1, \qquad |H| > H_s,$$

$$\frac{M}{B_s} = \pm \left(-1 + \frac{H}{H_c} \right), \qquad 0 \le |H| \le H_s,$$

$$H_s = 2H_c,$$

$$\mu \equiv \frac{B_s}{H_c}. \tag{9.21}$$

To make use of this approximation, consider a magnetic medium originally saturated in the negative direction to $-B_s$. Let the medium be thin enough so that only the x component of magnetization, M_x, is important. In this

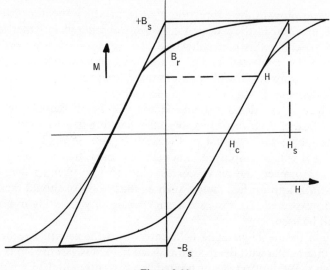

Figure 9.23

case (9.9) describes the head field. To normalize all quantities, let

$$y_0 \equiv y_n + 1,$$

$$H_0 \equiv y_0 H_s,$$

$$H_n \equiv \frac{H_g}{H_0}. \tag{9.22}$$

H_0 is the field strength required at the head gap ($x = y = 0$) to produce the saturation field H_s directly in front of the gap ($x = 0$) at a distance y_0. H_n is the dimensionless ratio of the actual field strength to H_0. Using these definitions and (9.9), Equation (9.21) may be written as

$$\frac{M_x}{B_s} = \frac{2H_n}{1 + (x_n/y_0)^2} - 1, \tag{9.23}$$

for a medium initially saturated to $-B_s$. If (9.23) exceeds 1, of course, M is limited to $+B_s$ by the saturation characteristic of Figure 9.23. The function (9.23) is plotted in Figure 9.24 for several values of H_n. It is symmetrical about $x = 0$. Notice the importance of saturation recording from two standpoints. One is the increased slope of M. This will show up later as an increase in readback signal, which is proportional to dM/dx. Another improvement achieved by saturation recording is the increased pulse definition and relative stability of position over variations in the recording field strength.

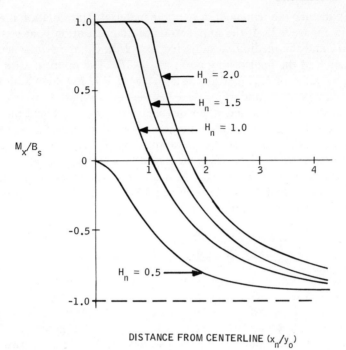

Figure 9.24

Although the magnetization curve of Figure 9.24 gives some indication of the pulse packing that may be achieved, recordings do not occur in splendid isolation in this manner, and the effects of adjacent recording along the same track must be considered. In particular, let the recording medium move to the right some normalized distance Δx_n, while maintaining the write head field at the relative strength $+H_n$. Clearly, all regions to the right of $x = 0$ will be subjected to decreasing, but positive, field strengths. Regions to the left will come under the head and be driven to $+B_s$. As $\Delta x_n \to \infty$, the field at the leading (right) edge of the recording of Figure 9.24 will decrease to zero, and magnetization will return to some value determined by a minor hysteresis loop. For the idealized linear characteristic of Figure 9.23, assume that squareness is perfect, even for minor loops, and that the return path for a point originally at H is along the horizontal dotted line (to $H = 0$).

Now suppose that at the point Δx, the field in the write head is suddenly switched from $+H_n$ to $-H_n$. Clearly, points directly under the head will be driven back to $-B_s$. Points more distant from the head will be subjected to some value of negative field strength. If sufficiently large in magnitude, this field can disturb the remanent state of the material. The situation is as illustrated in Figure 9.25. If the initial value of magnetization is at some high value M_1 and the medium is subjected to $-H$, a change in magnetization will occur, and the final resting point $M_2 \neq M_1$. The opposite case, for an initial low value of M_1, is also illustrated with $-H$ less than sufficient to reach the major loop, and $M_2 = M_1$. Let the field strength to which a point on the medium is subjected when the field $+H_n$ is first turned on be H_1, and let the field strength to which the same point is subjected at a later time (and distance) when the field $-H_n$ is applied be H_2. Then from Figure 9.25 it is clear that if $H_1 + H_2 < H_s$, there is no effect on the previous magnetization, whereas if $H_1 + H_2 > H_s$, the magnetization takes on a value independent of M_1 and entirely determined by H_2 (as if it had been previously saturated to $+B_s$). Therefore, we may write

$$H_1 < H_s - H_2, \qquad M \to M_1(H_1),$$
$$H_1 > H_s - H_2, \qquad M \to M_2(H_2), \qquad (9.24)$$

in which $M(H)$ is given by (9.23). However, note that Figure 9.24 is, in effect, a plot of $H(x)$ if we make the vertical scale run from 0 to H_s rather than from -1 to $+1$. Therefore, (9.24) can be evaluated graphically simply by inverting and superimposing upon itself the plot of Figure 9.24. This is done in Figure 9.26 and 9.27 for two values of H_n. Note the two major

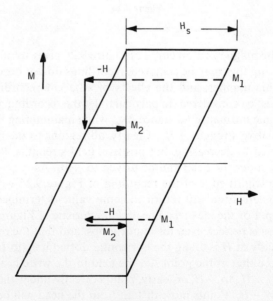

Figure 9.25

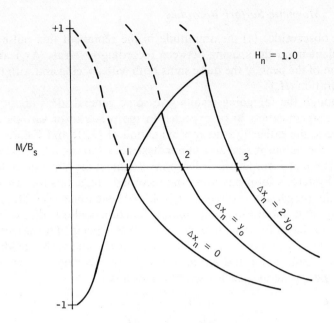

DISTANCE ALONG TRACK (x_n/y_0)

Figure 9.26

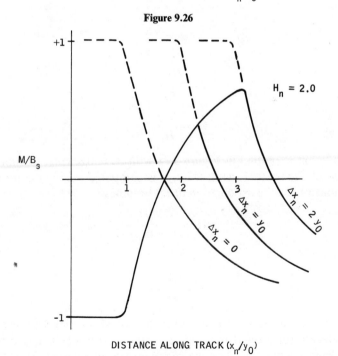

DISTANCE ALONG TRACK (x_n/y_0)

Figure 9.27

467

points observable: (a) the amplitude of the remanent flux pulse is highly dependent upon the spacing between recording reversals (Δx_n), and (b) the position of the peak of the pulse shifts both with spacing and with the degree of saturation (H_n).

Although the foregoing results give some general ideas about the form of the magnetization to be expected in the presence of various recording sequences, the rather drastic approximation of (9.21) and Figures 9.23 and 9.25 to the actual major and minor hysteresis loops casts some doubt on their degree of accuracy. In order to investigate what effect the actual shape of the hysteresis loop has upon the recorded magnetization, an improved approximating function is required. It should have some desirable properties; as follows: it should (a) be a smooth function without discontinuities in either its value or its derivatives, (b) be capable of being fit to measured points on the actual hysteresis loop, such as B_r and H_c, and (c) be capable of representing both major and minor loops. A great variety of functions could satisfy these general specifications. One such is

$$\frac{M_x(H)}{B_s} = \frac{1 - \gamma\, e^{-\alpha H}}{1 + \gamma\, e^{-\alpha H}},\tag{9.25}$$

in which

$$\gamma = \frac{1 + S_r}{1 - S_r},$$

$$S_r = \frac{B_r}{B_s},$$

$$\alpha = \frac{\ln \gamma}{H_c}.$$

This function is plotted in Figure 9.28 with $S_r = 0.8$, corresponding to the function of Figure 9.21, with which it is compared. Using the normalizing definitions (9.22), the exponent in (9.25) becomes,

$$\alpha H = \frac{b H_n}{1 + (x_n/y_0)^2},\tag{9.26}$$

in which

$$b = \frac{H_s}{H_c} \ln \gamma.$$

Again, H_n is the ratio of the actual field applied, to that field just required to produce saturation at the point directly in front of the gap centerline a normalized distance y_0. For the characteristic of Figure 9.21, $\gamma \cong 10$, and

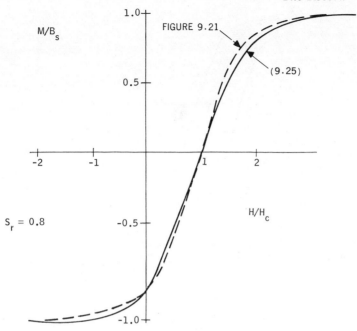

Figure 9.28

$b \cong 5$, implying $H_s/H_c \cong 2$. The function (9.25) is plotted with these parameters in Figure 9.29 for several values of H_n. Note that the lack of squareness changes the detailed shape of the recorded magnetization somewhat over the ideal approximation of Figure 9.24, but that the over-all form is not greatly different. Improvements in the squareness ratio will increase pulse definition and slope, as shown by the plot with $\gamma = 20$ ($S_r \cong 0.9$) in Figure 9.30.

In order to treat the effect of sequences of recorded pulses, again consider the record of Figure 9.29 or 9.30 moved to the right a distance Δx_n, and the applied field reversed (normalized $-H_n$). Again, the remanent magnetization due to the initial application of $+H_n$ will be subjected to some negative field. Let us assume that the final resting value of a point in the medium is the simple sum of its original value M_+, due to $+H_n$, and a disturbance ΔM. Thus,

$$M = M_+ - \Delta M, \tag{9.27}$$

in which M_+ is given by (9.25) and (9.26). The magnitude of ΔM will be a function both of M_+ and the value of the field. Suppose that (9.25) still describes the form of the behavior of the medium on a minor loop. Let $f(x_n)$ be (9.25) with x_n in (9.26), and $f(x_n + \Delta x_n)$ be (9.25) with $x_n + \Delta x_n$

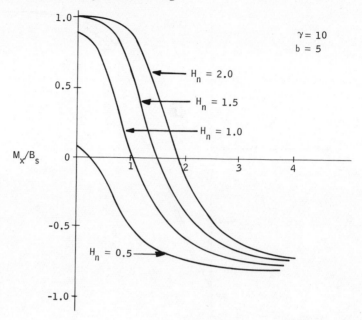

DISTANCE FROM CENTERLINE (x_n/y_0)

Figure 9.29

in (9.26). Then, scaling (9.25) appropriately, we obtain

$$\Delta M = (M_+ + B_s)\frac{1 + f(x_n + \Delta x_n)}{2}. \tag{9.28}$$

This implies that the shape of a minor hysteresis loop is the same as that of the major loop, with a change in scale factor. For $\gamma \gg 1$, as $x_n \rightarrow \infty$, $f(x_n + \Delta x_n)$ $\rightarrow -1$ and $\Delta M \rightarrow 0$; if $f(x_n + \Delta x_n) \rightarrow +1$, ΔM approaches the value $M_+ + B_s$, and by (9.27) the point is returned to $-B_s$. These relationships are illustrated in Figure 9.31, in which the minor loop as a function of reverse field is shown dotted. Noting that M_+ is given by (9.25) directly, we see that (9.27) becomes

$$\frac{M_x}{B_s} = \tfrac{1}{2}[f(x_n) - f(x_n + \Delta x_n) - f(x_n)f(x_n + \Delta x_n) - 1]. \tag{9.29}$$

This function is plotted in Figures 9.32 through 9.35 for two values of $\gamma(S_r \cong 0.8$ and $S_r \cong 0.9)$ and of H_n (saturation, and twice saturation), and for several values of Δx_n. Several conclusions can be drawn from these figures.

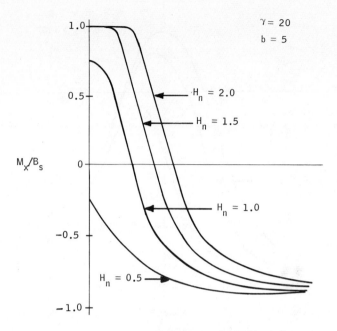

DISTANCE FROM CENTERLINE (x_n/y_0)

Figure 9.30

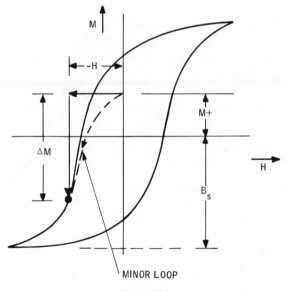

Figure 9.31

471

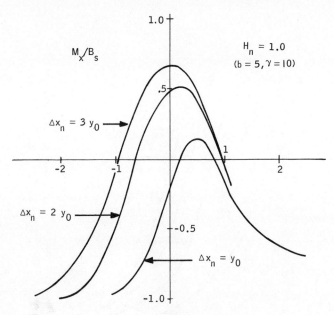

Figure 9.32

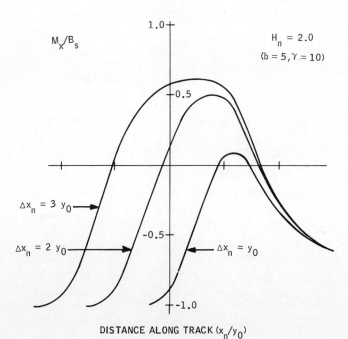

Figure 9.33

472

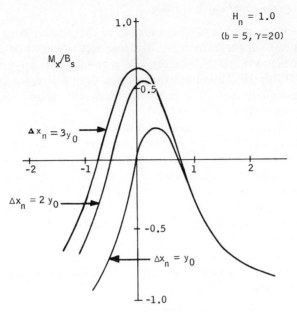

$H_n = 1.0$

$(b = 5, \gamma = 20)$

M_x/B_s

$\Delta x_n = 3y_0$

$\Delta x_n = 2\,y_0$

$\Delta x_n = y_0$

DISTANCE ALONG TRACK (x_n/y_0)

Figure 9.34

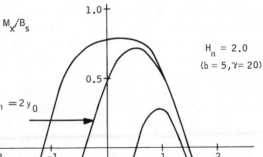
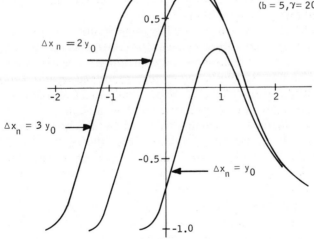

M_x/B_s

$H_n = 2.0$

$(b = 5, \gamma = 20)$

$\Delta x_n = 2\,y_0$

$\Delta x_n = 3\,y_0$

$\Delta x_n = y_0$

DISTANCE ALONG TRACK (x_n/y_0)

Figure 9.35

473

One is that increasing the squareness ratio of the medium beyond a certain point does not substantially affect the readback signal. For example, for $H_n = 2$, the increase in dM/dx for $\Delta x_n = 2$ is about 20 percent for an increase in S_r from 0.8 to 0.9. This simply means that even for very good recording media, the write-head field distribution ultimately establishes a limit on recorded pulse definition.

Another conclusion that can be drawn from Figures 9.32 through 9.35, is that although saturation recording is desirable, over saturation does not improve the recorded magnetization substantially. Thus $1.0 < H_n < 2.0$ is the range of operation most appropriate. Finally, it is clear that Δx_n must exceed 1 if the magnetization is to reverse at all, but that $\Delta x_n > 3$ does not greatly improve performance over $\Delta x_n = 2$. This is the reason for the general rule of thumb that has arisen that the pulse spacing should be "several" times the maximum head-to-surface spacing. Taking y to be given by the far side of the medium $S + D$, and noting that $\Delta x_n \cong 2y_0$ is sufficient to achieve good recording definition according to Figures 9.32 through 9.35, we obtain the required pulse spacing. Thus,

$$\Delta x = g_x \Delta x_n \cong 2g_x y_0 = 2(S + D + g_x). \qquad (9.30)$$

As might be expected, this same result can be based upon strictly geometrical arguments from the highly stylized hysteresis curve of Figure 9.23. One may ask, for example, what is the distance between the point at which a reverse field occurs such that $\Delta M \cong B_s - B_r$, a small degradation in remanent flux, and the point at which a reverse field occurs such that $H \cong H_s$, and the flux is completely reversed. For normal magnetic material parameters, the same approximation as (9.30) is obtained.

It should be apparent from the foregoing results that the region of interest is that between subsequent flux reversals. Further, to a reasonable approximation, the transition region constitutes a linear flux change from some effective flux B_e determined by remanence and small reverse fields to a similar effective flux $-B_e$. Therefore long sequences of pulse reversals can be immediately plotted, to a first order, by assuming that the length of the transition region, Δx_n, remains fixed, and that the flux outside this region is unaffected. When the pulse period T is such that $V_0 T < \Delta x$, a sequence-dependent flux level may result, and flux may never reach the effective remanence level B_e. Such a recording is illustrated for a sequence of time periods in Figure 9.36. The record-head field strength as a function of time is on the bottom line. The distance $V_0 T$, corresponding to the period T between reversals of record field strength, has been taken to be approximately $\Delta x/2$. The magnetization plots are based on the following rules. All portions of the record to the right of x_1 are unaffected by the condition of the field strength at the head. All portions of the record to the left of x_2 are assumed

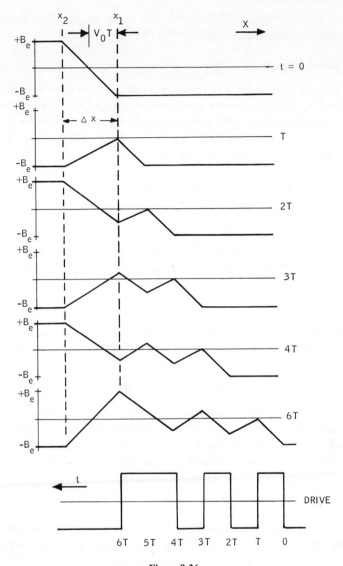

Figure 9.36

to be driven to the maximum possible value of magnetization $\pm B_e$ corresponding to the direction of record-head field strength. Magnetization in the interval Δx is obtained by connecting the values of magnetization at x_1 and x_2 by a straight line. It can be seen that both the magnitude and the rate of change of magnetization with distance are functions of the recorded configuration of information. This makes the recovery problem particularly difficult, because the information pattern itself is a variable.

The Read Signal

The foregoing estimates of recorded flux patterns form the basis on which corresponding estimates of the readback signal can be based. Some very simple considerations can give good order-of-magnitude values for the readback signal to be expected. First, assume that all recorded flux links the read winding. Then the sensed voltage must be

$$e = N_r \frac{d\phi}{dt} \times 10^{-8} \qquad (9.31)$$

in which N_r is the number of turns on the read winding.

To evaluate $\Delta\phi$ and Δt, consider the sequence of events corresponding to a change in magnetization passing under the head gap as depicted in Figure 9.37. In *a*, the flux from one record links the read winding (dotted lines) in a particular direction (arrow), and the flux from the adjacent record passes through the head piece without linking the winding. In *b*, the head is symmetrically placed, and no flux links the winding. In *c*, the roles of the two recordings are reversed, and the flux linking the winding is in the opposite sense. The change in flux linked by the read winding is therefore approximately

$$\Delta\phi = \Delta BA = 2B_e A = 2B_e DD_w. \qquad (9.32)$$

The time for making this flux transition, from *a* to *c* in Figure 9.37, is certainly no less than that corresponding to the distance, Δx between records; that is, $\Delta t \cong \Delta x / V_0$. Combining (9.30), (9.31), and (9.32), we obtain a reasonable estimate for resulting readback voltage as follows:

$$e \cong \frac{N_r V_0 B_e DD_w \times 10^{-8}}{S + D + g_x} \qquad (9.33)$$

The above estimate says nothing, however, about the voltage waveshape to be expected from a recorded magnetization transition. It implies, in fact, that the output is a pulse or square wave, beginning abruptly at one edge of the transition and ending abruptly at the over. Such behavior cannot, of course, be expected from the diffuse field nature of the flux linkages involved.

The key to a more accurate analysis of this problem is a simple one. It is to assume that during readback, the head, surrounding space, and recording medium are all linear. This may seem a surprising assumption in view of the nonlinear magnetic characteristics that have been of concern heretofore. It is reasonable, however, on the basis that readback signals

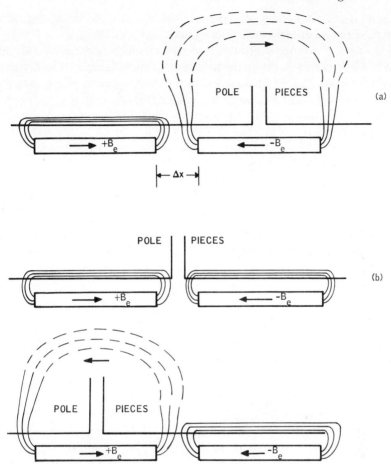

Figure 9.37

constitute very small perturbations upon the magnetic states of any portions of the recording medium or head. Thus, to the extent that minor hysteresis loops corresponding to very small field changes (fractions of the coercive force) are reversible and approximately linear, this assumption is an accurate one, closely approximated in practice.

With this principle of reciprocity, the head-recording linkages can be treated in a quite general manner. In a full treatment the coupling from the head to a point in space is described by a general vector weighting $\bar{H}(x, y, z)$. For present purposes, however, it is sufficient to treat these quantities in simple one-dimensional form. The treatment follows closely that of Hoagland

[2] and others who have used the reciprocity principle to advantage in analyzing a number of recording techniques and head designs.

Consider the arrangement of Figure 9.38 in which two current-carrying coils are linked in a linear space by a mutual inductance L. The first coil, carrying flux ϕ_1, can be identified with the read winding. The second coil replaces the permanent magnetization of an elemental volume of the recording medium by means of an equivalent magnetomotive force Ni_2, which in turn produces the equivalent flux ϕ_2, which would be observed at that point in the medium. For this pair of coils, the flux in one coil due to the current in the other is

$$\phi_2 = Li_1,$$

$$\phi_1 = Li_2. \tag{9.34}$$

A value for L can be obtained by taking the ratio of ϕ to i from either of the above equations. In particular, let

$$L = \frac{\phi_2}{i_1} \tag{9.35}$$

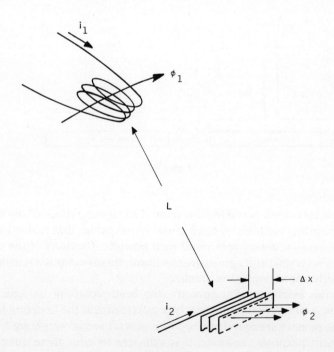

Figure 9.38

be the measure of flux produced at the volume element in the recording medium by a current i_1 in the head winding. Provided the medium is linear,

$$\phi_2 = B_2 A = \mu H_2 A, \tag{9.36}$$

in which A is the cross-sectional area of the volume element in the direction of ϕ_2, and H is the field strength produced at the volume element by the head current. However, H, as a function of position and head current, is already known from field plots or approximating functions. Therefore, for a fixed value of y, let

$$H_2 \equiv H_i(x)i_1, \tag{9.37}$$

in which $H_i(x)$ is the field strength per unit current produced by the head. Combining the above, we obtain

$$L = \mu A H_i(x). \tag{9.38}$$

Now calculate the equivalent current i_2 necessary to maintain a given flux density at the volume element. Considering i'_2 as a current sheet, the equivalent current per unit distance required to produce (or represent) the magnetization $M = B$ within the actual volume element is

$$i'_2 = \frac{H}{0.4\pi} = \frac{M(x)}{0.4\pi\mu}. \tag{9.39}$$

Using the second equation of (9.34), we can now write an expression for the incremental contribution to the total read-head flux, $\Delta\phi_1$, due to the incremental flux in a volume element of length Δx; that is, since $\phi_1 = Li_2$, the incremental contribution is

$$\Delta\phi = Li'_2 \, \Delta x = \frac{A H_i(x) M(x) \, \Delta x}{0.4\pi}, \tag{9.40}$$

in which (9.38) and (9.39) are used. Taking the limit of incremental contributions and summing, we obtain

$$\phi_1 = \frac{A}{0.4\pi} \int_{-\infty}^{\infty} H_i(x) M(x) \, dx. \tag{9.41}$$

The readback voltage is proportional to the derivative of (9.41). In order to account for the movement of the head relative to the medium, the reference coordinates for H and M must be chosen properly. This is illustrated in Figure 9.39. It can be seen from this figure that (9.41) takes on the form of the general convolution integral. Although variations in x are the most important, the form of (9.41) is not necessarily so restricted and can be applied over all space. Further, although the x component of magnetization

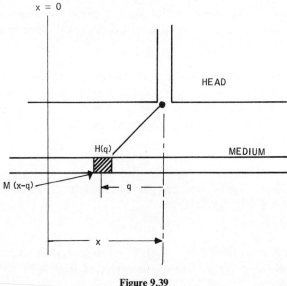

Figure 9.39

is usually the most important, in general, components may exist in any direction. Therefore, both M and H should be regarded as vector quantities. Dropping the constants from (9.41) and letting $H_i(x) = H(x)$ be the normalized field, we see that the readback voltage is of the general form

$$e(t) = e\left(\frac{x}{V_0}\right) = \frac{d}{dt} \int \overline{M}(x - q)\overline{H}(q) \; dq. \tag{9.42}$$

Small-read-head approximation. Evaluation of (9.42) has been accomplished only in special cases. In general we shall restrict the discussion to magnetization lying along a fixed value of y, and, frequently, to magnetization having only a single component M_x. Two main simplifying assumptions have been made in estimating (9.42). One is that the read and write heads are not the same. In particular, if the read-head gap is much smaller than the write-head gap and the head-to-surface spacing and thickness are much less than the read-head gap length, the read-head field-strength distribution relative to the magnetization distribution may be regarded as an impulse function; that is,

$$\overline{H}(q) = 0, \qquad q \neq 0,$$
$$\overline{H}(q) = 1, \qquad q = 0. \tag{9.43}$$

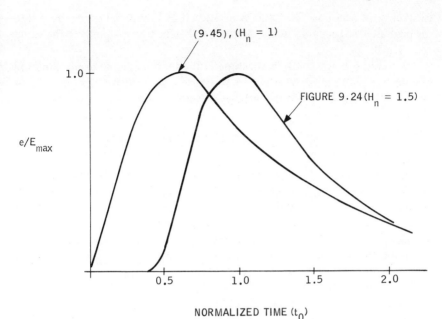

(9.45), ($H_n = 1$)

FIGURE 9.24 ($H_n = 1.5$)

e/E_{max}

| 0.5 | 1.0 | 1.5 | 2.0 |

NORMALIZED TIME (t_0)

Figure 9.40

Therefore, with x scaled to t by the factor V_0, (9.42) reduces to the form

$$e \sim \frac{d\overline{M}}{dx}. \tag{9.44}$$

This is particularly convenient because M as a function of x has already been determined for a number of cases. For example, using the simplest possible assumptions about the hysteresis loop, the function (9.23) describes the magnetization as a function of x. Normalizing time such that $t_0 \equiv V_0 t / g_x y_0$, and dropping all constants of proportionality, we obtain

$$e \sim \frac{dM_x}{dt_0} \sim \frac{t_0}{(1 + t_0^2)^2}. \tag{9.45}$$

This function is plotted in Figure 9.40, normalized to its peak value. Clearly the approximation is valid only for a write field strength up to saturation, because (9.23) does not take saturation into account. As H_n exceeds 1.0, dM/dx goes to zero within the saturation region of Figure 9.24, and the pulse peak shifts outward, away from the gap centerline. When this effect is taken into account, the result is as shown in Figure 9.40. For $H_n \le 1.0$, the peak of (9.45) occurs when $d^2 M/dt_0{}^2 = 0$, which corresponds to the value $t_0 = 1/\sqrt{3}$.

To obtain a better analytical estimate, the more accurate approximating function of (9.25) must be used. In this case, again dropping constants of proportionality, we obtain the voltage waveform as

$$ e \sim \frac{dM_x}{dt_0} \sim H_n \frac{t_0}{(1 + t_0{}^2)^2} \frac{\gamma \, e^{-bH_n/1 + t_0{}^2}}{(1 + \gamma \, e^{-bH_n/1 + t_0{}^2})^2}. \tag{9.46} $$

This function is plotted in Figure 9.41 for two values of H_n, and typical values of γ and b. It closely resembles readback signals actually encountered. Note that (9.46) is identical but negative for $t_0 < 0$. The position of the pulse peak in terms of the medium and head parameters can be obtained by setting $d^2 M/dt_0{}^2 = 0$. Using a different approximating function, Stein [12] has found that the pulse peak as a function of head-to-surface spacing occurs at approximately t_0 up to $y_n = 1$ and thereafter moves outward, away from the centerline, approximately in proportion to $y_n - 1$. Notice that the peak amplitude for $H_n = 2.0$ is less than that for $H_n = 1.0$. Barkouki and Stein

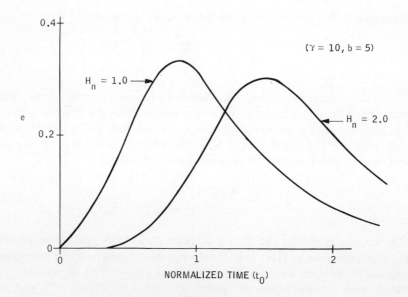

Figure 9.41

[10] have taken data showing that the peak amplitude occurs for H_n within the range $1.0 < H_n < 1.25$.

In principle, the waveshape of sequences of recordings can be obtained by differentiating (9.29) with $f(t_0)$ and $f(t_0 + \Delta t_0)$ replacing $f(x_n)$ and $f(x_n + \Delta x_n)$, respectively. Thus

$$e \sim \frac{dM_x}{dt_0} \sim f'(t_0)[1 - f(t_0 + \Delta t_0)] - f'(t_0 + \Delta t_0)[1 + f(t_0)]. \quad (9.47)$$

The function $f'(t_0)$ corresponds to (9.46). The resulting expressions are cumbersome, however, and a graphical approach is convenient. Note that the above functions have already been determined (if the time scale is shifted for $t_0 + \Delta t_0$) in Figures 9.32 through 9.35, and in Figure 9.41. Therefore values for the terms in (9.47) are immediately available. The results for two values of H_n are given in Figure 9.42. Note again the pulse position displacement and attenuation due to oversaturation recording. The negative peaks are the ones of interest, corresponding to the transition between $+H_n$

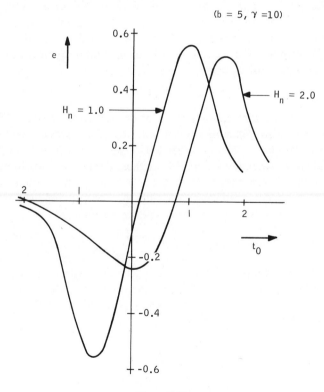

Figure 9.42

and $-H_n$ during recording; the positive peaks to the right correspond to the transition from the original magnetization at $-B_s$ that occurs when $+H_n$ is first turned on in the assumed model.

Short-flux-transition approximation. A second technique for evaluating (9.42) has been to assume that the flux transition of interest occupies only a small distance in space. This is a reasonable assumption, because the head field changes relatively slowly in space, and dq can be made as small as desired. In order to obtain solutions for particular cases, the derivative of the integral in (9.42) can be put in finite difference form. Thus, again dropping proportionality constants, we can write the output voltage as

$$e = H_i(x)\frac{1}{\Delta t}\left[\int_{-\infty}^{\infty} M_x(x)\,dx\bigg|_t - \int_{-\infty}^{\infty} M_x(x)\,dx\bigg|_{t+\Delta t}\right]. \qquad (9.48)$$

Consider now the classical case of a step change in magnetization from $-B_e$ to $+B_e$, as illustrated in Figure 9.43a. At any given time the step change is located a distance x from the head-gap centerline, at which point the head field would be $H_x(x)$. The quantity in brackets in (9.48) can be evaluated with the aid of Figure 9.43b. Note that the value of $M_x(x)$ at t and at $t + \Delta t$ is everywhere the same except in the region $\Delta x = v_0 \Delta t$. Thus the difference between the integrals is zero except in this region, and the quantity in brackets is simply $2B_e V_0 \Delta t$. Substituting into (9.48), we obtain

$$e = kH_i(V_0 t), \qquad (9.49)$$

in which k is a constant characteristic of the recording system. This is an important conclusion, meaning essentially that the readback voltage time waveform is identical to the normalized head field strength spatial waveform, for a step change in magnetization. We are therefore in a position immediately to plot output voltage versus time, because field strength versus distance has already been determined. Evaluating the constant of proportionality in (9.49), we can see that the readback voltage becomes

$$e = Ef(x_n, y_n) \equiv N^2 \frac{AB_e V_0}{g_x \times 10^8} f(x_n, y_n), \qquad (9.50)$$

in which $f(x_n)$ is a dimensionless function such as (9.8) corresponding to the relative field strength $H(x_n, y_n)/H_g$. Specific components of H are $H_x = H \cos\theta$ and $H_y = H \sin\theta$, the former being evaluated in (9.9). Let relative voltage and relative time be defined as

$$e_r \equiv \frac{y_0 e}{E},$$

$$t_r \equiv \frac{V_0/g_x}{y_0}t. \qquad (9.51)$$

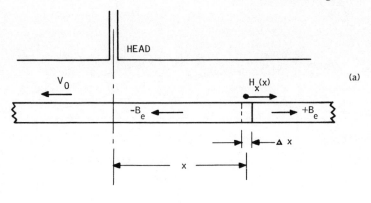

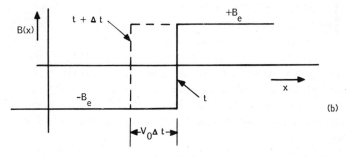

Figure 9.43

Then the contributions of the x and y components of magnetization to readback signal can be written in the form

$$e_r(x) = \frac{1}{1 + t_r^2},$$

$$e_r(y) = \frac{t_r}{1 + t_r^2}. \qquad (9.52)$$

These functions are plotted in normalized form in Figure 9.44. The total output voltage is simply the scalar sum $e_r(x) + e_r(y)$. The total output signals for a constant total step change of magnetization, $2B_e$, but various ratios of $B_e(x)$ to $B_e(y)$ are plotted in Figure 9.45. It should be apparent that other possible waveforms due to various combinations of plus and minus x and y components of magnetization can be obtained from Figures 9.44 and 9.45 by appropriate rotation about the x or y axes, or both. For example, reversing

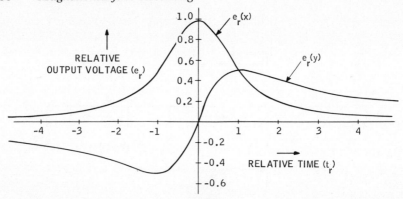

Figure 9.44

the direction of x magnetization but not that of the y would rotate Figure 9.45 about both x and y axes. Note the undesirable "overshoot" in the third quadrant of Figure 9.45 when the y component of magnetization is significant. This may have the effect of reducing the peak amplitude of adjacent recordings. The plots shown are drawn, however, for a symmetrically placed winding on the read head. It has been shown that overshoot may be partially compensated by nonsymmetrical placement of the read winding, resulting in a compensating nonsymmetrical read-head field distribution weighting function.

Consider now the output voltage resulting from a continuous series of alternating step changes in magnetization along the surface of the medium. By the principle of superposition the total relative output voltage is simply

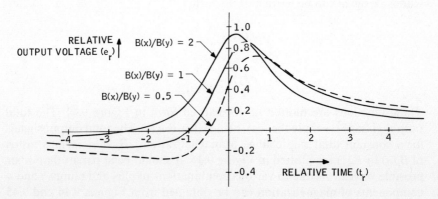

Figure 9.45

given by the sum of contributions from each of the step changes; that is,

$$e_r = \sum_{n=0}^{\infty} (-1)^n \, e_r(t_r + n \, \Delta t_r), \tag{9.53}$$

where Δt_r is relative time corresponding to relative pulse packing. This is equivalent to sliding the x- or y-component voltages of Figure 9.44 along the x axis an appropriate distance and then adding values algebraically. The result is the plot of Figure 9.46 for various values of the pulse-packing parameter P, defined as number of pulses per unit length $S + D \gg g_x$.

The preceding results are based on an assumed step change in magnetization. Other recording patterns are, however, of great practical interest. In particular, consider the case of an impulse recording of length J, as in Figure 9.47. This may be regarded as the superposition of two step changes in magnetization for each of which (9.49) holds true. The net relative output voltage is therefore

$$e_r \sim f(x_1) - f(x_2). \tag{9.54}$$

If J is sufficiently small, the difference between the two functions may be replaced by $J[df(x)/dx]$, and the output voltage is proportional to the derivative of the read-head field strength distribution. Carrying out this operation for the idealized small-gap case and rescaling e_r by J, we find that the resulting output voltages due to the two components of magnetization are

$$e_r(x) = -\frac{t_r}{(1 + t_r^2)^2}$$

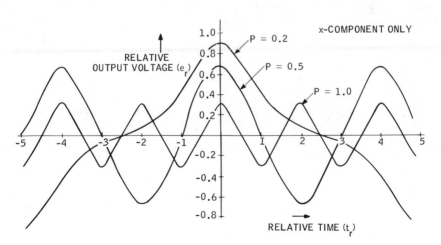

Figure 9.46

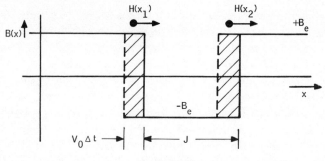

Figure 9.47

and

$$e_r(y) = \frac{1 - t_r^2}{(1 + t_r^2)^2}. \tag{9.55}$$

These functions are plotted in Figure 9.48.

It may appear from the foregoing estimates of the output signal that it is desirable to make N as large as possible in order to increase output voltage. If the signal-to-noise ratio is considered however, no improvement can be obtained by increasing N, because the noise level is also proportional to N^2. Noise, other than that implicit in the previous discussion of the attenuation caused by the presence of adjacent records in the same track, may be classified into three main categories. The first, due to stray fields linking the read-head winding, can be solved by appropriate magnetic shielding. The second type of noise is due to irregularities or inclusions of foreign material within

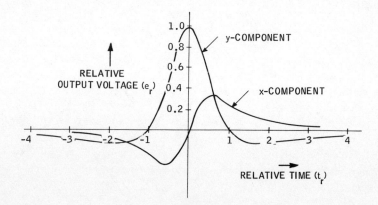

Figure 9.48

the recording medium itself. In general terms, it can be seen that the signal from noise of this type resembles that resulting from an impulse recording as described by equations such as (9.55). The noise to be expected from a given inclusion in the surface can be estimated by assuming that B_e is proportional to the portion of medium cross-sectional area occupied by the inclusion, and that J is proportional to its length. If a record is made precisely at the location of an inclusion, the readback signal voltage will be reduced by an amount corresponding to this estimated noise signal. This occurrence is termed a *drop-out* and may become serious if the inclusion is large compared to the record size. Clearly, as packing density is increased, the probability of readback errors of this type is similarly increased. In practice, all media contain large numbers of inclusions that have random distributions with respect to both size and location, and the amount of noise expected from this source therefore depends primarily upon the quality of the manufacturing process.

A third type of noise is a result of records in adjacent tracks. The worst case corresponds to a step change in magnetization on an adjacent track of opposite polarity to that on the given track. If the read-head field strength is assumed to decrease inversely with distance, to a first approximation the signal-to-noise ratio due to this factor alone is given by $D_t/(S + D)$. For the case of an impulse recording, readback voltage drops off inversely as the square of distance, and the signal-to-noise ratio will be correspondingly improved. Generally speaking, these estimates of signal-to-noise ratio are conservative, because the field strengths from a series of alternating recordings will tend to cancel as distance from the track increases. The major limitation upon track spacing is therefore primarily dependent upon the recording process as previously discussed.

Read-Write Systems

Having considered individual phenomena in surface recording, let us now consider the process from an over-all viewpoint. In this section attention is focused on the characteristics of a complete set of recorded information. In particular, we are interested in the various recording patterns that have been employed, their characteristic readback signals, their advantages and disadvantages with respect to noise and tolerance of the various parameters, and the circuits required to produce and recover the necessary information.

In order to evaluate qualitatively a given recording pattern, it is common practice to sketch flux distribution and readback voltage signals based upon three simplifying assumptions: (a) that the medium directly under the head-gap centerline at the instant of recording is always driven to saturation, (b) that the flux "spreads" a distance from this centerline by an amount appropriate to the particular packing density under consideration, and (c) that (9.44) holds approximately, and therefore the output voltage is given directly by the derivative of the flux distribution. The use of these assumptions is illustrated in the following discussion.

There are four major methods of surface recording that can be distinguished although each is subject to a number of individual variations. The first of these is known as the *return-to-zero* (*RZ*) or pulse method of recording. It most nearly resembles the impulse recording discussed in the preceding section and is illustrated in Figure 9.49. In Figure 9.49*a* it is assumed that the flux spread in the recording process is just sufficient to fill one record cell. If the readback voltage is sampled shortly after the beginning of each digit cell-time, as indicated by the arrow, the polarity of the voltage is an unambiguous indication of the previously recorded information, negative corresponding to 0, and positive to 1. An equally valid but reversed polarity indication of the recorded information is also available in the latter half of the cell period, as indicated by the dotted arrow in cell period 1. For packing densities equal to, or less than that illustrated, the information in each cell is essentially independent of all other information, and the recording in a single cell may be readily reversed without producing any other effect, as illustrated by the dotted waveform in cell period 6. In the presence of noise it is common practice to select a threshold voltage above whose positive value at the sampling times indicated the output is regarded as a 1 and below whose negative value, it is a 0. The fact that the peak output voltage occurs slightly before the center of the cell period permits an interesting feature of this recording system. It is possible to detect the information in a cell, and, if the delays in the circuitry are sufficiently short, to make a decision whether or not to change that information prior to recording time. Although this feature has the potential of reducing the amount of logic or storage circuitry required in certain special systems, it possesses severe limitations with respect to timing and reliability, particularly in high-packing-density media, and it is not in wide use.

The flux plot of Figure 9.49 has been drawn assuming zero magnetization of the surface medium between cells. A totally demagnetized surface is difficult to produce and maintain, however, and if for any reason the magnetization of the medium is not zero, then the output voltages produced by different recordings will not be equal. This difficulty can be overcome by recording upon a medium previously saturated in one direction or the other throughout its entire length. For a negatively saturated medium, for example, the 0 recording will produce little or no flux change, whereas a 1 will produce a change from $-B_e$ to $+B_e$. This usually results, depending upon the packing density, in a higher output voltage.

As the packing density is increased with the same record field distribution, the flux density and output voltage approach that shown in Figure 9.49*b*.

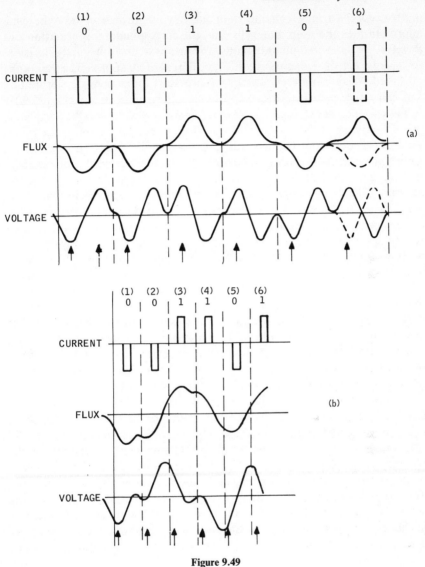

Figure 9.49

If sources of noise remain constant, necessitating approximately the same threshold level, uncertainties will begin to arise regarding the recorded information as indicated by the output voltage level at sampling time in cells 2 and 4. This may be overcome by reducing noise and hence threshold level, or by making an appropriate logic decision. Note that for the pattern illustrated, cells from which uncertain output voltages arise contain the same information as the preceding cells, and the corresponding information can therefore be recovered unambiguously by means of an appropriate decision. As packing density is increased, this type of recording approaches that of the next method, and a different criterion is necessary for resolving uncertainties.

A *non-return-to-zero* (*NRZ*) recording pattern is illustrated in Figure 9.50. In this case the flux change is always $2B_e$ in magnitude. A readback signal is present only when there is a transition from one binary level to the other. Three levels of output voltage must therefore be distinguished: (a) positive (above the threshold level), indicating a 1 in the next cell; (b) negative, indicating a 0 in the next cell; and (c) zero (or within the threshold region), indicating no change from the information in the present cell to that of the next cell. As in the case of *RZ* recording, the spread of the record field can produce an output indication prior to the time it is necessary to write in a cell. In *NRZ* recording it is very difficult to change an isolated bit satisfactorily without recording the entire track or a portion of it set off by appropriate spacing.

As packing density is increased, the waveform approaches that illustrated in Figure 9.50*b*. Notice that the flux level of an isolated bit may be somewhat reduced, (e.g., cell 5). Unwanted noise due to spreading also begins to occur in cells further removed from a change in flux, as shown by the output voltage at sampling time in cells 1 and 3. A logic decision can also be used to recover this waveform if it is observed that a positive output indicates a 1 in the following cell only if it is not followed by a more positive output the next cell-sampling time. Thus the positive output in cell 1 is followed by a more positive output in cell 2, and the latter is therefore the only valid indication of a 1 in the succeeding period, cell 3. The reliability of this method rapidly deteriorates as the packing density is further increased. Note that for the recovery system described, the reading circuit is required to have a knowledge of the previous bit in order to make a decision regarding the succeeding bit. This additional memory requirement may be transferred to the recording circuitry by a variant of the system that calls for the write current to remain unchanged when a 1 is recorded but to change value when a 0 is recorded. Thus a series of 1's results in no output and a series of 0's in alternating outputs. Such a scheme, of course, makes it impossible to change a single bit without recording the entire sequence of information.

A third method of recording is the Manchester or phase modulation method illustrated in Figure 9.51. In this type of recording the write current always changes polarity at the center of the cell, the direction of change distinguishing between a 1 and a 0. Thus flux always changes polarity at the middle of the cell and may or may not change polarity between cells. The read circuit is only required to distinguish between a positive and a negative

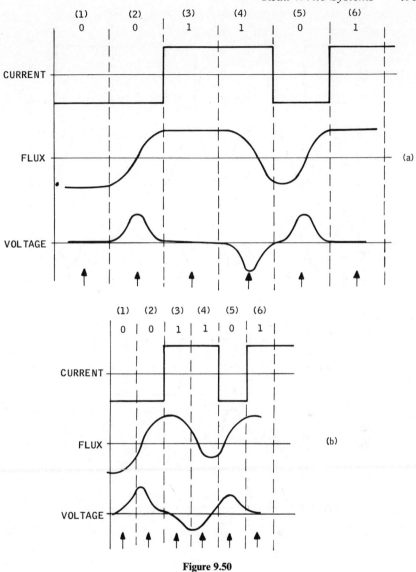

Figure 9.50

polarity output voltage to determine the contents of the cell. As packing density is increased, the waveforms of Figure 9.51*b* are encountered. It is apparent that polarity is still a valid indication of the contents of the cell, but that the threshold level is approached because of flux attenuation caused by the spread of the record-head field. It is possible under this

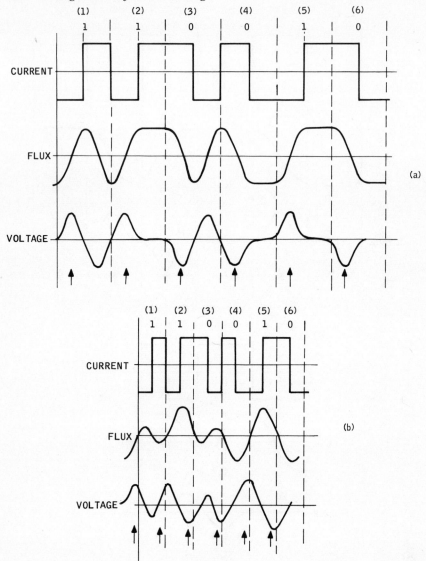

Figure 9.51

recording scheme to reverse the information in a single isolated cell, but this can be done reliably only at the lower packing densities. In practice the Manchester recording system has proven to be highly reliable and easy to recover and is widely used in digital computer recording systems.

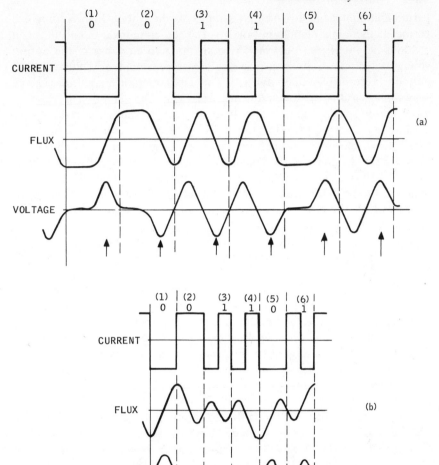

Figure 9.52

A fourth method which combines some of the features of both the *NRZ* and the Manchester systems is frequency modulation recording as illustrated in Figure 9.52. In this method the same polarity of current is passed through the recording head for the entire cell period in the case of a 0, but the polarity of the current is reversed at the center of the cell for a 1. The write current is always reversed at the beginning of the cell period. The rule for the recovery of information is based upon sampling the voltage near the end of the cell period and comparing it with that sampled at the end of the preceding cell period. If there is a change in polarity, the cell contains a 0, and if not, it contains a 1. It is also possible to count the zero crossings of the output voltage waveform. Cells with two zero crossings contain a 1; those having only one zero crossing contain a 0. This latter method may, however, be somewhat sensitive to low-frequency noise that can temporarily raise or lower the dc level of the output voltage. It is not possible to reverse the recording in a single cell without re-recording the entire sequence of information. The waveforms resulting from higher packing densities are indicated in Figure 9.52b. In this case the rule for recovery may be modified slightly to say that only the signals whose sampling voltages lie outside the threshold region are to be considered in deciding whether there has been a change in polarity. Thus the change in signal output from the indeterminate voltage of cell 3 to the negative sampled voltage of cell 4 is not indicative of a 0. However, the change from a negative voltage greater than the threshold level in cell 4 to a positive voltage greater than the threshold level in cell 5 does indicate a 0.

With regard to the minimum peak signal level, we have seen that the read-head voltage is directly proportional to V_0, B_e, D_h, D, and $1/S$, and is also a function of $H(x)$, which in turn depends on g_x, S, and D, and on the geometry of the head and read winding. In most systems containing a quantity of separate heads and a large recording-medium surface area the dimensions of each one of these factors may be regarded as making up an independent random variable possessing a particular probability density function. In a drum system, for example, V_0 is usually confined within quite narrow limits but may vary within these limits due to hunting of the drum drive motor. The value of B_e will, of course, depend on the uniformity of the magnetic characteristics of the medium as well as on the uniformity of recording. The values of D_h and g_x are primarily functions of the head manufacturing process. In the case of drum memory systems, variations in the value of S are due to differences in mechanical placement from head to head and to drum eccentricity. In the case of magnetic tape recording systems, a phenomenon known as "flutter" may cause the tape to vary in spacing from the head during the length of the record. Superimposed upon this effect may be considerable variations in V_0 due to the intermittent nature of the medium motion.

It is highly desirable to make an evaluation of variables such as the above, with confidence limits appropriate to the given system reliability. In this way the probability of a failure in any element of a large system may be approximately evaluated. It should be noted that in a large storage system it may not be feasible to test experimentally all possible configurations of information in all possible cell locations, and that a theoretical approach may be the only way to evaluate system reliability.

Read, Write, and Selection Circuits

Pulse amplifiers for producing the desired current waveform in the write-head winding constitute a straightforward design problem. The primary requirement upon the write amplifier is that it be capable of passing a current of magnitude I_w in both directions through the head winding. It is therefore common practice to center-tap the head winding and provide a means of switching the current through one leg or the other, as shown in Figure 9.53a. The effective number of turns on the write winding is half the total number of turns on the head, because half of the winding is always open-circuited. The inductance of this portion of the winding will be defined as L_w. The rise of write current versus time is primarily a single exponential with time constant $\tau = L_w/R_1$, as in Figure 9.53b. If the write current is just sufficient to saturate the entire thickness of the medium, the final value of $i_w(t)$ for $S + D \gg g_x$ is

$$I_w \cong \frac{2.5H_s(S + D)}{N_w}. \qquad (9.56)$$

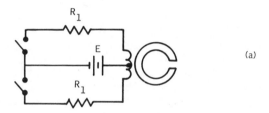

(a)

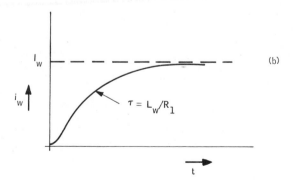

(b)

$$\tau = L_w/R_1$$

Figure 9.53

The above description fits a standard push-pull pulse amplifier, shown in simplified form in Figure 9.54. The damping resistor R_2 is selected for critical damping, based upon measured or calculated L_w and measured values of winding distributed capacitance C_w, or head resonant frequency. The drive transistor is normally driven to saturation, or, in an equivalent emitter-follower arrangement, to some standard voltage. The turn-ON time of the transistor is normally negligible compared to the recording frequency.

The head-winding inductance is primarily a function of the gap reluctance. Therefore

$$L_w \cong N_w{}^2 \frac{0.2\pi g_y g_z}{g_x \times 10^8}. \tag{9.57}$$

Neglecting initial damping effects, the first-order time constant of the circuit, $\tau = L_w/R_1$, should be several times less than the period T of the write cycle. Define $k \equiv T/\tau \gg 1$, and let $R_1 = kL_w/T$. After the initial transient, the requirement is that $I_w = E/R_1$. Therefore, using (9.56) and (9.57), the supply voltage must be at least

$$E > \frac{\pi k N_w H_s g_y g_z (S + D)}{2 g_x T \times 10^8}. \tag{9.58}$$

With head geometry, recording frequency, and the factor k selected, the only remaining variable under the control of the circuit designer is N_w. Let the transistor maximum derated values of collector voltage and current be denoted by E_s and I_s. It is a requirement imposed on the selection of transistors that $E_s \geq 2E$, because at the initial instant when E appears across one leg of the write-head winding it will also appear across the other leg. It is of course necessary that $I_s \geq I_w$. With these inequalities, the range of

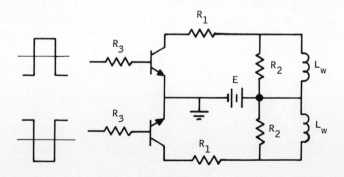

Figure 9.54

permissible values of N_w becomes

$$\frac{H_s(S + D)}{0.4I_s} \leq N_w \leq \frac{g_x TE_s \times 10^8}{\pi k H_s g_y g_z (S + D)}. \tag{9.59}$$

In the limiting case, therefore, a circuit is realizable only if the product of rated collector voltage and current is at least given by

$$E_s I_s \geq \frac{\pi k H_s{}^2 (S + D)^2 g_y g_z}{0.4 g_x T \times 10^8}. \tag{9.60}$$

With this realizability condition met, values of all of the circuit parameters, including the number of turns on the write head, may be readily determined.

The design of a stable, reliable, high-gain read amplifier for recovering the information represented by the read-head signals is, next to the design of the head itself, the most exacting task facing the engineer developing a surface-recording memory unit. It will be found, however, that all of the character-istics required of a read-head amplifier are qualitatively identical to those enumerated for the magnetic-core matrix-memory sense amplifier.

The over-all transfer characteristic required for a read amplifier for any type of recording system includes a "dead zone" in the neighborhood of zero input voltage, thus establishing a threshold, a region of high gain ideally approaching infinity, and a region of saturation or clipping corresponding to a plus or minus input signal level. This transfer characteristic resembles in every respect that specified for the core memory sense amplifier. As in that amplifier, it is also frequently desirable to insert a transformer between the signal source (the head) and the amplifier input. In a surface-recording system a transformer fulfills several desirable functions as follows (a) if the frequency response of the transformer is selected to encompass primarily only those frequencies present in the recording itself, it serves as a filter for some of the undesirable noise; (b) it can usually be designed to match the input impedance of the amplifier; (c) in a system containing a large number of heads that are not in fact identical, a transformer may be used to tune or change the phase of individual heads in such a way as to make them appear more nearly identical as far as the read amplifier is concerned; and (d) a transformer is sometimes desirable for purposes of dc decoupling. Much the same factors governing input transformer design for the matrix-memory sense amplifier also apply to the surface-recording system read-amplifier transformer. The major quantitative difference lies in the magnitude of the source impedance represented by the read-head inductance as compared to the single-turn sense winding in the matrix case. An appropriately high value of transformer magnetizing inductance must therefore be designed.

The problem of input signal limiting and amplifier recovery may similarly require consideration in the surface-recording system design. This is true particularly if the same head is used alternatively for writing and reading. A quantitative difference exists, however, in that the magnitude of the signal induced on the read winding when the write amplifier is active will usually be considerably larger than the noise induced on the sense winding by half-select drive currents in the core-matrix memory case. Except in special cases, this serious problem is mitigated by the fact that a considerably longer recovery time is usually allowable in a surface-recording system. Required recovery times commonly range from several bit times to a complete drum revolution of several milliseconds, as compared to the recovery time of a few microseconds required of a matrix-memory sense amplifier.

The above considerations lead to the generalized block diagram of a read amplifier shown in Figure 9.55. Generally, both positive and negative signals must be amplified. Since one-quadrant amplifiers are usually more convenient, the coupling network (e.g., transformer) is considered to produce both the read signal and its negative. Signal limiting is accomplished by any of a variety of standard clipping circuits. Gain can be provided by amplifiers that are commonly operated in a linear region. They must, however, be followed by logic or decision circuitry that is compatible with logic-level signals employed in the remainder of the system, because outputs are commonly employed to drive standard gates. The design of this circuitry does not represent any substantial departure from that of standard decision devices previously discussed. The major care to be exercised lies in the design of the amplifier to meet the three criteria of stability, frequency response, and gain.

In any surface-recording memory system involving a large number of heads it is economically desirable to design the system in such a way that the number of read and write circuits is less than the number of heads. This can, of course, be done if a means is found to connect any one of a number

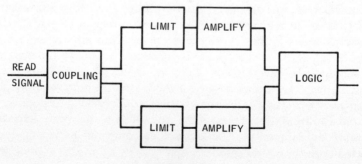

Figure 9.55

of heads to either a single read or a single write circuit. Such a system necessarily implies at least two assumptions: (a) that reference is required to the stored information only at relatively infrequent intervals, and (b) that it is not necessary to regenerate the stored information. If such conditions cannot be met, a recirculating line employing one read amplifier and one write amplifier per line must be used. The network for accomplishing this many-to-one connective operation is referred to as the *selection system*.

In some cases, electromechanical switching systems are feasible. Such systems are used for those items of equipment in which rapid memory access is not an overriding requirement and in which low cost is essential. Reductions not only in numbers of read-write circuits but in numbers of heads has been achieved by providing for a variety of relative mechanical motions between head and recording surface. In general, electromechanical selection can consist of one or a combination of (a) mechanical switching between read-write circuits and heads and (b) mechanical switching between heads and recording surface locations. The trade-off is invariably a struggle between cost and speed. The design of electromechanical systems is indeed a challenging problem for the circuit engineer; but it is a subject beyond the scope of this text.

All-electronic selection systems are confined to making connections between a set of read-write heads and a single or smaller set of read-write amplifiers. In general, these systems may be described as requiring the association of some addressing logic with each head. Because the simplest possible logic is a coincident (two-input) decision, selection systems are frequently arranged on an xy basis. For example, a two-input AND gate is effectively associated with each head, and appropriate xy logic signals are applied to the matrix for selection. Because it is undesirable to provide even a simple amplifier at each head, it is also common practice to make the logic selection signals compatible with read or write signal requirements.

A schematic system embodying these concepts is illustrated in Figure 9.56. Row selections are represented by transistors driven by I_1 or I_2; column selections are represented by transistors driven by I_3 or I_4. The system operates as follows. Let addressing currents I_2 and I_3 flow, driving transistors b, c, and d to saturation. Current can now flow from ground through transistor b, through both legs of head 3, through transistors c or d, and thence to the negative supply E_1 through resistors R_1. Because both diodes in the legs of head 3 are forward-biased, any signal appearing across head 3 will also appear across R_1, and thus at the input to the read amplifier. If the bias source E_2 is more negative than E_1, the diodes in the legs of head 1 are reverse-biased and cannot transmit head 1 signals to the read bus. Similarly, although the center-tap of head 4 is driven positive to 0 V by transistor b, its signal cannot appear on the read bus if transistors e and f are cut off by

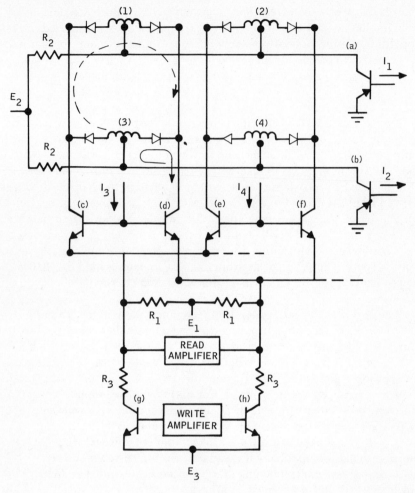

Figure 9.56

the absence of address signal I_4. The value of R_1 is generally selected to provide a small symmetrical biasing current in the legs of the head to be read. Because approximately equal and opposite currents flow in the two legs of the head, the net field produced is nearly zero and does not affect the recorded magnetization.

In order to write information on the medium, one or the other write-amplifier transistors g or h is driven into saturation. This places a voltage, E_3 in series with R_3 across one leg or the other of the head. In Figure 9.56, transistor h is assumed saturated, in which case current flows in the direction indicated by the solid line through head 3. Notice that if E_2 is more negative than E_3, all diodes except those connected directly to head 3 remain cut off, thereby preventing back currents such as that illustrated by the dotted line through head 1. When writing in this manner, we see that the negative voltage E_3 corresponds directly to the drive signal illustrated in Figure 9.54. Note that before turn-ON of the write amplifier, the load seen by transistor c or d consists of R_1 in series with half the head, a diode, and transistor b. The saturated transistor b, the forward-biased diode, and the head resistance may be regarded as negligible compared to R_1. When one of the write-amplifier transistors is turned on, the effective supply voltage for the transistor c or d switches to the value E_3, and the load resistor to approximately R_3, assuming $R_3 \ll R_1$. Note that I_3 must be sufficient to insure that the selection transistors do not come out of saturation for values of collector current equal to I_w. Note further that the selection transistors are operated at all times in regions of very low dissipation. If extremely fast selection switching speeds are not required, low-frequency transistors are satisfactory for the selection matrix, and only the write-amplifier transistors need possess fast response times in order to switch head currents rapidly during the recording cycle. The design of the above circuits follows in a straightforward manner from previous considerations of decision elements and head drive signals.

REFERENCES

[1] A. S. Hoagland, "Magnetic Drum Recording of Digital Data," *AIEE Trans. Commun. Electron.*, 14, 381–385 (September 1954).

[2] A. S. Hoagland, "Magnetic Data Recording Theory: Head Design," *AIEE Trans. Commun. Electron.*, 27, 506–512 (November 1956).

[3] A. S. Hoagland, *Digital Magnetic Recording*, Wiley, New York, 1963.

[4] R. M. Bozorth and D. M. Chapin, "Demagnetizing Factors of Rods," *J. Appl. Phys.*, 15, 320–326 (May 1942).

[5] D. E. Wooldridge, "Signal and Noise Levels in Magnetic Tape Recording," *AIEE Trans. Commun. Electron*, 65 (June 1946).

[6] D. E. Wooldridge, "Remanent Flux Curves for Magnetic Recording Tapes," *Minnesota Mining and Manufacturing Co.*, Bulletin No. 20 (January 1953).

[7] D. G. N. Hunter and D. S. Ridler, "The Recording of Digital Information on Magnetic Drums," *Electr. Engr.*, 29, 356, 490–496 (October 1957).

[8] W. E. Stewart, *Magnetic Recording Techniques*, McGraw-Hill, New York, 1958.

[9] J. J. Miyata and R. R. Hartel, "The Recording and Reproduction of Signals on Magnetic Medium Using Saturation-type Recording," *IRE Trans. Electron. Computers*, EC-8, 2, 159–169 (June 1959).

[10] M. F. Barkouki and I. Stein, "Theoretical and Experimental Evaluation of RZ and NRZ Recording Characteristics," *IRE Trans. Electron. Computers*, EC-12, 2, 92–100 (April 1963).

[11] S. J. Begun, *Magnetic Recording*, Rinehart, New York, 1958.

[12] I. Stein, "Generalized Pulse Recording," *IEEE Trans. Electron. Computers*, **EC-12**, 2, 77–92 (April 1963).

[13] E. Hopner, "High-density Binary Recording Using Nonsaturation Techniques," *IEEE Trans. Electron. Computers*, **EC-13**, 3, 255–261 (June 1964).

[14] L. F. Shew, "High-density Magnetic Head Design for Noncontact Recording," *IEEE Trans. Electron. Computers*, **EC-11**, 6, 764–773 (December 1962).

[15] J. E. Gillis, "A Technique for Achieving High Bit Packing Density on Magnetic Tape," *IEEE Trans. Electron. Computers*, **EC-13**, 2, 112–117 (April 1964).

[16] J. Geurst, "The Reciprocity Principle in the Theory of Magnetic Recording," *Proc. IEEE*, **51**, 11, 1573–1577 (November 1963).

EXERCISES

9.1 The output voltage during readback resulting from a step change in write current in a magnetic drum recording system is shown in Figure P.9.1. Draw waveforms of write current, recorded flux, and readback voltage for the binary sequence . . . 10110 . . . for the following types of recording.
(a) *RZ*
(b) *NRZ*
(c) Phase modulation

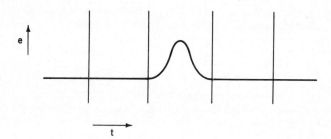

Figure P.9.1

9.2 Improvements are contemplated in a magnetic drum recording system. The present system works satisfactorily, and write current is just sufficient to saturate the magnetic material farthest from the gap. After each of the following effects, list the possible changes in the head, surface, recording current, surface velocity, etc., that would bring about the effect to a first approximation.
(a) Double permissible packing density along track.
(b) Double permissible storage capacity of drum.
(c) Double the readback signal.
(d) Cut the required write current in half.
(e) Cut the readback signal in half.

9.3 The parameters of a drum recording system are
Head-to-surface spacing: 1 mil (10^{-3} in.),
Medium thickness: negligible,
Head-gap length: negligible,
Surface velocity: 1,000 in/s.

The hysteresis loop of the recording medium, measured under recording system conditions, is given in Figure P.9.2.
(a) Find the write drive (ampere-turns) required for saturation recording.
(b) What is the minimum length of the transition on the recording from $+B_r$ to $-B_r$ for a step change in write signal?
(c) If the cross section of a recording is such that the flux corresponding to a change from $+B_r$ to $-B_r$ is 0.3 Mx, what is the peak readback signal voltage per turn of read-head winding?

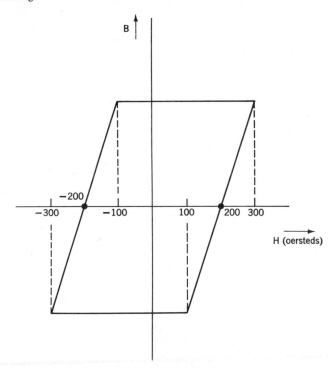

Figure P.9.2

9.4 A modified *NRZ* recording system is to be tried. The system involves recording a 1 in a given cell if the binary information to be recorded *changes* from the preceding cell to the given cell. Otherwise a 0 is recorded.
(a) Sketch waveforms of write current, recorded flux, and readback voltage for this system.
A 3,600-rpm., 8-in.-diameter, 10-in.-long drum is to be used. The effective flux retained by a record on the medium is 200 G. The read-head winding is limited by frequency response to 100 turns. The head-to-drum spacing equals the recording medium thickness. A minimum readback voltage of 100 mV is necessary for reliable operation. The bit rate required is 250 kc (0.4 μs/bit).
(b) Taking the track-to-track separation to be equal to the track width, what is the capacity of the drum in bits?

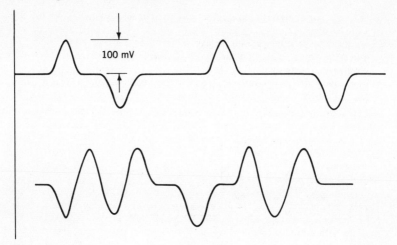

Figure P.9.3

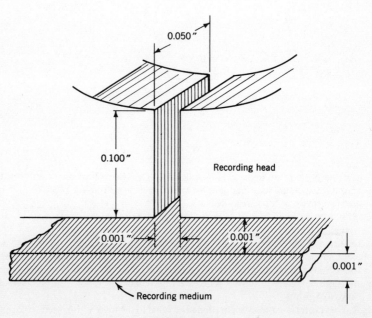

Figure P.9.4

9.5 Waveforms observed at the output of a read head in a drum system for two different recording schemes are shown in Figure P.9.3.

(a) Reconstruct the binary pulse sequence.

(b) It is decided to increase packing density by cutting head-to-surface spacing from 2 to 1 mil, and recording medium thickness from 1 to 0.5 mil, without changing head construction. The head-gap length (g_x) is 0.1 mil. Approximately what will be the new peak output voltage for the *NRZ* waveform (recorded under the new conditions)?

9.6 In a surface recording system, the peak readback signal for a step change in magnetization pattern is 100 mV when the head is in contact with the moving magnetic medium. The spacing between head pole pieces is 0.5 mils.

(a) What is the approximate peak readback signal when the head is 1 mil from the magnetic medium?

(b) At this spacing, what is the width in microseconds of the readback pulse between half-amplitude points if the recording surface velocity is 1,000 in./s?

9.7 The geometry of a recording head in the vicinity of the recording gap is illustrated in Figure P.9.4. The saturation field strength of the recording medium is $H_s = 1,000$ Oe.

(a) If the write amplifier cannot supply more than 100 mA, what should be the number of turns on the write head?

(b) What is the inductance of the write head?

(c) What should be the effective source resistance of the write amplifier if the drum of Problem 9.3 is used with a recording packing density of 150 bits/in. and *NRZ*-type recording?

Index